in situ
HYBRIDIZATION
in ELECTRON
MICROSCOPY

Methods in Visualization

Series Editor: Gérard Morel

in situ HYBRIDIZATION in ELECTRON MICROSCOPY

Gérard Morel, Ph.D., D.Sc.
Annie Cavalier, Eng.
Lynda Williams, Ph.D.

CRC Press
Taylor & Francis Group
Boca Raton London New York

CRC Press is an imprint of the
Taylor & Francis Group, an **informa** business

CRC Press
Taylor & Francis Group
6000 Broken Sound Parkway NW, Suite 300
Boca Raton, FL 33487-2742

© 2001 by CRC Press LLC
CRC Press is an imprint of Taylor & Francis Group, an Informa business

First issued in paperback 2019

No claim to original U.S. Government works

ISBN 13: 978-0-367-45537-8 (pbk)
ISBN 13: 978-0-8493-0044-8 (hbk)

Visit the Taylor & Francis Web site at
http://www.taylorandfrancis.com

and the CRC Press Web site at
http://www.crcpress.com

Library of Congress Cataloging-in-Publication Data

Morel, Gérard.
   In situ hybridization in electron microscopy / Gérard Morel, Annie Cavalier, Lynda Williams.
      p. cm. — (Methods in visualization)
   ISBN 0-8493-0044-4
   1. In situ hybridization. 2. Microscopy. I. Cavalier, Annie. II. Williams, Lynda. III.
   Title. IV. Series.

QH452.8 M67 2000
572.8'4—dc21   00-050750

Library of Congress Card Number 00-050750

SERIES PREFACE

Visualizing molecules inside organisms, tissues, or cells continues to be an exciting challenge for cell biologists. With new discoveries in physics, chemistry, immunology, pharmacology, molecular biology, analytical methods, etc., limits and possibilities are expanded, not only for older visualizing methods (photonic and electronic microscopy), but also for more recent methods (confocal and scanning tunneling microscopy). These visualization techniques have gained so much in specificity and sensitivity that many researchers are considering expansion from in-tube to *in situ* experiments. The application potentials are expanding not only in pathology applications but also in more restricted applications such as tridimensional structural analysis or functional genomics.

This series addresses the need for information in this field by presenting theoretical and technical information on a wide variety of related subjects: *in situ* techniques, visualization of structures, localization and interaction of molecules, functional dynamism *in vitro* or *in vivo*.

The tasks involved in developing these methods often deter researchers and students from using them. To overcome this, the techniques are presented with supporting materials such as governing principles, sample preparation, data analysis, and carefully selected protocols. Additionally, at every step we insert guidelines, comments, and pointers on ways to increase sensitivity and specificity, as well as to reduce background noise. Consistent throughout this series is an original two-column presentation with conceptual schematics, synthesizing tables, and useful comments that help the user to quickly locate protocols and identify limits of specific protocols within the parameter being investigated.

The titles in this series are written by experts who provide to both newcomers and seasoned researchers a theoretical and practical approach to cellular biology and empower them with tools to develop or optimize protocols and to visualize their results. The series is useful to the experienced histologist as well as to the student confronting identification or analytical expression problems. It provides technical clues that could only be available through long-time research experience.

Gérard Morel, Ph.D.
Series Editor

ACKNOWLEDGMENTS

The authors would particularly like to thank Doctors Judy Brangeon, Francine Puvion-Dutilleul, Dominique Le Guellec, Jean-Guy Fournier, George Pelletier, Marc Thiry, and Alain Trembleau for the illustrations used in this book. The authors also thank various colleagues, too numerous to mention, who have helped with this project.

The work was carried out in the framework of the European "Leonardo da Vinci" project (Grant F/96/2/0958/PI/II.I.1.c/FPC), in association with Claude Bernard-Lyon 1 University (*http:// brise.ujf-grenoble.fr/LEONARDO*).

THE AUTHORS

Gérard Morel, Ph.D., D.Sc., is a Research Director at the National Center of Scientific Research (CNRS), at University Claude Bernard–Lyon 1, France. Gérard Morel obtained his M.S. and Ph.D. degrees in 1973 and 1976, respectively, from the Department of Physiology of University Claude Bernard–Lyon 1. He was appointed Histology Assistant at the same university in 1974 and became Doctor of Science in 1980. He was appointed by CNRS in 1981 and became a Research Director in 1989. He is a member of the American Endocrine Society, International Society of Neuroendocrinology, Society of Neuroscience, American Society of Cell Biology, Société Française des Microscopies, Société de Biologie Cellulaire de France, and Société de Neuroendocrinologie Expérimentale, and he has been the recipient of research grants from the European Community, INSERM (National Institute of Health and Medical Research), La Ligue Contre le Cancer, l'ARC (Association de Recherche contre le Cancer), University Claude Bernard, and private industry. Dr. Morel's current research interests include the internalization and trafficking of ligands and receptor molecules, particularly peptide receptors, the regulation of gene expression, and paracrine interactions. He is an expert in the submicroscopic detection of low levels of gene expression.

Annie Cavalier is a Biology Engineer at CNRS at the University of Rennes 1, France. She was appointed a technical assistant in cytology in 1969. By continually improving her knowledge of cell and molecular biology with regular participation in state-of-the-art training courses, she has been appointed a CNRS Engineer (1993) and has graduated from University Claude Bernard–Lyon 1. She is a member of the Société Française des Microscopies and was actively involved in the organization of the 13th International Congress of Electron Microscopy (Paris, July 1994) and the first Congress of the Société Française des Microscopies (Rennes, June 1996). Annie Cavalier has participated in writing many scientific papers, chapters, and two books. Her current interests are the structure and function of water channels (aquaporins) and glycerol facilitators. Due to her unique technical skills, she collaborates with many scientific groups.

Lynda Williams, B.Sc., Ph.D., is a Principal Scientific Officer at the Rowett Research Institute in Aberdeen where she has worked since 1985. She obtained her B.Sc. from Greenwich University in 1977 and her Ph.D. from Queen Elizabeth College in 1982. She was a post-doctoral fellow at the Centre de Recherches en Endocrinologie Moleculaire at Université Laval, Quebec, before moving to Scotland. She is a member of the British Neuroendocrine Group and the Society for Endocrinology and has been the recipient of grants from NATO, the British Heart Foundation, and the European Community, among others. Her current research interests include energy balance and appetite control, particularly the signaling and transport of leptin. Previous research interests include localization of melatonin receptors. Her expertise is in autoradiography at both the light and electron microscope.

CONTENTS

GENERAL INTRODUCTION

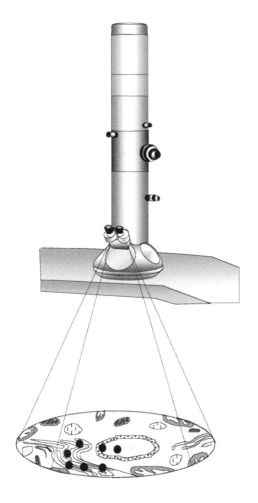

The aim of this book is to demonstrate to the researcher the scope that electron microscope *in situ* hybridization can offer in characterizing and visualizing specific nucleic acid sequences in tissues and at the subcellular level.

Advances in biotechnology and the development of new resins have allowed this technique to be applied to all types of biological samples (organs, biopsies, and cells).

In this book, the different types of probes are described, followed by details of the three techniques that may be used: hybridization on ultrathin sections of tissue embedded in hydrophilic resin (post-embedding method), hybridization prior to embedding (pre-embedding method), and hybridization on ultrathin sections of frozen tissue (frozen tissue method). For each technique, the different stages are described in detail: the preparation of tissue, pretreatment, hybridization, and visualization of the hybridization products.

The arrangement of the text in two columns allows the reader to find useful comments next to the theory and practical details alongside illustrations for each stage in the protocol. The technical details are presented in a way that is designed to help researchers perfect their own protocol for any probe and tissue type. The summary tables provide the criteria for choosing the probe type and technique to be used with all the necessary solutions detailed in the appendices. A glossary and detailed index efficiently aid the search for information.

HISTORICAL PERSPECTIVE

The first reports of light microscopy in *in situ* hybridization were published simultaneously in 1969 by John et al. and Gall and Pardue.

➯ John, H., Birnstiel, M.L., and Jones K.W., RNA–DNA hybrids at cytological levels. *Nature*, 223, 582, 1969.
➯ Gall, J.G. and Pardue, M.L., Molecular hybridization of radioactive DNA to the DNA of cytological preparations. *Proc. Natl. Acad. Sci.*, 63, 378, 1969.

In 1971, Jacob et al. published studies using *in situ* hybridization at the electron microscope using tritium labeled probes for ribosomal RNA. The hybrids were visualized by autoradiography after a very long exposure time.

⇨ Jacob, J., Todd, K., Binstiel, M.L., and Bird, A., Molecular hybridization of ^{3}H-labeled ribosomal RNA in ultrathin sections prepared for electron microscopy. *Biochim. Biophys. Acta*, 228, 761, 1971.
⇨ Only radioactive labels were available (essentially ^{3}H).

The applications of *in situ* hybridization were limited, up until the 1980s, due to methodological constraints.

⇨ The only embedding material available was epoxy resin.

In the 1980s, a great deal of progress was made in molecular genetics (cloning and characterization of genes); the necessity of localizing these genes led to the rapid development of *in situ* hybridization techniques.

⇨ Oligonucleotide synthesis

New methods of labeling (radioactive and non-radioactive) and visualization systems (enzymatic and colloidal gold) contributed to the development of *in situ* hybridization.

⇨ Appearance of antigenic labels

The development of ultrastructural *in situ* hybridization started in 1986 with:

- The use of radioactive probes, allowing the visualization of nucleic acids with great sensitivity

⇨ Limited by autoradiography:
- Resolution
- Duration of the exposure

- The appearance of new acrylic resin embedding medium

⇨ For example, Methacrylates, Lowicryl, LR White, Unicryl, etc.

- The development of cryo-ultramicrotomy techniques

⇨ Development of new cryo-ultramicrotomes
⇨ Methods of staining ultrathin frozen tissue

The recent availability of non-radioactive precursors has helped considerably with the development of the methodology to improve the resolution and reduce the exposure time.

⇨ Several hours in comparison with weeks or months for ultrastructural autoradiography

Precursors are not as sensitive as radioactive probes, which give optimal subcellular resolution. However, the incorporation of nucleotides coupled to antigens is the most used labeling system. The hybrids are visualized by an immunocytochemical reaction using antibodies conjugated to colloidal gold. The resolution obtained is directly related to the size of the colloidal gold particle used.

⇨ For example, (biotin, digoxigenin, or fluorescein) UTP or dUTP; BrdU can also be used
⇨ Direct or indirect reaction

⇨ 5 to 15 nm in diameter

THE THREE TECHNIQUES

The techniques used for *in situ* hybridization at the electron microscope level are well established.

There are three techniques for the preparation of biological tissue and the hybridization reaction:

1. Post-embedding technique
2. Pre-embedding technique
3. Frozen tissue technique

Each of these three techniques has advantages and disadvantages. The choice of technique depends on the relative importance of the subcellular morphology or the sensitivity required.

➯ Each of the methods has some disadvantages.

➯ On sections of acrylic resin
➯ On thick sections (vibratome sections)
➯ On ultrathin sections of frozen tissue

Post-Embedding Technique

This technique provides a compromise between ultrastructural preservation and resolution.

➯ *See* Chapter 4.

➯ For highly expressed nucleic acids, embedding in hydrophilic resin is satisfactory. The development of the new resins (i.e., Lowicryl, LR White, Unicryl) allows the precise identification of structures but limits the accessibility of the probe to the targets at the surface of the section.

Pre-Embedding Technique

The embedding in epoxy resin after *in situ* hybridization, depending on the fixation used, gives good morphology. The probe penetrates into the "thick" section to form hybrids; an immunocytochemical reaction is then used to visualize the reaction (immunoglobulin, enzyme, or colloidal gold). The diffusion of these into the thickness of the section is a problem and limits the efficiency of this method.

➯ *See* Chapter 5.

➯ Nucleic acids are very hydrophilic and diffuse easily.
➯ The use of radioactive labels prevents this problem but requires long exposure (i.e., ultrastructural autoradiography).

Frozen Tissue Technique

The great advantage of this method is its high sensitivity. Freezing tissue avoids the need for embedding and maintains water in the tissue, allowing the molecular structure of the nucleic acid to be preserved.

➯ *See* Chapter 6.

➯ This approach is most appropriate in the case of nucleic acids with low levels of expression.

➯ Freezing and staining are very tricky stages.

The objective of this book is to encourage interest in *in situ* hybridization at the level of the electron microscope and to present the different methods in theory and practical approaches, their possibilities, advantages, and disadvantages.

➯ This book provides a clear description of both principles and practice.

ABBREVIATIONS

AEC	⇌ 3-amino-9-ethylcarbazole
ATP	⇌ adenosine triphosphate
BCIP	⇌ 5-bromo-4-chloro-3-indolyl phosphate
bp	⇌ base pair
Bq	⇌ becquerel
cDNA	⇌ complementary deoxyribonucleic acid
Ci	⇌ curie
CTP	⇌ cytosine triphosphate
cpm	⇌ counts per minute
cRNA	⇌ complementary ribonucleic acid
DNA	⇌ deoxyribonucleic acid
DAB	⇌ 3′ diaminobenzidine tetrahydrochloride
dATP	⇌ deoxyadenosine triphosphate
dCTP	⇌ deoxycytosine triphosphate
DEPC	⇌ diethylpyrocarbonate
dGDP	⇌ deoxyguanosine-5′-diphosphate
dGMP	⇌ deoxyguanosine-5′-monophosphate
dGTP	⇌ deooxyguanosine 5′-triphosphate
DNase	⇌ deoxyribonuclease
dNTP	⇌ deoxynucleoside triphosphate
DTT	⇌ dithiothreitol
dUTP	⇌ deoxyuridine-5′-triphosphate
EDTA	⇌ ethylene diamine tetra-acetic acid
Fab	⇌ immunoglobulin fragment obtained by proteolysis (papain)
(Fab')2	⇌ immunoglobulin fragment obtained by proteolysis (pepsin)
Fc	⇌ immunoglobulin fragment obtained by proteolysis (papain)
GTP	⇌ guanosine triphosphate
IgG	⇌ immunoglobulin G
kBq	⇌ kilobecquerel
kDa	⇌ kilodalton
mRNA	⇌ messenger ribonucleic acid
MW	⇌ molecular weight
NBT	⇌ nitro blue tetrazolium
NTP	⇌ nucleoside triphosphate
oligo(dT)	⇌ oligo-deoxythymidine
PBS	⇌ phosphate buffered saline
PCR	⇌ polymerase chain reaction
PF	⇌ paraformaldehyde
RNA	⇌ ribonucleic acid
RNase	⇌ ribonuclease
rRNA	⇌ ribosomal ribonucleic acid
RT	⇌ room temperature

SSC	⇌ standard saline citrate
TdT	⇌ terminal deoxytransferase
Tm	⇌ melting temperature
tRNA	⇌ transfer ribonucleic acid
UDP	⇌ uridine-5′-diphosphate
UTP	⇌ uridine-5′-triphosphate
$^v/_v$	⇌ volume/volume
$^w/_v$	⇌ weight/volume

Chapter 1

Probes

CONTENTS

DEFINITION

A probe consists of a labeled nucleic acid (DNA or RNA) whose nucleotide sequence is complementary to the target nucleic acid.

➥ The nucleotide sequence of a probe is anti-sense matching the sense sequence of the target.

➥ The label used must match the visualization method used (*see* Chapter 5).

Probes are one of the following types:

- Double-stranded DNA
- Single-stranded DNA
- Synthetic DNA
- RNA

➥ Generally cDNA
➥ Generally obtained by PCR
➥ Oligonucleotides
➥ Generally cRNA

Target nucleic acid sequences include:

- DNA in interphase nuclei
- Chromosomal DNA
- Ribosomal RNA
- Transfer RNA
- Messenger RNA
- Viral RNA, etc.

➥ rRNA
➥ tRNA
➥ mRNA

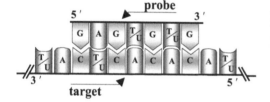

➥ Principles of probe hybridization to target nucleic acid:
  - The target is oriented $3' \rightarrow 5'$.
  - The probe is oriented $5' \rightarrow 3'$, i.e., anti-sense.

Figure 1.1 Probe hybridization.

1.1 TYPES OF PROBES

There are four types of probes, depending on the type of nucleic acid (DNA or RNA) and how it is generated.

➥ Probe (labeled polynucleotides)

1.1.1 Double-Stranded DNA Probes

These probes are double-stranded fragments:

- Of genomic DNA corresponding to a previously cloned sequence

- Complementary to RNA

➥ These sequences may contain regions that are not expressed in the cell (introns). They can be used in viral or genome research.

➥ DNA complementary to the sequence of messenger RNA (complementary DNA: cDNA) is obtained, the sequence of which is published in the literature or in a data bank.

• Amplified by PCR

⇝ A probe can be amplified (in an exponential fashion) and the sequence verified from published data.

The DNA is inserted into a plasmid (e.g., pBR322, pGEM-13...)

⇝ The plasmids are then amplified using bacteria; DNA is extracted and purified on a gel before being cut with restriction enzymes to isolate the insert.

❏ *Advantages*
• The probes are very stable.
• Amplification using a bacterial culture

⇝ Stability of double-stranded DNA probes
⇝ Allows the storage of a renewable stock
⇝ Straightforward method

• The labeling protocol is well established.
• Labeling at the same time as PCR synthesis

⇝ Using random priming (*see* Section 1.3.1)
⇝ *See* Section 1.3.2.
⇝ Limits the PCR yield

❏ *Disadvantages*
• Obtaining a plasmid with the appropriate insert from the originator
• The probe must be amplified.

⇝ Sometimes difficult

⇝ Reagents and equipment are necessary for amplification.

• Re-hybridization of double-stranded probes

⇝ Denaturation prior to use is required (*see* Section 1.6.2).
⇝ There is a reduction in the *in situ* hybridization signal.

1.1.2 Single-Stranded DNA Probes

These probes are:
• Fragments amplified by PCR

⇝ From a known sequence, a probe can be produced increasing exponentially in quantity in the process (*see* Section 1.3.2).

• Single-stranded DNA fragments inserted into a bacteriophage vector (e.g., M13 and its derivatives)

⇝ It is necessary to amplify the plasmids with the aid of bacteria (see above).

❏ *Advantages*
• The probes are very stable.
• Available in large amounts
• There is no need to denature.
• There is no hybridization between probes.

⇝ DNA probes
⇝ After amplification
⇝ Except in the case of palindromic sequences
⇝ There may be palindromic sequences which lead to intra-chain hybridization.

• Labeling at the same time as PCR synthesis

⇝ Limited by PCR yield

❏ *Disadvantages*
• Obtaining the sequence from the originator
• The probe must be amplified.

⇝ Sometimes difficult
⇝ It is necessary to have reagents and equipment for amplification.

• Relatively weak specific activity

⇝ Labeling techniques limited to 5′ end labeling or 3′ extension

1.1.3 Oligonucleotide Probes

These probes are of single-stranded synthesized DNA whose sequence is complementary to that of target nucleic acid.

The criteria for construction are:

- A complementary sequence to the target nucleic acid sequence

- A sequence comprising a coding region

- A sequence of between 20 and 50 nucleotides
- A G + C content of ≤ 50–55%

- An absence of palindromic sequences

- Compare the sequence to that in a gene bank

↝ The sequences are published in the literature or in the various data banks.
↝ They are accessible commercially.
↝ The choice of sequence is essential.

↝ The published cDNA generally corresponds to the mRNA converted into cDNA. The probe sequence must be the complementary and anti-sense sequence to this cDNA (*see* Figure 1.2).
↝ Especially for mRNA research
↝ The sequence includes only exons.
↝ A sequence which includes 2 exons is best as it limits detection to genomic sequences).
↝ An oligonucleotide is generally specific if it is more than 18 nucleotides long.
↝ Nonspecific interactions occur if there is more than 60% G + C.
↝ These lead to intra-chain or inter-chain loops, which inhibit hybridization.
↝ This must be checked to prevent the use of incorrect sequences.

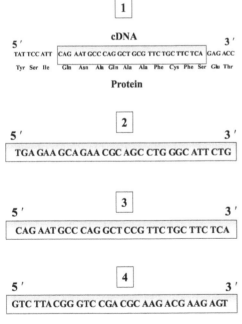

1 = **Example of oligonucleotide probe construction** specific for rat growth hormone (rGH) from a complementary DNA (cDNA) sequence.

2 = **Anti-sense probe**
- The complementary sequence is anti-sense to the cDNA

3 = **Sense probe**
- Negative control probe

4 = **Non-sense probe**
- Negative control probe

Figure 1.2 Probe and control sequence construction.

❏ *Advantages*

- Very accurate specific sequence

- Very stable probe

↝ Where there is a high degree of homology between genes coding for related proteins, a short and very specific sequence must be selected.
↝ DNA probes

- Available in large amounts

- Well established procedure for labeling

- Synthesis of labeled probe

- Synthesis of sense probe
- Synthesis of non-sense probe

- Quantification of the signal

- Simultaneous use of multiple oligonucleotides

↬ Easy to obtain commercially in large quantities
↬ 3′ extension is the most widely used method for *in situ* hybridization (*see* Section 1.3.4).
↬ The incorporation of a labeled nucleotide is carried out during synthesis (for antigenic nucleotides, the efficiency of labeling cannot be tested) (*see* Section 1.5.2).
↬ Synthesis is easy and produces excellent negative controls: sense and non-sense probes cannot hybridize to the target, thus no signal is produced (*see* Figure 1.2).
↬ An oligonucleotide hybridizes to only one target nucleic acid.
↬ Signal amplification
↬ Quantification is difficult since the efficiency of hybridization is not constant.

❑ *Disadvantages*

- The determination of *in situ* hybridization and washing temperatures is difficult.
- Simultaneous use of multiple oligonucleotides

- Generally labeled by 3′ extension

- Efficiency of 3′ extension is limited for antigenic nucleotides.

↬ Parameters must be determined experimentally.
↬ The hybridization and washing conditions must be compatible.
↬ The precise control of extension length is difficult.
↬ The addition of unlabeled nucleotides can help.

1.1.4   RNA Probes

These probes consist of single-stranded RNA (riboprobes) resulting from the transcription of a DNA insert subcloned into a plasmid (*see* Figure 1.3).

The riboprobes are obtained by *in vitro* transcription from one strand of the insert (both strands can be transcribed into RNA). The plasmid is first linearized using a restriction enzyme site downstream from the insert in relation to the polymerase used.

↬ The DNA insert (double or single-stranded) is contained in a circular plasmid including binding sites for RNA polymerase (promoters), which allow the transcription of the DNA insert into RNA.
↬ If transcription is initiated from the opposite and complementary strand to the mRNA, an anti-sense probe is obtained.

↬ The promoters for RNA polymerases are generally SP6, T3, or T7.
↬ Cleavages (1) and (2) are obtained using restriction enzymes characteristic for the vector or plasmid.
(**P1**) promoter 1
(**P2**) promoter 2

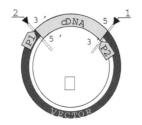

Figure 1.3   Insert in a plasmid.

❑ *Advantages*

- Synthesis of the whole length of the probe

  ➦ Usually cut into fragments of 200 to 300 nucleotides

- Labeling during synthesis

  ➦ This synthesis of the RNA by *in vitro* transcription allows the incorporation of labeled nucleotides. The whole probe is labeled.

- High specific activity

  ➦ This depends on the number of labeled nucleotides used, and the ratio of labeled to unlabeled nucleotides.

- Very stable hybrids

  ➦ The stability of the hybrids is in increasing order: DNA–DNA, DNA–RNA, RNA–RNA.

- Sense probe

  ➦ If the transcription is carried out using the strand coding for the mRNA, the probe obtained is called sense and serves as the negative control.

❑ *Disadvantages*

- Difficult preparation

  ➦ RNA polymerases are very labile enzymes which must be treated with great care.

- Limited storage

  ➦ These probes are very unstable.

1.1.5   Choice of Probe

Table 1.1 Criteria for Selecting the Type of Probe to Use

Criteria of Choice	Probes			
	Double-Stranded DNA	Single-Stranded DNA	Oligonucleotide	RNA
Availability	++	++	+++	+
Storage	+++	+++	+++	+
Stability	+++	+++	+++	+
Specific activity	++	++	++	+++
Manipulation	+++	+++	+++	+
Efficiency	+	++	++	+++
Controls	+	++	++	+++

1.2   PROBE LABELS

Labels are nucleotide triphosphates either carrying radioactive isotope(s) or modified by the addition of an antigenic molecule.

➦ The storage conditions and use are determined by the manufacturer.

➦ Antigenic nucleotide

The type of label used determines the type of detection method:

➦ The sensitivity of the detection method also depends on the type of label.

- Autoradiography is used for radioactive probes.
- Immunocytochemistry is used if the probe label is antigenic.

↬ This method is the most sensitive and allows quantification.
↬ A rapid method, which requires neither specific equipment nor facilities.

The labeling is carried out:

- After obtaining a fragment of DNA

↬ There are three methods which can be used:
- The substitution of an existing nucleotide by a labeled nucleotide
- The addition of labeled nucleotides
- The addition of molecular labels

- During the synthesis of single-stranded DNA, oligonucleotide, or RNA probes

↬ The labeled nucleotide is incorporated:
- PCR amplification (*see* Sections 1.3.2 and 1.3.3) or introduction into the extension of the oligonucleotide (*see* Section 1.3.4) (i.e., labeled deoxyribonucleotide triphosphates)
- During *in vitro* transcription of RNA (*see* Section 1.3.5) (i.e., labeled ribonucleotide triphosphates)

1.2.1 Radioactive Labels

↬ This is less commonly used for electron microscopy.

These are radioactive isotopes that emit β^- rays. Their position in the nucleotide structure depends on the type of isotope and the technique used for incorporation.

↬ The radioactive markers used must have a sufficiently high energy.

1.2.1.1 Types

The isotopes that may be used for *in situ* hybridization at the electron microscope level are ^{35}S and ^{33}P.

↬ The choice of isotope depends on whether the research aim is sensitivity or resolution of detection.

Isotope characteristics are:

- Half-life

↬ Time required for half the radioactive atoms to decay

- Emission energy

↬ Energy liberated when the emission of rays is evaluated in MeV (Mega-electron volts)

- Resolution

↬ Distance between the localization of a silver grain and the source of the emission (a function of the energy of the isotope and of autoradiography)

- Sensitivity

↬ Number of labeled molecules that can be detected (signal/background noise ratio)

- Autoradiographic efficiency

↬ Number of silver grains obtained per β^- emission

1.2.1.1.1 ^{35}S

The characteristics of ^{35}S are:

- Half-life: 87.4 days

- Emission energy: 0.167 MeV
- Resolution: 460 nm
- Sensitivity:
  Specific activity ≈ 3000 Ci/mmol
- Autoradiographic efficiency:
  0.5 grain/β⁻ emission

❑ *Advantages*

- Less energetic isotope

- An excellent compromise

❑ *Disadvantages*

- Modification of the nucleotide (substitution of an oxygen molecule in the phosphate group)
- Risk of oxidation

- Difficult to use

↝ It remains the most commonly used label.

↝ Long enough for probe use before radiolysis occurs
↝ Emission of β⁻ rays less energetic than ^{33}P
↝ Allows subcellular resolution
↝ Sensitivity similar to ^{33}P

↝ Greater than that of ^{33}P

↝ The use of ^{35}S does not require extensive radioprotection.
↝ Sensitivity, efficiency

↝ Chemical bond less stable (*see* Figure 1.4)

↝ Necessitates the use of protective chemicals (i.e., DTT, mercaptoethanol)
↝ ^{35}S can lead to chemical modifications in the molecule, causing background.

1.2.1.1.2 ^{33}P

Its characteristics are:

- Half-life: 25.4 days
- Emission energy: 0.25 MeV
- Resolution: ≈ 500 nm
- Sensitivity
- Autoradiographic efficiency

❑ *Advantages*

- Weak emission energy
- Short half-life
- No modification of the nucleotide

❑ *Disadvantages*

- Less used for electron microscopy
- Cost
- Short half-life

↝ Short exposure time
↝ Close to that of ^{35}S
↝ Good resolution
↝ Close to that of ^{35}S
↝ Close to that of ^{35}S

↝ No problems with radioprotection
↝ Short exposure times
↝ Physiological label

↝ The resolution has not yet been determined.
↝ Still high
↝ Difficult to store; must be used rapidly.

1.2.1.2   Position of the Label

Only labels in the α position are used for electron microscopy.

↝ The following labeling methods are used:

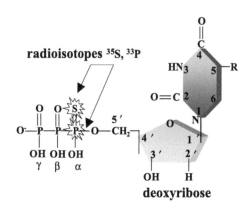

deoxyribose

- Random primer extension (*see* Section 1.3.1)
- PCR (polymerase chain reaction) (*see* Section 1.3.2)
- 3′ extension (*see* Section 1.3.4)
- *In vitro* transcription (*see* Section 1.3.5)

Figure 1.4 Structure of a deoxyribonucleotide triphosphate labeled by substitution of ^{35}S, ^{32}P, or ^{33}P.

POSITION α

Only phosphates in the α position are included in the phosphodiester bond, thus it is the only label which remains in the polynucleotides.

⇨ Used in most labeling techniques
⇨ This phosphate group allows a phosphodiester bond to be formed during polymerization (*see* Figure 1.10).

1.2.1.3 Advantages

- Sensitivity: the best results are obtained with less energetic emitters, i.e., less penetrating radiation.

- Control of the labeling process

- Quantitative estimation

⇨ Radioactive probes are more sensitive than antigenic probes.
⇨ The autoradiographic signal remains located at the source of emission.
⇨ Allows the determination of the specific activity of the probe (*see* Section 1.5.1)
⇨ The density of the signal is proportional to the radiation emitted, and thus to the number of hybrids formed.

1.2.1.4 Disadvantages

- Resolution
- Exposure time

- Quantification is generally difficult.

- Use of radioisotopes requires precautions.
- Cost of manipulations is high.

- Lifespan of the radioisotope is limited.
- Radiolysis

⇨ Less than that obtained with antigenic probes
⇨ After hybridization, autoradiography is carried out.
⇨ Necessitates an exposure time of between one day and several months
⇨ Quantification is determined by the number of copies of nucleic acid in each cell. It is difficult to carry out in practice (1 mRNA molecule does not correspond to 1 grain of silver). Generally, only a relative estimation is obtained.
⇨ Radioprotection
⇨ Cost of the labeled nucleotides and the emulsions; labeling of the probes
⇨ Requires planning
⇨ Difficult to evaluate

1.2.2   Antigenic Labels

1.2.2.1   Types

These are nucleotide triphosphates of nitrogenous bases to which an antigenic molecule was added.

The labels most often used are biotin, digoxigenin, and fluorescein (FITC).

☞ Antibodies are commercially available which form antigen–antibody complexes after hybridization.

☞ These molecules are added by the substitution of the group –CH3 of a thymidine, transforming it into uracil (*see* Figure 1.5).

☞ The substitution of a methyl in position 5 allows the addition of an antigenic molecule while transforming the dTTP into dUTP-X antigen.

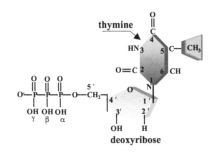

Figure 1.5 Structure of thymidine triphosphate (dTTP).

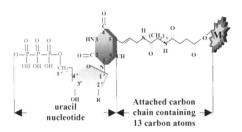

- Antigen X-UTP (R = OH)
- Antigen X-dUTP (R = H)

X represents the number of carbon atoms of the antigenic base, usually between 6 and 21. (**M**) antigenic label (i.e., biotin, digoxigenin, etc.)

Figure 1.6 Labeled nucleotide structure.

1.2.2.1.1   Biotin

Biotin is a vitamin (vitamin H) of low molecular weight (MW = 244) (*see* Figure 1.7).

☞ It is found in a number of animal tissues.

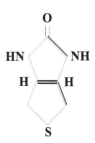

Figure 1.7 Structure of biotin.

1. Affinity

- Avidin
- Streptavidin

☞ $K_d \approx 10^{15}$ M^{-1}

☞ Glycoprotein extract of egg white

☞ Glycoprotein extract from *Streptomyces avidinii* culture medium

2. Detection

- Avidin or multiple avidins
- Streptavidin
- Anti-biotin IgG (mono or polyclonal)

❏ *Advantages*
- The biotinylated nucleotides exist in the forms of biotin X-dUTP and biotin X-UTP.

❏ *Disadvantages*

- Efficiency of incorporation

- Biotin is present in certain tissues (e.g., muscle, liver, kidney, heart, etc.).

➥ These molecules can be conjugated to:

- An enzyme (usually peroxidase)
- Colloidal gold

➥ The length of the chain means that the antigenic group is exposed at the edge of the hybrid.

➥ Very short and very long carbon chains limit hybridization.
➥ Endogenous biotin must be inhibited (using blocking agents) (*see* Appendix B 6.1). Treatment with proteases and checking the absence of labeling from endogenous biotins.

1.2.2.1.2 DIGOXIGENIN

Digoxigenin is a molecule extracted from foxglove (*Digitalis purpurea* and *Digitalis lanata*) of MW ≈ 300–400 (*see* Figure 1.8).

➥ Antigenic molecule not present in animal tissues

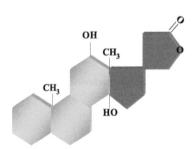

Figure 1.8 Structure of digoxigenin.

Detection:

- Fab fragments
- Anti-digoxigenin, IgG (mono or polyclonal)

❏ *Advantages*
- Molecule of plant origin
- It is conjugated to dUTP or UTP in the forms of digoxigenin X-UTP and digoxigenin X-dUTP.

- Specific detection in animal tissue

➥ A molecule which can be conjugated to:

- An enzyme (usually peroxidase)
- Colloidal gold

➥ No endogenous digoxigenin in animal tissues
➥ X represents the carbon chain, comprising between 6 and 21 atoms.
➥ The digoxigenin is attached to carbon 5 of uracil via the attached carbon chain by an enzymatic reaction.

❏ *Disadvantages*

- The structure of this molecule is related to that of steroids.
- Toxicity

↪ This molecule could bind to steroid receptors.
↪ Must not be inhaled

1.2.2.1.3   FLUORESCEIN

Fluorescein (FITC) is a fluorochrome which emits photons upon UV excitation, but is mostly used as an antigenic molecule (*see* Figure 1.9).

↪ Fluorescence allows direct visualization of the label.
↪ Seldom used in direct detection. This property is used in controlling probe labeling (*see* Section 1.5) and for cytogenetics.

Figure 1.9 Structure of fluorescein.

❏ *Advantages*

- Control in probe labeling

- Antigenic probe

❏ *Disadvantages*

- Direct use difficult for mRNA
- Sensitivity

↪ The property of fluorescence allows direct visualization of the quality of labeling.
↪ Antibodies are commercially available for visualizing the antigen and thus the hybrid.

↪ With the exception of cytogenetics
↪ Not detectable without amplification

1.2.2.2   Advantages

- The resolution obtained with non-radioactive probes is better than with radioactive probes.
- Autoradiography is avoided.
- Rapid results
- Considerable number of different labels

- Numerous detection systems
- High stability of the probes

↪ More reliable localization of the hybrids and more precise identification of subcellular structures
↪ No special equipment needed for protocols
↪ Hybrid detection time is short.
↪ Many probes carrying different labels may be used simultaneously.

↪ Usable for up to a year

1.2.2.3 Disadvantages

• Less sensitive than radioactive labels	⇨ There must be a sufficient amount of antigen present to allow detection of the hybrids by a visible immunohistological reaction.
• Efficiency of incorporation of the labeled nucleotide	⇨ The efficiency of incorporation of the polymerase varies according to the length of the nucleic acid chain (a problem of steric hindrance).
• Control of probe labeling	⇨ Requires a considerable quantity of probe and a specific detection system (*see* Chapter 5, Section 5.8). ⇨ Specific activity not quantified with precision
• Quantification	⇨ Difficult

1.2.3 Choice of Label

Table 1.2 Criteria for Choosing the Probe Label

Chosen Criteria	Labels	
	Radioactive	Antigenic
Biohazard	Contamination	Biological
Limitations	Location/equipment	Sometimes needs to be screened from light
Cost	Needs to be renewed frequently	Lower
Availability	Periodically	Constant
Storage of label	Short	Long
Stability	Short	Long
Efficiency of labeling	Good	Limited
Efficiency of detection	High	At the threshold of detection
Specific activity	High	Weak
Duration of the protocol	Long	Rapid
Storage of the probe	Short	Long
Sensitivity	High	Limited
Resolution	At the tissue level	Cellular
Quantification	Possible	Very difficult, or impossible

1.3 LABELING TECHNIQUES

Table 1.3 Where to Find Discussions of the Different Labeling Techniques

Technique	Probes			
	Double-Stranded DNA	**Single-Stranded DNA**	**Oligonucleotide**	**RNA**
Random priming	§1.3.1	§1.3.1		
PCR	§1.3.2	§1.3.2		
Asymmetric PCR	§1.3.3	§1.3.3		
3′ extension	§1.3.4	§1.3.4	§1.3.4	
In vitro transcription				§1.3.5

1.3.1 Random Priming

1.3.1.1 Principles

- Random labeling by **duplication** of DNA
- Double-stranded DNA from the insert isolated from the plasmid

↝ By enzymatic digestion, then purification by preparative electrophoresis, migration in an agarose gel, and extraction of the required fragment.

The different steps are:

- The double-stranded DNA is denatured.

↝ The insert is denatured to separate the two strands (by heat: 95°C for 5 min).

- It is incubated in the presence of a mixture of synthetic polynucleotides: **the primers**

↝ The polynucleotides that serve as primers are generally hexanucleotides, but can be nonanucleotides in random sequences.

The Klenow enzyme attaches itself to the 3′ hydroxylated ends of the primers and incorporates unlabeled/labeled nucleotides; these bind by complementarity to the DNA strand which serves as the template, thus allowing the synthesis of the complementary strands.

↝ The Klenow enzyme is the fragment of DNA polymerase I after digestion with subtilisin (disappearance of the DNase activity). This polymerase forms double-stranded DNA from a single-strand and a primer.

↝ The incorporation of the nucleotides carrying antigens is slower than native nucleotides.

The α phosphate of the nucleotide to be incorporated attaches itself to the 3′ hydroxyl of the deoxyribose of the preceding nucleotide to form a diester bond (*see* Figure 1.10).

There is no degradation of the new strand.

↝ The label must be carried on an α phosphate (radioactive nucleotide) (*see* Figure 1.4) or on a base (nonradioactive nucleotide) (*see* Figure 1.6).

↝ The Klenow polymerase lacks exonuclease activity.

All the newly synthesized DNA is labeled and corresponds in total copy number to the original.

↝ The quantity obtained seems to be larger if the reaction time is extended.

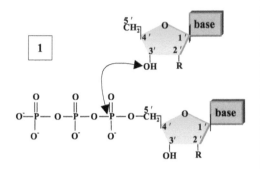

1 = The α phosphate of the nucleotide attaches to the 3′ hydroxyl of the deoxyribose

2 = Formation of a diester bond

Figure 1.10 Formation of a diester bond.

❑ *Advantages*
- Specific activity:
  1. The specific activity of the labeled probe is increased if the insert is purified and no contaminating DNA is present.

      ➱ The specific activity corresponds to the number of isotopes or antigens incorporated in relation to the mass of the probe.
      ➱ The specific activity is ≈ 5–10 times greater than that of nick translation.

  2. Several labeled nucleotides can be used.

      ➱ The use of 1 to 4 labeled nucleotides adds proportionally to the specific activity of the probe, but increases radiolysis for radioactive nucleotides. Generally, a single labeled nucleotide is used, or sometimes two.
      ➱ The ratio of antigenic labeled to unlabeled nucleotides has to be high in order to facilitate their incorporation (a ratio of 1/4 is generally advised).
      ➱ The percentage of labeled nucleotide incorporated is identical in each synthesis.

- Uniform labeling of DNA

- Only a small amount of DNA is necessary.

      ➱ Labeling can be carried out with a quantity of DNA ≥ 10 ng.

❑ *Disadvantages*

- Preparation of the DNA is longer than with the nick translation method.
- Smaller DNA fragment than with nick translation

      ➱ It is necessary to use the insert rather than the entire plasmid.
      ➱ The generated fragments are of the order of 100 to 200 nucleotides in length.

1.3.1.2 Summary of Different Steps

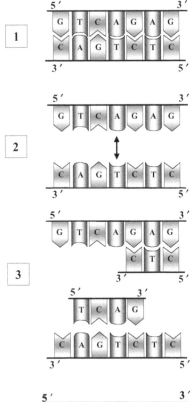

1 = **Separation of the insert from the plasmid**
Labeling is carried out after separation of the insert from the plasmid (vector) by enzymatic ligation and agarose gels.

2 = **Denaturation**
The DNA is double-stranded; denaturation (separation of the double strands) is an indispensable step for the hybridization of primers.

3 = **Hybridization of the primers**
Hybridization with oligonucleotides whose sequences contain all the possible combinations: the type of primer (hexa- or nonanucleotide) dictates the length of the newly-synthesized strands. The label must be on the α phosphate or the base.

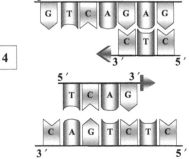

4 = **Binding of the Klenow enzyme**
Binds to the 3' hydroxylated ends, which then incorporate the labeled or non-labeled precursors

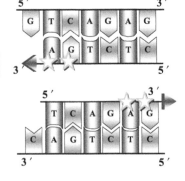

5 = **Incorporation of nucleotides labeled with dGTP or dATP ☆ and unlabeled nucleotides**
- Synthesis incorporates the labeled nucleotides.
- The length of the primers influences the length of the newly formed strands.
- Polymerization is effective in the 5' → 3' sense

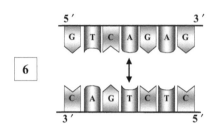

6 = Denaturation
- Before use, the fragments must be denatured.
- The labeled fragments are between 100 and 200 nucleotides in length.

Figure 1.11 Labeling of DNA by random priming.

1.3.1.3 Equipment/Reagents/Solutions

EQUIPMENT
- Centrifuge
- −20°C or −80°C freezer

- Vortex mixer

- Water bath or heating block   100°C
    37°C

➾ Greater than or equal to 14,000 *g*
➾ Storage of products and precipitation of the probe
➾ Use carefully. Do not disperse the products on the wall of the tube.
➾ Denaturation of DNA
➾ Incubation of the enzyme

REAGENTS
- Dithiothreitol (DTT)
- EDTA
- Hexanucleotides

➾ **Only used for *in situ* hybridization**

➾ Ethylene diamine tetra-acetic acid
➾ Or nonanucleotides

SOLUTIONS

- Buffer 10X or reaction buffer containing hexanucleotides 10X
- Dithiothreitol 1 *M*
- DNA to be labeled

➾ **All the solutions are prepared using RNase-free reagents in a sterile container** (*see* Appendix A1).
➾ Generally supplied with the enzyme
➾ Provided with the labeling kit
➾ *See* Appendix B2.11.
➾ The DNA is between 400 bp and several kbp.
➾ The typical final concentration for labeling is 1 mg/mL (100 ng to 3 mg).

- dNTP (dATP, dTTP, dCTP, dGTP) 10X in 500 m*M* Tris–HCl buffer containing 50 m*M* MgCl$_2$/1 m*M* DTT/BSA 0.5 mg/mL; pH 7.8

➾ Storage at −20°C
➾ DNase-free BSA

   1. Radioactive labeling

   2. Antigenic labeling

➾ 0.5 m*M* radioactive or non-radioactive dNTPs (dATP, dTTP, dCTP, dGTP)
➾ 1 m*M* (dATP, dCTP, dGTP); 0.65 m*M* (dTTP) + 0.25 m*M* labeled dUTP

- 500 m*M* EDTA; pH 8.0
- Klenow enzyme 2 U/µL

➾ *See* Appendix B2.13.
➾ Storage at −20°C in glycerol 1:1 ($^v/_v$)

- Labels
  1. Antigenic labels
     - 1 m*M* biotin 11-dUTP
     - 1 m*M* digoxigenin 11-dUTP
     - 1 m*M* fluorescein 12-dUTP ⇨ Storage in aliquots at –20°C
  2. Radioactive labels
     - α ^{35}S/dCTP, α ^{33}P/dCTP
     - α ^{35}S/dATP, α ^{33}P/dATP ⇨ Storage in aliquots at –80°C
- Probe ⇨ Concentration 10 to 50 ng
- Sterile water ⇨ *See* Appendix B1.1.

1.3.1.4   Protocol for Radioactive Probes

DENATURATION

1. Place the following reagents in a sterile Eppendorf tube in the given order: ⇨ The labeling uses single-stranded DNA.
   - Distilled sterile water   9 μL
   - Probe   2 μL   ⇨ Typical final concentration for labeling is 0.5 μg/μL
2. Denaturation   **10 min at 100°C**   ⇨ 92°C or above for several minutes to allow separation of the two strands of DNA
3. Place immediately in   **10 min at –10°C**   ⇨ To prevent re-association at the time of cooling, the solution is placed very rapidly into a bath below 0°C (the two strands remain separated).
   - Ice/NaCl, or
   - Ice/ethanol

DUPLICATION BY THE KLENOW ENZYME

1. Add:
   - 0.5 m*M* dNTP (dCTP, dGTP, dTTP)   3 μL   ⇨ 1 μL of each unlabeled nucleotide
   - Probe   1 μL   ⇨ 25 ng of denatured, double-stranded DNA
   - Labeled dATP (50 μCi)   5 μL   ⇨ Another labeled dNTP can be used
   - Hexanucleotides 10X   2 μL   ⇨ Hexa- or nonanucleotide primers
   - Klenow enzyme   1 μL (+ 1 μL)   ⇨ Final concentration: 2 units, but must be modified according to the quantity of probe to be labeled
   ⇨ Due to the instability of the newly synthesized strands, the addition of more of the enzyme increases the quantity of DNA produced. It should be added at the last moment.
   - Sterile water   to 20 μL
2. Incubate: **60 min at 37°C**   ⇨ The reaction can be continued for up to 20 h.
   ⇨ The enzyme is destroyed after 90 min of incubation. Add more if necessary.
3. Add:
   - EDTA (200 m*M*)   2 μL   ⇨ Stops the reaction
   ⇨ The same effect can be obtained by heating for 10 min at 65°C.

⇨ Pass through a Sephadex G50 column or Nuctrap push column (not always necessary) (*see* Section 1.4.2)

❑ *Next stage:*

- Purification of the DNA by ethanol precipitation

⇨ *See* Section 1.4.1.

1.3.1.5   Protocol for Antigenic Probes

1. Denaturation

⇨ *See* Section 1.6.2.

2. Duplication by the Klenow enzyme:

Mix the unlabeled nucleotides:

⇨ Prepare a solution of unlabeled nucleotides by the addition of an equal volume of each nucleotide.

- dNTP mixture   2 μL

⇨ Final concentration (25 nM dATP, dCTP, dGTP, 16 nM dTTP)

- Labeling buffer   2 μL

⇨ Initial concentration 10X

- Labeled dNTPs   1 μL

⇨ (Biotin, digoxigenin, fluorescein) — dUTP used at a final concentration of 9 nM

- Hexanucleotides 10X   2 μL

⇨ Hexa- or nonanucleotide primers

- Klenow enzyme   1 μL
                                 (+ 1 μL)

⇨ Final concentration: 2 units

⇨ Due to the instability of the newly synthesized strands, it is possible to add more enzyme to increase the quantity of DNA formed. This should be added at the last moment.

- Sterile water   **To a final volume of 20 μL**

Incubate.   **60–90 min at 37°C**

⇨ The reaction can be continued for up to 20 h.

⇨ The enzyme is destroyed after 90 min of incubation. Add more if necessary.

⇨ During polymerization, one nucleotide in every 20–25 is labeled. The incubation time is therefore longer (given the problem of steric hindrance).

Add:

- EDTA 200 mM 2 μL

⇨ The same effect can be obtained by heating for 10 min at 65°C.

❑ *Next stage:*

- Purification of the DNA by ethanol precipitation

⇨ *See* Section 1.4.1

⇨ Do not separate on a Sephadex column because steric hindrance occurs.

1.3.2 PCR Amplification

1.3.2.1 Principles

To amplify a specific sequence (e.g., a probe) in a mixture of DNA.

↬ Polymerase chain reaction

The probe and its complementary strand are present in a mixture of nucleic acid (e.g., plasmid DNA, *see* Figure 1.3). Two primers define this sequence: the first is «sense» at the 5′ extremity to the probe and the second is «antisense» at the 3′ extremity to the probe sequence (*see* Figure 1.12). After hybridization of these oligonucleotides, duplication is achieved by the action of a polymerase, which extends the oligonucleotides by using the cDNA as a template.

A large copy number is obtained by a series of cycles comprising a primer hybridization stage, followed by a stage of elongation of this primer, and finally a denaturation stage which separates the newly formed strand from the original template strand.

↬ The quantity obtained is proportional to the number of cycles used (2^n, where n is the number of cycles).

↬ Primers (*see* Glossary)

↬ The *Taq* enzyme withstands the temperature of denaturation.

❑ *Advantages*

- Production of a large amount of probe
- No pretreatment of the stored DNA

- Method of amplification of a specific sequence
- Labeling during the course of synthesis

- Newly synthesized probes can be up to several hundred bases long
- High specific activity

- High specificity of the probe

- DNA probe

↬ According to the principles of the method
↬ It is possible to carry out the amplification and labeling at the same time.
↬ The specificity of the sequence can be checked in a database.
↬ Labeling may reduce the efficiency of the amplification.
↬ The longer the sequence, the less important the amplification.
↬ Several types of labeled nucleotides can be incorporated into the newly synthesized sequence.
↬ Due to the choice of the sequence amplified
↬ Great stability of DNA probes

❑ *Disadvantages*

- Requires special equipment
- Very stringent reaction conditions
- Considerable risk of contamination

↬ This is generally available.
↬ Conditions must be carefully controlled.
↬ Loss of probe specificity

1.3.2.2 Summary of Different Steps

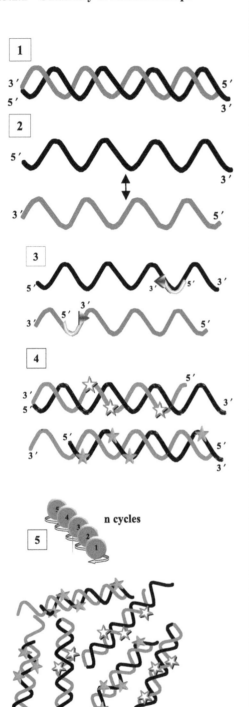

1 = **Genomic DNA**

2 = **Denaturation**
Genomic DNA is denatured.

3 = **Addition of primers**
The primer specific to each end of the two strands of the sequence to be amplified is hybridized.

4 = **Action of *Taq* polymerase**
The enzyme extends the primer by duplication of the DNA strand (end of the first cycle).

5 = **Amplification**
The following cycles lead to the amplification of the target sequence and, therefore, the probe. The labeled nucleotides are incorporated at the time of synthesis of the newly formed strand.

The quantity of the probe obtained is proportional to the number of cycles.

Figure 1.12 Labeling by symmetrical PCR.

1.3.2.3   Equipment/Reagents/Solutions

1. Equipment
   * Liquid PCR machine
     ⇝ Standard equipment

2. Reagents
   * KCl
     ⇝ Molecular biology grade
   * Labeled dNTP
     – dUTP X-antigen
       ⇝ Storage at –20°C
     – Radioactive dNTP
       ⇝ Storage at 4°C or –80°C
   * Mineral oil
     ⇝ For PCR
   * MgCl$_2$
     ⇝ Molecular biology grade
   * Sense and anti-sense primers
     ⇝ Storage in aliquots at –20°C
   * *Taq* polymerase enzyme
     ⇝ Storage at –20°C
   * Tris
     ⇝ Molecular biology grade
   * Unlabeled dNTPs
     ⇝ Storage at –20°C

3. Solutions

   ⇝ **All the solutions are prepared using DNase-free reagents in a sterile container** (*see* Appendix A2).

   * Buffer (10X)
     ⇝ 100 mM Tris - HCl/500 mM KCl; pH 8.3
     ⇝ Storage in aliquots at –20°C
   * 2 mM unlabeled dNTPs
     ⇝ Storage at –20°C
   * Labeled dNTPs
     ⇝ Storage at –20°C
     – 0.7 mM dUTP X-antigen
       ⇝ Addition of 1.3 mM dTTP
     – Radioactive dNTP
   * 25 mM MgCl$_2$
     ⇝ *See* Appendix B2.8.
   * Sense and anti-sense primers
     ⇝ 0.1 to 1 µM (stored in aliquots at –20°C)
   * Sterile water
     ⇝ *See* Appendix B1.1
   * *Taq* polymerase enzyme
     ⇝ 5 U/µL (storage at –20°C)

1.3.2.4   Reaction Mixture for Radioactive Probes

1. Place the following reagents in a sterile Eppendorf tube:

• Linearized DNA	≈ 50 ng	⇝ To be determined
• Primers	250 nmoles	⇝ Of each primer
• 3 unlabeled dNTPs	x µL	⇝ 2 mM
• 4th labeled dNTP	x µL	⇝ ≈ 50 µCi
• 4th unlabeled dNTP	x µL	⇝ ≈ 1 mM
• MgCl$_2$	2–10 µL	⇝ To be determined
• Buffer (10X)	5 µL	
• *Taq* polymerase	1.5 U	⇝ To be determined (0.5 to 2.5 U)
		⇝ Volume according to the concentration
• H$_2$O	to 50 µL	

2. Mix and centrifuge.
3. Cover with oil.   100 µL
4. Place in the thermocycler.

1.3.2.5 Reaction Mixture for Antigenic Probes

1. Place the following reagents in a sterile Eppendorf tube:

• Linearized DNA	≈ **50 ng**	➭ To be determined
• Primers	**250 nmoles**	
• dATP, dGTP, dCTP	**x μL**	➭ 2 mM
• Labeled dUTP	**x μL**	➭ 0.7 mM
• Unlabeled dTTP	**x μL**	➭ 1.3 mM
• MgCl$_2$	**2–10 μL**	➭ To be determined
• Buffer (10X)	**5 μL**	
• *Taq* polymerase	**1.5 U**	➭ To be determined (0.5 to 2.5 U)
		➭ Volume according to the concentration
• H$_2$O	**to 50 μL**	

2. Mix and centrifuge.
3. Cover with oil. **100 μL**
4. Place in the thermocycler.

1.3.2.6 PCR Protocol

1. First cycle

• Denaturation	**7 min at 94°C**	
• Hybridization	**1 min at 60°C**	➭ This temperature varies according to the primer.
• Elongation	**1 min at 72°C**	➭ The time can be extended if the probe to be synthesized is long.

2. Following cycles (*n* cycles) — ➭ The number depends on the required quantity of probe: $Q = q \times 2^n$, where q is the quantity of probe at the start and n is the number of cycles.

• Denaturation	**1 min at 94°C**	➭ For large numbers of cycles, it is sometimes necessary to reduce this time to preserve the efficiency of the enzyme.
• Hybridization	**1 min at 60°C**	➭ This temperature varies according to the primer.
• Elongation	**1 min at 72°C**	➭ After 10–20 cycles, this time is generally increased to compensate for the loss in efficiency of the enzyme.

3. Last cycle

• Denaturation	**1 min at 94°C**	

• Hybridization	**1 min** **at 60°C**	
• Elongation	**10 min** **at 72°C**	↬ Time is increased to complete the extension of the newly formed strands.

❏ *Next stage:*
  • Precipitation with ethanol

↬ *See* Section 1.4.1.

1.3.3 Asymmetric PCR Amplification

↬ Polymerase chain reaction

The aim is to make up for the principal drawback of classical PCR by obtaining the probe in the form of two complementary strands, i.e., by generating the anti-sense sequence of the target nucleic acid.

1.3.3.1 Principles

The use of a single primer ensures the synthesis of the probe to be used. After denaturation of the two strands of DNA (first step), the primer hybridizes to the target DNA strand (second step); then, during the course of the third step, elongation (i.e., probe synthesis) takes place. In each cycle, an anti-sense copy of the target nucleic acid is synthesized. The number of copies obtained is n (where n represents the number of cycles).

↬ This relationship is no longer exponential, but arithmetical.

❏ *Advantages*

  • Production of the probe alone
  • Single-stranded DNA probe
  • Labeling during the course of synthesis

  • High specific activity

  • No pretreatment of the stored DNA

  • Newly synthesized probe can reach several hundred bases.

↬ A feature of the method
↬ High stability of DNA probes
↬ The label is incorporated into the structure of the probe.
↬ One or more labeled nucleotides may be incorporated.
↬ The amplification may be carried out immediately prior to use.
↬ Limited by the efficiency of the enzyme *Taq* polymerase

❏ *Disadvantages*

  • Weak yield

  • Special equipment is required.
  • Very stringent reaction conditions
  • Considerable risk of contamination

↬ The quantity of the probe obtained is limited.
↬ Generally available
↬ Conditions must be carefully controlled.
↬ Loss of probe specificity

1.3.3.2 Summary of Different Steps

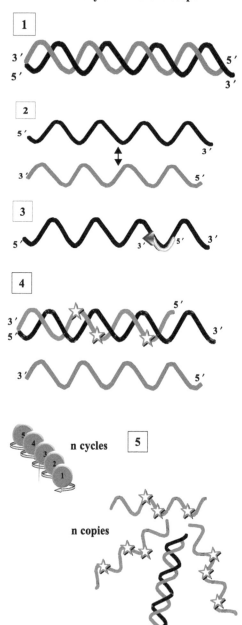

1 = **Genomic DNA**

2 = **Denaturation**
Genomic DNA is denatured

3 = **Addition of primer ⇒**
The primer is specific to the end of the strand that is complementary to the sequence to be detected.

4 = **Action of *Taq* polymerase**
The enzyme extends the primer by duplication of the DNA strand (end of the 1st cycle).

5 = **Amplification**
The following cycles lead to the amplification of the probe. The labeled nucleotides are incorporated at the time of synthesis of the newly formed strand.

The quantity of the probe obtained is proportional to the number of cycles.

Figure 1.13 Labeling by asymmetric PCR.

1.3.3.3 Equipment/Reagents/Solutions
1. Equipment
   • Liquid PCR machine
2. Reagents
   • Anti-sense primer

↪ Standard for PCR

↪ Only the anti-sense primer allows the synthesis of the sequence that is complementary to the target.
↪ Storage in aliquots at –20°C

• KCl	⇝ Molecular biology grade
• Labeled dNTP	
– dUTP X-antigen	⇝ Storage at –20°C
– Radioactive dNTP	⇝ Storage at 4°C or –80°C
• MgCl$_2$	⇝ Molecular biology grade
• Mineral oil	⇝ For PCR (stored at –20°C)
• *Taq* polymerase enzyme	⇝ Storage at –20°C
• Tris	⇝ Molecular biology grade
• Unlabeled dNTPs	⇝ Storage at –20°C

3. Solutions	⇝ **All the solutions are prepared using DNase-free reagents in a sterile container** (*see* Appendix A2).
• Anti-sense primer	⇝ 0.1 to 1 μ*M* (storage in aliquots at –20°C)
• Buffer (10X)	⇝ 100 m*M* Tris–HCl/500 m*M* KCl; pH 8.3
	⇝ Storage in aliquots at –20°C
• Labeled dNTP	⇝ Stored at –20°C
– 0.7 m*M* dUTP X-antigen	⇝ Addition of 1.3 m*M* dTTP
– Radioactive dNTP	
• 25 m*M* MgCl$_2$	⇝ *See* Appendix B2.8.
• Sterile water	⇝ *See* Appendix B1.1.
• *Taq* polymerase enzyme	⇝ 5 U/μL (stored at –20°C)
• 2 m*M* unlabeled dNTPs	⇝ Stored at –20°C

1.3.3.4   Reaction Mixture for Radioactive Probes

1. Place the following reagents in a sterile Eppendorf tube:

• Linearized DNA	≈ **50 ng**	⇝ To be determined
• Primer	**250 nmoles**	⇝ Only the anti-sense primer
		⇝ A very weak concentration of the sense primer (< 50 to 100 times)
• 3 unlabeled dNTPs	x μL	⇝ 2 m*M*
• 4th labeled dNTP	x μL	⇝ ≈ 50 μCi
• 4th unlabeled dNTP	x μL	⇝ ≈ 1 m*M*
• MgCl$_2$	2–10 μL	⇝ To be determined
• Buffer (10X)	5 μL	
• *Taq* polymerase	1.5 U	⇝ To be determined (0.5 to 2.5 U)
		⇝ Volume according to the concentration
• H$_2$O	to 50 μL	

2. Mix and centrifuge.
3. Cover with oil.   100 μL
4. Place in the thermocycler.

1.3.3.5   Reaction Mixture for Antigenic Probes

1. Place the following reagents in a sterile Eppendorf tube:

• Linearized DNA	≈ **50 ng**	⇝ To be determined
• Primer	**250 nmoles**	⇝ Only the anti-sense primer

- dATP, dGTP, dCTP   **x µL**   ↪ 2 mM
- Labeled dUTP   **x µL**   ↪ 0.7 mM
- Unlabeled dTTP   **x µL**   ↪ 1.3 mM
- MgCl$_2$ **2–10 µL**   ↪ To be determined
- Buffer (10X)   **5 µL**
- *Taq* polymerase   **1.5 U**   ↪ To be determined (0.5 to 2.5 U)
         ↪ Volume according to the concentration
- H$_2$O   **to 50 µL**

2. Mix and centrifuge.
3. Cover with oil.   **100 µL**
4. Place in the thermocycler.

1.3.3.6   Protocol for Asymmetric PCR

1. First cycle

- Denaturation   **7 min at 94°C**   ↪ The time can be reduced.
- Hybridization   **1 min at 60°C**   ↪ This temperature must be determined according to the primer sequence.
- Elongation   **1 min at 72°C**   ↪ The time can be increased according to the length of the probe to be synthesized.

2. Following cycles   ↪ The number depends on the required quantity of the probe: Q = q × n, where q is the quantity of the probe at the start and n is the number of cycles.

- Denaturation   **2 min at 94°C**   ↪ For large numbers of cycles, it is sometimes necessary to reduce this time so as to preserve the efficiency of the enzyme.
- Hybridization   **1 min at 60°C**   ↪ This temperature varies according to the primer.
- Elongation   **1 min at 72°C**   ↪ After 10–20 cycles, this time is generally increased so as to compensate for the loss in the efficiency of the enzyme.

3. Last cycle   ↪ **Particularly important**

- Denaturation   **2 min at 94°C**
- Hybridization   **1 min at 60°C**
- Elongation   **10 min at 72°C**   ↪ Time is increased to complete the extension of the newly formed strands.

❑ *Next stage:*
- Precipitation with ethanol   ↪ *See* Section 1.4.1.

1.3.4 3′ Extension

1.3.4.1 Principles

The labeling of an oligonucleotide of short length (less than 50 mers) is carried out by the addition at the 3′ end of labeled nucleotides using an enzyme, terminal deoxytransferase (TdT). This requires a free 3′ OH and nucleotide triphosphates.

In the presence of cobalt ions, the enzyme TdT catalyzes the polymerization of a labeled deoxyribonucleotide.

The extension poly (T) is to be avoided.

The labeled nucleotides are carriers of: radioactive isotopes ^{35}S or ^{33}P or an antigenic molecule.

❏ *Advantages*

- In the case of strong homologies

- Can be used to label breaks in genomic DNA
- Single-stranded probe
- Little DNA required
- Small fragments
- Sensitivity

- Rapid method
- Radioactive labeling

- Nonradioactive labeling

- Quantification

❏ *Disadvantage*
Weak signal using antigenic labeling

⇝ Method applicable to oligonucleotides

⇝ The methods of random primer extension (*see* Section 1.3.1) and PCR (*see* Section 1.3.2) cannot be used with these probes.
⇝ It is possible to add unlabeled nucleotides so as to have a longer extension in the case of antigenic nucleotides.

⇝ Cobalt is the cofactor of TdT.

⇝ Presence of a poly (A) end in messenger RNA of eukaryotic cells, which would form hybrids
⇝ α (^{35}S or ^{33}P)-dATP
⇝ Such as
  - Biotin X-dUTP
  - Digoxigenin X-dUTP
  - Fluorescein X-dUTP

⇝ The synthesis of oligonucleotides for specific regions
⇝ This is a simple and rapid method for visualizing deletions in a heterogeneous tissue.
⇝ No reassociation possible with the probe
⇝ Greater than or equal to 10 pM
⇝ Easy penetration of the tissue
⇝ The sensitivity of detection can be improved to the level of detecting mRNA by simultaneously using several labeled oligonucleotides to detect the same target nucleic acid.
⇝ No manipulation by cloning
⇝ Determined by specific activity, which can be very high
⇝ It is possible to obtain large quantities of DNA for storage.
⇝ Each hybrid corresponds to a target nucleic acid.

⇝ Each oligonucleotide corresponds to a target nucleic acid. If the expression is weak, the signal is weak. It can be increased by using several probes for the same target sequence.

1.3.4.2 Summary of Different Steps

1 = **Synthesized oligonucleotide probe**
The 3′ end must be hydroxylated.

2 = **Enzymatic action**
Terminal deoxytransferase (TdT) attaches itself to the 3′ end containing a free OH.

3 = **Addition of labeled dNTPs (dATP or dUTP) ☆ and unlabeled dATP and polymerization**
The TdT enzyme catalyzes the polymerization of the radioactive and antigenic deoxynucleotide triphosphates.

Figure 1.14   Labeling by 3′ extension.

1.3.4.3 Equipment/Reagents/Solutions

1. Equipment

- Centrifuge
- Vortex
- Water bath/incubator at 37°C

➦ Greater than or equal to 14,000 g

➦ Enzyme reaction

2. Reagents

➦ Molecular biology grade
➦ Only to be used for *in situ* hybridization

- Antigenic dUTP labels
  - Biotin X-dUTP
  - Digoxigenin X-dUTP
  - Fluorescein X-dUTP
- Cobalt chloride (CoCl₂)
- EDTA
- Oligonucleotides

➦ Labeling kit

➦ Choose 25 to 45 nucleotides (mers); a probe of 30 nucleotides is a good average.
➦ Storage at −20°C

- Potassium cacodylate
- Radioactive dNTP labels

➦ Essentially $\alpha(^{35}S$ or $^{33}P)$-dATP; other radioisotopes are not often used. The use of other dNTPs leads to non-specific signals.

- Terminal deoxytransferase (TdT)

➦ Enzyme, buffer, and CoCl₂ available in kit form

3. Solutions

➦ **All the solutions are prepared using DNase- and RNase-free reagents in a sterile container** (*see* Appendix A2).
➦ Storage at −20°C

- Antigenic deoxynucleotides
  - 1 mM biotin X-dUTP
  - 1 mM digoxigenin X-dUTP
  - 1 mM fluorescein X-dUTP
- CoCl₂ 25 mM

➦ Storage at −20°C (*see* Appendix B2.6.2)

- 500 mM EDTA; pH 8.0
- Labeling (tailing) buffer (5X)

- Oligonucleotides (2–100 pM/µL)
- Radioactive deoxynucleotides
  – α(^{35}S or ^{33}P) - dATP

- Terminal deoxytransferase (TdT)
  (50 U/µL)

- Water

↝ *See* Appendix B2.13.
↝ 1 M potassium cacodylate/125 mM Tris–HCl/1.25 mg/mL BSA; pH 6.6
↝ Storage at –20°C
↝ For labeling
↝ The radioisotopes (^{35}S, ^{33}P) are present in position α (*see* Figure 1.4) for labeling by 3′ extension.
↝ Stored at –80°C or 4°C in aliquots
↝ Specific activity ≈ 3000 Ci/mM
↝ 200 mM potassium cacodylate/1 mM EDTA/200 mM KCl/0.2 mg/mL BSA/50% glycerol ($^v/_v$); pH 6.5
↝ Storage at –20°C. The presence of glycerol allows the pipetting of the enzyme without warming the whole tube. The enzyme is labile.
↝ Treated with DEPC (*see* Appendix B1.2)

1.3.4.4 Protocol for Radioactive Probes

1. Reaction mixture
   Place the following reagents in a sterile tube in the order indicated:
   - Sterile water   **15.5 µL**
   - Oligonucleotides
     (10 pM or 100 pM)   **10 µL**
   - Reaction buffer (5X)   **10 µL**
   - 25 mM CoCl$_2$ **10 µL**

   - 10 pM labeled dATP   **3 µL**

   - TdT (50 U/µL)   **1.5 µL**

2. Incubation   **60 min
                                                              at 37°C**

3. Stopping the reaction
   Add:
   EDTA 10 mM **2 µL**

❑ *Next stage:*
   - Precipitation with ethanol

↝ Labeled oligonucleotides (*see* Figure 1.4): α(^{35}S or ^{33}P)-dATP

↝ Eppendorf type

↝ For a final volume of 50 µL

↝ According to requirements

↝ If a red precipitate forms, check the radioactive nucleotide.

↝ ≈ 50 µCi in the case of the α^{35}S - dATP probe
↝ Labile enzyme; must not be brought up to room temperature
↝ The labeling is carried out at 37°C (minimum 40 min). It is not necessary to prolong the incubation time, nor to increase the amount of enzyme, because the extension is generally very long (check the specific activity).
↝ Optional step, which can be replaced by a water bath at 4°C, or a heater (10 min at 65°C)

↝ *See* Section 1.4.1.

1.3.4.5 Protocol for Antigenic Probes

1. Reaction mixture
   Place the following reagents in a sterile tube, in the order indicated:

↝ Labeled oligonucleotides (*see* Figure 1.6): (Biotin, digoxigenin, fluorescein)-dUTP

↝ Eppendorf type

• Sterile water	**4 μL**	➾ For a final volume of 20 μL
• Oligonucleotides		➾ The oligonucleotides are diluted in sterile
(10 p*M* or 100 p*M*)	**5 μL**	water and stored at –20°C.
• Reaction buffer (5X)	**4 μL**	
• 25 m*M* CoCl$_2$	**4 μL**	
• 1 m*M* labeled dUTP	**1 μL**	➾ Only 1 to 2 labeled nucleotides are added at the 3′ end.
• 10 m*M* dATP	**1 μL**	➾ The addition of unlabeled dATP allows the incorporation of several labeled dUTPs.
• TdT (50 U/μL)	**1 μL**	➾ The enzyme terminal deoxynucleotide transferase catalyzes the polymerization of nucleotides from the 3′ end.
		➾ Labile reagent; must not be reheated

2. Incubation **60 min at 37°C**

➾ It is not necessary to prolong the incubation time or to increase the amount of enzyme, as the extension remains limited.

3. Stopping the reaction
   Add:
   10 m*M* EDTA **2 μL**

➾ Optional step, which can be replaced by a water bath at 4°C, or a heater (10 min at 65°C).

❑ *Next stage:*
   • Precipitation

➾ The oligonucleotides are purified by ethanol precipitation (*see* Section 1.4.1).

1.3.5   *In Vitro* Transcription

➾ Transcribes DNA into an RNA probe

1.3.5.1   Principles

RNA probes are obtained from an *in vitro* transcription system and are labeled during synthesis.

The transcription vector (plasmid) into which the DNA is inserted must possess the RNA promoters situated on both sides of the cloning site.

➾ The most frequently used promoters for RNA polymerase are SP6, T3, and T7.

For transcription, the plasmid is linearized by cutting between the insert and the promoter not in use.

Depending on which promoter is used, transcription will produce sense or anti-sense probes.

➾ This step presupposes that the orientation of the cDNA in the vector can be read from the restriction map (*see* Glossary).

Transcription begins at the transcription initiation site (the site of insertion of the promoter). The growing strand extends in the 5′ → 3′ direction. Each nucleotide put into place is complementary to the corresponding nucleotide in the insert.

➾ Labeling is achieved by the action of RNA polymerase, which incorporates successive labeled or unlabeled nucleotides during the course of RNA synthesis.
➾ The newly synthesized RNA is labeled.

Transcription can be performed according to the polymerase used, so that either sense mRNA (sense probe) or anti-sense probes (from the opposite strand) can be produced.

↪ The single-stranded RNA formed is complementary to the DNA strand that serves as its template.

• Transcription with polymerase 1

↪ Polymerase 1 makes labeled copies of the same sequence as the target nucleic acid. The synthesis of the RNA strand serves as a control for determining background labeling (no specific hybridization with the target RNA) [*see* Figure 1.15].

• Transcription with polymerase 2

↪ Polymerase 2 makes labeled copies complementary to the target nucleic acid (hybridization with the target RNA). This synthesis of the anti-sense RNA is the probe (*see* Figure 1.15).

When the enzyme arrives at the site of cutting, it uncouples from the template and retranscribes the insert.

↪ Depending on the stability of the enzymes and the strength of the promoter, several cycles of initiation on the same DNA template can considerably amplify the amount of RNA transcribed in relation to the DNA template.
↪ The RNA obtained is single-stranded.

❏ *Advantages*

• Produces an RNA probe
• The probes obtained are single-stranded.

↪ RNA–RNA hybrids are the most stable.
↪ There is no hybridization in the reaction mixture.

• The amount of labeled probe is greater than the amount of template.
• The specific activity is high.

↪ Synthesis of labeled transcripts (amplification 10 X with labeled molecules)
↪ Possibility of using several labeled nucleotides

• The probe is large.
• It is possible to obtain both sense and anti-sense probes.

↪ Copies the whole insert
↪ Easily controlled

❏ *Disadvantages*

• Insertion of the cDNA into a vector to obtain transcription
• Lability of the probes
• Restriction enzymes are used to linearize the plasmid.
• Probes are ligated to obtain smaller fragments.

↪ Construction of this is very important.

↪ Avoid all risk of contamination by RNases.
↪ Restriction enzymes have a limited lifespan and are very labile.
↪ The original probes are the same length as the DNA template.

1.3.5.2 Summary of Different Steps

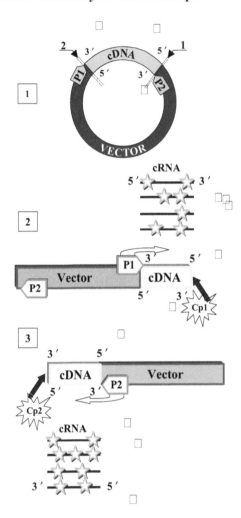

1 = Insertion into a plasmid
- The target cDNA is inserted into a plasmid equipped with two promoters, P1 and P2, recognized by two different RNA polymerases and situated on either side and on each strand of the insert.
- Two cutting sites are present between the insert and each promoter. They make possible the linearization of the plasmid by means of a cut between the insert and the promoter not in use.

(P1) promoter 1
(P2) promoter 2

2/3 = Synthesis of RNA probes by *in vitro* transcription in the presence of labeled nucleotides:

2 = Synthesis of sense RNA probes
The sense transcripts are obtained by the action of the polymerase attaching itself to the promoter P1 (sense probe) (i.e., a copy of the target sequence).

(Cp1) cut 1

3 = Creation of complementary RNA probes
The anti-sense transcripts are obtained by the action of the polymerase attaching itself to the promoter P2 (anti-sense probe) (i.e., complementary to the target sequence).

(Cp2) cut 2

Figure 1.15 *In vitro* transcription.

1.3.5.3 Equipment/Reagents/Solutions

1. Equipment

- Centrifuge
- Electrophoresis tank
- Power pack
- Vacuum jar or Speedvac
- Vortex mixer
- Water bath at 37°C

⮕ Greater than or equal to 14,000 *g*

⮕ Drying the probe

⮕ Enzyme reaction

2. Reagents

⮕ Molecular biology grade, RNase-free, only to be used for labeling

- Agarose
- Antigenic ribonucleotides
  - Biotin X-UTP
  - Digoxigenin X-UTP
  - Fluorescein X-UTP

⮕ *See* Figure 1.6.

- Chloroform
- Cobalt chloride
- Dithiothreitol (DTT)
- EDTA
- Isoamyl alcohol
- Phenol
- Plasmid containing the insert to be transcribed   ⇝ *See* Figure 1.3.
- Radioactive ribonucleotides   ⇝ Labeled in position α (*see* Figure 1.5)
- Transfer RNA
- Tris

3. Solutions   ⇝ **All the solutions are prepared using RNase-free reagents in a sterile container** (*see* Appendix A2).

- Antigenic ribonucleotides   ⇝ Storage at –20°C
- Buffers   ⇝ Labeling kit
  - Transcription buffer (10X)   ⇝ 400 mM Tris–HCl/60 mM MgCl$_2$/100 mM DTT/20 mM spermidine; pH 8.0 (all present in the labeling kit)
  - TBE buffer (5X)   ⇝ 450 mM Tris/450 mM boric acid/10 mM EDTA
  - TE buffer (10X)   ⇝ *See* Appendix B3.6 (100 mM Tris/10 mM EDTA)

- Chloroform/isoamyl alcohol
  - Chloroform (49 vol)
  - Isoamyl alcohol (1 vol)
- DNase I   ⇝ *See* Appendix B2.12.
- Dithiothreitol (DTT) 1 M ⇝ *See* Appendix B2.11.
- 500 mM EDTA; pH 8.0   ⇝ *See* Appendix B2.13.
- 70%, 100% ethanol
- Radioactive ribonucleotides   ⇝ Storage at –80°C
- 10 mM ribonucleotides (rNTP mix)
- ATP, GTP, CTP   ⇝ Dilute each nucleotide in water. The rNTP mix is neutralized at pH 7.0.
     ⇝ Storage at –20°C
- RNA polymerases   ⇝ To be determined depending to the promoter used, which is generally T3, T7, or SP6
- 10% SDS
- Sterile water   ⇝ DEPC-treated (*see* Appendix B1.2)
- Transcription kit   ⇝ Labeling kit. Storage at –20°C
- Transfer RNA (tRNA) (10 mg/mL)   ⇝ *See* Appendix B2.5.

1.3.5.4 Protocol for Radioactive Probes

⇝ **Wearing gloves is essential when handling riboprobes** (to avoid degradation of the RNA by exogenous RNases).

1. Linearization of the plasmid
   - Linearized plasmid   **≥ 100 μL**   ⇝ The plasmid must first be precipitated, then
        ≈ 20–50 μg   purified on a gel, and finally linearized. It is taken up in a minimum volume.

Denature:	**5–10 min at 92°C**	

2. Purification
   Add:
   • Phenol/chloroform | **vol/vol** | ⇝ Vortex for 1 min (formation of an emulsion).
   Centrifuge: | **≥ 14,000 *g* for 5 min** | ⇝ Recover the supernatant (the DNA is in the upper aqueous phase).

3. Precipitation | | ⇝ *See* Section 1.4.
4. *In vitro* transcription | | ⇝ Riboprobe labeled with ^{35}S
   Place the following reagents in a sterile Eppendorf tube, on ice, in the indicated order:

   • Linearized plasmid (1 μg/μL) | **1 μL** | ⇝ Can be replaced by PCR fragments containing a promoter

   • Labeled nucleotide | **2 μL** | ⇝ Final concentration (α^{35}S - UTP/1 μ*M*)
   • Sterile H$_2$O | **to 20 μL** | ⇝ *See* Appendix B1.1.
   • Transcription buffer (10X) | **2 μL** |
   • DTT (100 m*M*) | **2 μL** | ⇝ Final concentration 10 m*M*
   • RNasin (20 U/μL) | **1 μL** | ⇝ Final concentration 1 U/μL
   | | ⇝ Inhibitor of RNase, prevents the degradation of synthesized RNA. This reagent does not withstand high temperatures. Add at the last moment.

   • NTP mix | **2 μL** | ⇝ Final concentration 0.4 m*M*
     – 2.5 m*M* ATP | | ⇝ First prepare a mixture of all the constituents, apart from DNA and RNA polymerase, and neutralize at pH 7.0. Add labeled UTP.
     – 2.5 m*M* GTP | |
     – 2.5 m*M* CTP | |
   • RNA polymerase (20 U/μL) | **2 μL** | ⇝ The polymerases (SP6, T3, or T7) allow the synthesis of the anti-sense and sense riboprobes. The latter serves as a control for the determination of background labeling.

   Incubate: | **60 min at 37°C** | ⇝ To avoid RNA degradation, do not extend the incubation time.

5. Digestion of the DNA
   Add:
   • Molecular biology grade DNase I | **2 μL** | ⇝ Concentration:1 U/μg
   Incubate: | **15 min at 37°C** | ⇝ To specifically eliminate the strands of DNA (elimination of the template)

   Add:
   • 100 m*M* EDTA | **2 μL** | ⇝ The enzyme is inhibited by EDTA (final concentration: 4 m*M*).

6. Extraction of proteins | | ⇝ Extraction with phenol and precipitation in alcohol

   • Add:
     – tRNA (10 mg/mL) | **1 μL** | ⇝ Final concentration: 400 μg/mL
   • Denature: | **5 min at 95°C** | ⇝ Or 2 min at 100°C

– 10% SDS	**4 µL**	➸ *See* Appendix B3.6.
– TE buffer	**to 100 µL**	
• Add:		
– Phenol (1 vol)	**100 µL**	➸ Vortex 1 min.
• Centrifuge:	**≥14,000 g**	➸ Extract the upper aqueous phase
	5 min	
– Chloroform/isoamyl alcohol	**100 µL**	➸ Chloroform/isoamyl alcohol (49:1 $^v/_v$)
• Centrifuge:	**≥ 4000 g**	
	10 min	

❑ *Next stage:*
• Purification, precipitation   ➸ *See* Section 1.4.

1.3.5.5   Protocol for Antigenic Probes

The reaction conditions are identical to those described for radioactive riboprobes.

➸ **Gloves are essential when handling ribo-probes** (to avoid degradation of the RNA by exogenous RNases).

Steps 1, 2, and 3 are identical to those for labeling using radioactive isotopes.

➸ *See* Section 1.3.4.4.

Step 4: *In vitro* transcription

Place the following reagents in a sterile Eppendorf tube, on ice, in the indicated order:

• Linearized plasmid (1 µg/µL)	1 µL	➸ Can be replaced by PCR fragments containing a promoter
• Labeled nucleotide	2 µL	➸ 3.5 m*M* (biotin, digoxigenin, fluorescein)-UTP
• Sterile H$_2$O	to 20 µL	➸ *See* Appendix B1.1.
• Transcription buffer (10X)	2 µL	
• RNasin(20 U/µL)	1 µL	➸ Final concentration: 1 U/µL ➸ Inhibitor of RNases to prevent RNA degradation. This reagent does not withstand high temperatures. Add at the last moment.
• NTP mix - 10 m*M* ATP - 10 m*M* GTP - 10 m*M* CTP - 6.5 m*M* UTP	2 µL	➸ Final concentration: 0.4 m*M* ➸ First prepare a mixture of all the constituents, apart from the DNA and RNA polymerase, and neutralize at pH 7.0. Add labeled UTP.
• RNA polymerase (20 U/mL)	2 µL	➸ The polymerases (SP6, T3, or T7) allow the synthesis of the anti-sense and sense riboprobes. The latter serves as a control for the determination of background labeling. ➸ All the constituents must be at the bottom of the tube.

Mix and centrifuge:

Incubate: **2 h**
                                              at 37°C

➸ To avoid RNA degradation, do not extend the incubation time.
➸ *See* Section 1.3.4.4.

Steps 5 and 6 are identical to those for transcription using a radiolabeled nucleotide.

❑ *Next stage:*
- Purification, precipitation

➭ *See* Section 1.4.

1.4   PRECIPITATION TECHNIQUES

Purification of DNA can be achieved:

- By ethanol precipitation

➭ The labeled probe is precipitated in ethanol to remove the free nucleotides and to concentrate the probe.

- On a Sephadex G-50 column

➭ Only the nucleotides are removed. If the volume is very large, it is necessary to precipitate the probe.

1.4.1   Ethanol Precipitation

1.4.1.1   Principles
Nucleic acids are water-soluble molecules which precipitate in an alcoholic solution in the presence of salts.

1.4.1.2   Equipment/Reagents/Solutions
EQUIPMENT

- Centrifuge
- Freezer (–20°C or –80°C)

- Vacuum jar or Speedvac
- Vortex

➭ Greater than or equal to 14,000 g
➭ Storage of reagents and precipitation products
➭ Drying the probe

REAGENTS

➭ Molecular biology grade, to be used only for precipitation.

- Ammonium acetate
- Sodium acetate or lithium chloride
- Transfer RNA

SOLUTIONS

➭ All the solutions are prepared using RNase- and DNase-free reagents in a sterile container (*see* Appendix A2).

- 7.5 M ammonium acetate; pH 5.5
- Ethanol (70% and 100%)
- 4 M lithium chloride
- 3 M sodium acetate; pH 5.2
- Transfer RNA (tRNA) (10 mg/mL)

➭ *See* Appendix B2.1.
➭ Stored at –20°C
➭ Stored at –20°C (*see* Appendix B2.7)
➭ *See* Appendix B2.2.
➭ *See* Appendix B2.5.

1.4.1.3   Protocol
1. Reaction mixture
   On ice, add the reagents in the following order:

• tRNA (10 mg/mL)	**2 µL**	↪ **Optional.** Facilitates the precipitation of oligonucleotide probes.
• 7.5 *M* ammonium acetate,	**1/5 of the final volume**	↪ Ammonium acetate can be replaced by 3 *M* sodium acetate or 4 *M* lithium chloride.
• 3 *M* sodium acetate, or • 4 *M* lithium chloride		In the latter case, it is necessary to wash the pellet with 70% ethanol (at –20°C) after precipitation and recentrifugation.
• Ethanol (100%)	**≈ 2–3 vol**	↪ Ethanol stored at –20°C
		↪ Volume of the reaction solution
Vortex, centrifuge:		↪ No reagent may remain on the sides of the tube.

2. Incubation   ↪ Precipitation of the probe

- Incubate: **60 min at –80°C**   ↪ Minimum 30 min at –80°C

     or overnight at –20°C   ↪ Minimum 2 h at –20°C

- Centrifuge: **>14,000 *g* ≥15 min at 4°C**   ↪ Orient the tube so as to find out the position of the pellet and remove the supernatant more easily.
↪ The speed of the centrifuge is essential in obtaining a pellet. In the case of labeling using a radioactive nucleotide, the supernatant contains all the free radioactive dNTP and must be removed.

3. Washing

- Remove the supernatant
- Wash the precipitate
   – 70% ethanol   **50 µL**   ↪ Elimination of salts
- Centrifuge   **>14,000 *g* ≥15 min at 4°C**   ↪ Orient the tube so as to facilitate the observation of the position of the pellet and the removal of the supernatant.
- Remove the supernatant
- Dry

1.4.2   Purification

1.4.2.1   Principles

The nucleic acids are separated from the free nucleotides and labeling reagents by passage through a column.

1.4.2.2   Equipment/Solutions

EQUIPMENT

- Centrifuge   ↪ Greater than or equal to 14,000 *g*
- Sephadex G-50 column
- Vacuum jar or Speedvac   ↪ Drying the probe
- Vortex

SOLUTIONS

➭ Solutions are prepared using RNase- and DNase-free reagents in a sterile container (*see* Appendix A2).

• TE buffer

➭ *See* Appendix B3.6.

1.4.2.3   Protocol
1. Wash the column with the buffer.
2. Drop on the labeled solution.
3. Separate the constituents.

➭ By pressure ("Push column") or centrifugation

4. Wash the column.
5. Elute the probe.
6. Precipitate the probe if necessary.

➭ *See* Section 1.4.1.

1.4.3   Drying

The pellet is dried by:

➭ Eliminates all trace of volatile alcohol

➭ Important step

• Vacuum jar, or	**30 min**
• Speedvac	**5–10 min**

➭ The probes can be stored dry or solubilized in a buffer (e.g., TE, *see* Appendix B3.6), or in sterile water, and stored at –20°C.

1.5   CONTROLS

➭ Additional (but very important) step

1.5.1   Radioactive Probes

1.5.1.1   Counting the Incorporated Radioactivity
Add:

• TE (10X) or DEPC-treated water	**80 μL**
• DTT (100 m*M*)	**20 μL**

➭ *See* Appendices B3.6 and B1.2.
➭ Final concentration: 20 m*M*
➭ Radioactivity incorporated into the probe/ radioactivity due to the nucleotide introduced into the labeling tube
➭ If the percentage of incorporation is ≤ 50%, the enzyme activity is too weak, or the nucleotide is degraded.

Check the incorporation of radioactive nucleotides by counting in scintillation fluid:

• Original solution	**1 μL**

1.5.1.2   Calculation of Specific Activity
Radioactivity incorporated/mass of the probe or molarity of the probe

➭ The specific activity of a probe is the best indication of labeling reproducibility and allows a comparison of activities of different probe types: cDNA, cRNA, or oligonucleotide.

- Labeled probe   **1 µL**
- Scintillation fluid   **>1 mL**

↝ Too high a specific activity is not always a guarantee of a good *in situ* hybridization signal. In the case of an oligonucleotide, a high specific activity corresponds to a 3′ extension much longer than the oligonucleotide itself.

In the case of cDNA or cRNA, it may lead to rapid radiolysis and fragmentation of the probes.

1.5.1.3 Verification of Labeling on an Analytical Gel of Alkaline Agarose 0.8% in Sterile TBE

↝ Verification of the labeling and of the size of the fragments obtained

EQUIPMENT

- Autoradiography cassette
- Cellophane paper
- Electrophoresis power pack
- Electrophoresis tank
- Gel dryer

SOLUTIONS

- Alkaline agarose buffer
- TBE buffer (5X)

↝ 50 mM NaOH/1 mM EDTA
↝ 450 mM Tris/450 mM boric acid/10 mM EDTA

PROTOCOL

1. Deposit:
   - Sample   **0.5 µL/well**
2. Allow to migrate (5–75 mA).   **30–45 min, 90 V**
3. Dry the gel.
4. Expose the gel covered in cellophane to autoradiographic film.   **6–8 h**

↝ About 500,000 cpm/well
↝ The sample and electrophoresis buffers must be freshly prepared and sterilized.
↝ Gel dryer on Whatman paper
↝ Use Kodak XAR film
↝ Visualization of the bands corresponding to the radioactive probes after exposure of the film

1.5.2   Antigenic Probes

1.5.2.1   Controls

Labeling is achieved by creating a range of dilutions on a nylon membrane. After drying, the labeling of the probe is visualized (*see* Figure 1.16).

↝ A labeling standard is provided in certain kits.
- Dilutions by a factor of 10: (1, 1:10, 1:100, 1:1000)

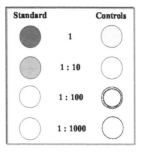

Figure 1.16 Controls for antigenic probes.

1.5.2.2 Biotin and Digoxigenin Detection

If the label is biotin or digoxigenin, this is revealed by immunoenzymatic detection of the antigenic molecule.

↪ The anti-digoxigenin or anti-biotin antibodies used can be conjugated to alkaline phosphatase. In this case, the chromogen used is NBT/BCIP (nitro blue tetrazolium/5-bromo-4-chloro-3-indolyl phosphate) (*see* Chapter 5, Section 5.2.2).

1.5.2.3 Visualization of Fluorescein

If the label is fluorescein or any other fluorescent label, it is possible to place an aliquot on a nylon membrane or filter paper and observe the fluorescence under a microscope. A signal is evidence of probe labeling.

↪ It is sometimes necessary to wash the membrane in order to remove unincorporated nucleotides if precipitation has not been carried out.

↪ This control is quick and can be carried out before the precipitation of the probe.

1.5.2.4 Disadvantages

- Uses a relatively large amount of probe (up to 10 ng)
- Does not allow the determination of the specific activity of the probe

1.6 STORAGE/USE

1.6.1 Storage

↪ Using a probe at the time of preparation removes all the risks inherent in storage.

Storage depends upon the nature of the probe and label:

PROBES

- Antigenic probe

- Radioactive probe

↪ DNA probes (cDNA and oligonucleotides) keep for much longer than cRNA probes.

↪ Radioactive labels lead to radiolysis of nucleic acids.

LABELS

- Radioactive label

↪ Storage is dependent on half-life:

Isotope	Half-Life	Storage
^{33}P	25.4 days	1 week
^{35}S	87.4 days	up to 1 month

- Antigenic label

↪ Antigenic labels are very stable. When dissolved in TE buffer at −20°C, they are generally stable for several months.

Storage methods:
• Dry

⇝ Just before use reconstitute the desired amount of probe in the hybridization buffer.

• In sterile water or buffer

⇝ Storage possible at –20°C for several months

• In hybridization buffer

⇝ Storage possible at –20°C for several months

1.6.2   Use

In the case of double-stranded probes, principally DNA, it is necessary to denature the probe:

⇝ Long RNAs must also be denatured (*see* Chapter 3). This step is carried out in a hybridization buffer.

1. Denaturing	**10 min**
	at 92°C
	or
	5 min
	at 96°C

⇝ If a probe is labeled with ^{35}S, DTT (10 mM) must be added after denaturation, since the probe will be heat sensitive (*see* Figure 1.17).

2. Place on ice.

⇝ Prevents the two complementary strands from re-hybridizing.

3. Place the probe on the sections immediately.

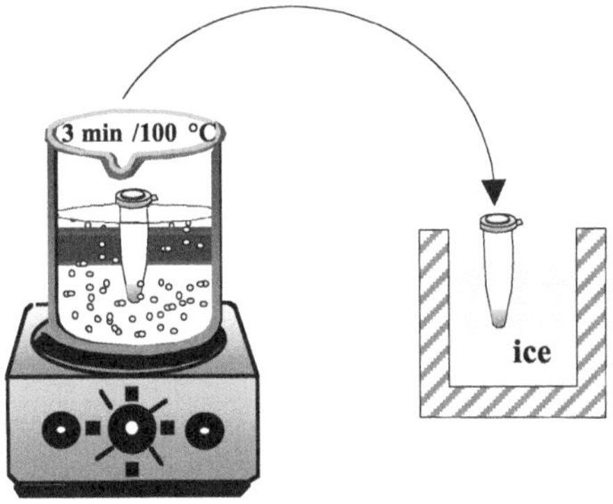

Figure 1.17 Denaturation of the probe.

Next stage:
• Hybridization

  – Post-embedding method
  – Pre-embedding method
  – Frozen tissue method

⇝ *See* Chapter 4, Section 4.6.
⇝ *See* Chapter 5, Section 5.5.
⇝ *See* Chapter 6, Section 6.9.

Chapter 2

Principles of Methodology

CONTENTS

CHOICE OF TECHNIQUE

Each technique has potential applications. The principal criteria to take into consideration when choosing which technique to use are:

- Sensitivity
- Resolution
- Morphology
- Simplicity of the technique
- Multiple labeling

The techniques currently in use are:

- Post-embedding technique
- Pre-embedding technique
- Frozen tissue technique

Each has advantages and disadvantages. None of the techniques meets all of the criteria.

The aim in each case is the visualization of the nucleic acid sequence in question at the level of the electron microscope.

Whatever technique is used, the first stage is fixation (*see* Chapter 3), which is common to all three techniques.

➭ It also depends on the materials available.

➭ *See* Chapter 4
➭ *See* Chapter 5
➭ *See* Chapter 6

➭ For electron microscopy, the tissue sections must not be thicker than 100 nm.

➭ Indispensable
➭ Fixation immobilizes the cellular constituents (lipids, proteins, nucleic acids).

$a < 100$ nm

Figure 2.1 Diagram of a cell section for transmission electron microscopy.

2.1 POST-EMBEDDING TECHNIQUE

2.1.1 Principles

In situ hybridization is carried out on tissue sections embedded in acrylic resin.

These resins were developed for cytology and are made up of a mixture of acrylate — methacrylate of low viscosity, allowing rapid tissue infiltration.

The «hydrophilic» nature of these resins allows some water to be left in the tissue, conserving the molecular structure, which aids in preserving antigenicity.

➭ Post-embedding method

➭ The most recently developed resins give improved detection of nucleic acids.

Nucleic acids are hydrophilic molecules but can be partially preserved *in situ* when embedded in resin.

⤳ Nucleic acids present within the resin are inaccessible after resin polymerization.

After tissue fixation, several stages of tissue preparation must be carried out prior to ultrathin sectioning:

- Dehydration
- Impregnation with resin
- Embedding in resin

- Polymerization

⤳ Usually in alcohol
⤳ At a low temperature, if possible
⤳ To allow a biological specimen to become hard enough to allow ultrathin sectioning
⤳ At a low temperature

Hybridization is carried out on sections following pretreatment. Visualization of hybrids labeled with antigenic groups is carried out by immunocytochemistry followed by staining prior to observation.

DEHYDRATION

After fixation, tissue is progressively dehydrated in increasing concentrations of solvents. The degree of dehydration varies depending on the resin used and the protocol.

⤳ Fixation generally uses a light fixative (*see* Chapter 3, Section 3.1.1.2.1). Biological samples are fragile and sensitive to dehydration and embedding. During fixation, dehydration, embedding, and polymerization, many cellular constituents are lost or denatured. This is considerably reduced at low temperature.

This stage may be carried out at room temperature or at a low temperature depending on the type of resin used.

⤳ Water contained in the tissue is replaced by organic solvent (cryosubstitution).

IMPREGNATION

Impregnation is carried out in a mixture of ethanol–resin with increasing concentrations of acrylic resin followed by pure resin. This allows the elimination of all solvents before embedding. The duration of this stage depends on the size of the sample and may be prolonged without damaging the ultrastructure.

⤳ By a mixture of solvent and embedding medium
⤳ All of these stages are carried out at a low temperature.
⤳ The different viscosity of acrylic resins means that the infiltration time varies and also depends on the sample size.
⤳ Solvent inhibits polymerization.

EMBEDDING

Embedding involves the replacement of all the solvent by resin, positioning of the sample at the bottom of the mold, and filling with resin.

⤳ The absence of air is indispensable for complete polymerization of acrylic resin.

POLYMERIZATION

Depending on the resin used, this is carried out under UV light at a low temperature ($-20°C$ to $-35°C$) for several days or by heat.

⤳ Hardening of resin
⤳ *See* Chapter 4, Section 4.3.

ULTRAMICROTOMY
After the resin is polymerized, the tissue is sectioned using an ultramicrotome.

⇝ Cutting sections of embedded tissue

In situ hybridization may be carried out on sections of variable thickness:

⇝ *See* Chapter 4, Section 4.4.3.

- Sections between 0.5 to 1 µm in thickness are called semithin.

⇝ These are generally sections of large areas of tissue for light microscopy.
⇝ Not often used for this technique

- Sections between 80 to 100 nm in thickness are called ultrathin.

⇝ For subcellular observations
⇝ The thickness of sections is limited due to the weak penetration of electrons.

PRETREATMENTS
This is limited by the lack of penetration of reagents into the section.

⇝ Indispensable — this permits the sites at the surface of the section to become accessible.

HYBRIDIZATION
The probe is usually labeled with an antigenic molecule. It is possible to use a radioactive isotope.

⇝ For example, biotin, digoxigenin, or fluorescein (*See* Chapter 1, Section 1.2.2).
⇝ For example, ^{35}S or ^{33}P

This is a key step carried out directly on ultrathin sections floating on a drop of reaction medium.

⇝ With no penetration of the probe into the section
⇝ It is best if the conditions for hybridization are optimized to remove nonspecific hybrids at a later time.

The post-hybridization washes must be carefully controlled.

⇝ Should be stringent, but it varies for each hybridization.

VISUALIZATION
Depends on the type of labeling used:

- Autoradiography for radioactive labels
- Immunocytochemistry for antigenic labels

⇝ *See* Chapter 1, Section 1.2.1.
⇝ *See* Chapter 1, Section 1.2.2.

Since the hybrids are only present on the surface of the section, the preservation of hydrogen bonds is important. To detect it, classical methods for visualizing an extremely labile antigen are used.

⇝ The salt concentration at this stage is very important in maintaining the stability of the hybrids.
⇝ Stabilization of the complex (i.e., fixation) is often necessary.

STAINING
Staining is necessary to visualize cellular structures.

⇝ Counting colloidal gold particles or silver grains (autoradiography)

OBSERVATION
Quantification of the signal is possible.

2.1.2 Summary of Different Stages

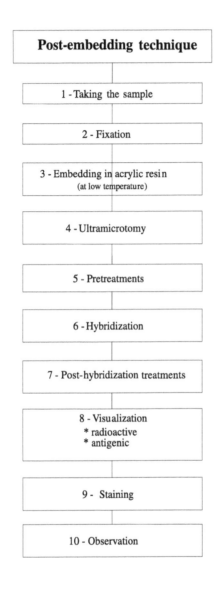

⇀ This method consists of embedding biological material in acrylic resin (e.g., Lowicryl, LR White, Unicryl, etc.) (*see* Chapter 4).

1: *See* Chapter 3, Section 3.2.1.

2: *See* Chapter 3, Section 3.2.4.

3: *See* Chapter 4, Section 4.3.

4: *See* Chapter 4, Section 4.4.
Ultrathin sections (80 to 100 nm)
This technique can be used on semithin sections of tissue embedded in epoxy resin.

5: *See* Chapter 4, Section 4. 5.

6: *See* Chapter 4, Section 4.6.

7: *See* Chapter 4, Section 4.7.

8: *See* Chapters 4 and 5.
Radioactive hybrids (*see* Chapter 5, Section 5.8.2)
Antigenic hybrids (*see* Chapter 4, Section 4.8)

9: *See* Chapter 4, Section 4.9.

10: *See* Examples of Results.

Figure 2.2 Principles of post-embedding technique.

2.1.3 Advantages/Disadvantages

2.1.3.1 Advantages

• Simple technique

⇀ Similar to immunocytochemistry.

• Good preservation of ultrastructure	↝ Generally better than that obtained with the frozen tissue technique
	↝ This approach is a good compromise between preservation of nucleic acids and morphology.
• High resolution	↝ Achieved by using colloidal gold for detection
• Multiple labeling	↝ Colloidal gold of different diameters can be used for different probes (5–15 nm).
• Preservation of antigens	↝ By embedding at a low temperature, cellular constituents may be conserved throughout dehydration and impregnation.
• Amplification possible	↝ During the visualization process
• Storage is easy:	↝ It is better to store the blocks rather than the sections.
– Embedded tissue blocks can be kept for an unlimited period of time.	
• Quantification is easy.	↝ Using antibodies conjugated to colloidal gold

2.1.3.2   Disadvantages

• Very toxic products used	↝ Particularly Lowicryl resins. Safety precautions are necessary during all stages of preparation. Inhalation of the toxic vapors must be avoided (use a fume hood and wear a mask and gloves).
	↝ LR White resins are less toxic.
• Detection is limited to the surface of the section.	↝ Sensitivity is reduced since only the surface of the section is accessible to the probe (*see* Chapter 4, Figure 4.1)
• Quantity of the probe used is important.	↝ The concentration of the labeled probe is ≈ 20 times higher than for the same detection in frozen tissue.
• Low sensitivity	↝ This is a disadvantage if the number of target molecules is low.
• Not compatible with all tissues	↝ Certain tissues are very difficult to impregnate.
• Variations in the hardness of the blocks	↝ Problems in cutting ultrathin sections
• Weak adherence between the resin and the biological material	↝ Problems in cutting sections
• Instability of the resin under the electron beam	↝ It is necessary to coat the grid with a film prior to mounting ultrathin sections (*see* Appendix A4).
• Multiple labeling techniques are tricky.	

2.2   PRE-EMBEDDING TECHNIQUE

2.2.1   Principles

In situ hybridization is carried out prior to embedding on sections 50 to 200 μm thick, which are then embedded in resin and ultrathin sections are cut.

⇝ Pre-embedding method

After fixation, delaying the procedure and storing the tissue at any stage is limited:

⇝ *See* Chapter 3, Section 3.2.4.

- Cutting thick sections
- Pretreatment
- Hybridization
- Post-hybridization treatments
- Visualization
- Embedding
- Staining
- Observation

If the probes used are radioactively labeled, the thick sections are embedded in epoxy resin. Autoradiography is carried out on ultrathin sections.

If the probes used are labeled with an antigen, the thick sections are embedded in:

- Epoxy resin, if visualization takes place prior to embedding, or
- Hydrophilic resin, if visualization takes place on the ultrathin sections.

Tissues are fixed before the following stages are carried out:

1. Thick sections

   Thick sections are between 50 and 100 μm; the penetration of the probes is not a problem, but the penetration of the antibodies is limited.

⇝ Sections are cut on a vibratome.

2. Pretreatment:

There are two aims:

- To increase the access of the probe to the target nucleic acid

  ↪ Deproteinization and to reduce nonspecific labeling
  ↪ It is necessary to make a compromise between the sensitivity of the technique and the morphology of the cell.

- To make the tissue permeable to the reagents used for visualization

  ↪ Permeabilization

3. Hybridization:

This is carried out on a thick section prior to embedding. The conditions used depend on the pretreatments carried out.

↪ The efficiency of the stage is dependent on carefully controlled parameters.

4. Post-hybridization treatments:

This mainly consists of washes, conserving all specific hybrids while removing any probe that has nonspecifically bound to the tissue.

↪ This depends on the type of tissue and the type of probe used.

5. Visualization:

↪ Depends on the type of labeling used

- If this precedes embedding, antigenic labels are visualized by an enzyme.

  ↪ Usually peroxidase

- If this occurs after embedding:
  - Antigenic labels are visualized by colloidal gold
  - Radioactive labels are visualized by autoradiography

    ↪ For example, ^{33}P or ^{35}S

6. Embedding:

- If the visualization process is complete or the label is radioactive, embedding is in an epoxy resin

  ↪ The substrate of the peroxidase reaction is made opaque to electrons.
  ↪ β^- rays only penetrate a short distance into resin (< 100 nm).

- If the visualization process will be carried out on ultrathin sections, embedding is in a hydrophilic resin.

7. Staining:

The intensity of the staining depends on the protocol used.

↪ Ultrathin sections are embedded in epoxy or hydrophilic resin.

8. Observation

2.2.2 Summary of Different Stages

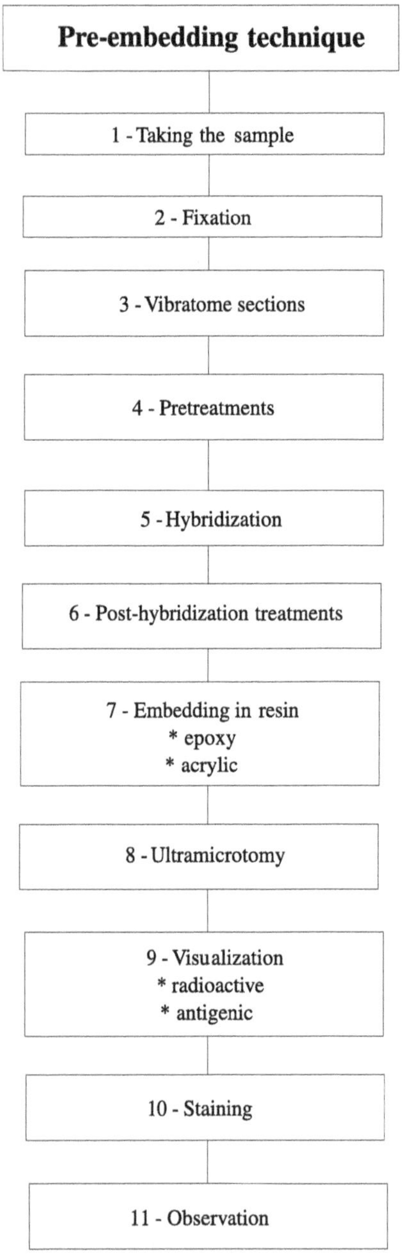
Pre-embedding technique
1 - Taking the sample
2 - Fixation
3 - Vibratome sections
4 - Pretreatments
5 - Hybridization
6 - Post-hybridization treatments
7 - Embedding in resin
* epoxy
* acrylic
8 - Ultramicrotomy
9 - Visualization
* radioactive
* antigenic
10 - Staining
11 - Observation

➪ On thick vibratome sections

➪ *See* Chapter 5.

1: *See* Chapter 3, Section 3.2.1.

2: *See* Chapter 3, Section 3.2.4.

3: *See* Chapter 5, Section 5.3.
Thick sections (50–200 nm)

4: *See* Chapter 5, Section 5.4.

5: *See* Chapter 5, Section 5.5.

6: *See* Chapter 5, Section 5.6.

7: *See* Chapters 4 and 5.
- In epoxy resin (*see* Chapter 5, Section 5.7.5.1). Visualization of antigenic hybrids is possible prior to embedding in epoxy resin (*see* Chapter 5, Section 5.8.3.1).
- In acrylic resin (*see* Chapter 4, Section 4.3)

8: *See* Chapter 4, Section 4.4.
- Ultrathin sections (80–100 nm)
- Semithin sections (1 µm)

9: *See* Chapter 5, Section 5.8.
- Radioactive hybrids (*see* Chapter 5, Section 5.8.2).
- Antigenic hybrids (*see* Chapter 5, Section 5.8.3).

10: *See* Chapter 5, Section 5.9.

11: *See* Examples of Results.

Figure 2.3 Principles of the pre-embedding technique.

2.2.3 Advantages/Disadvantages

2.2.3.1 Advantages

- Good ultrastructure
  - ↪ Better than that obtained with frozen tissue or tissue embedded in acrylic resin

- Good sensitivity
  - ↪ The lack of embedding results in good preservation of nucleic acids.

- Multiple labeling possible
  - ↪ *In situ* hybridization and immunocyto-chemistry.
  - ↪ Double labeling involving hybridization/hybridization is very tricky.

2.2.3.2 Disadvantages

- Difficult to store unembedded material
  - ↪ Tissue can be stored immediately after it is taken.

- Difficult to quantify
  - ↪ Use of antigenic labels limits the possibility of quantification.

- This technique gives the lowest resolution.
  - ↪ 1-nm-diameter colloidal gold is the only size that penetrates the tissue sufficiently, but it requires silver latensification to be easily visible.

- Antigenic labeled probes require tissue permeabilization.
  - ↪ The morphology is damaged by the use of detergent in the incubation medium.
- Visualization by autoradiography
  - ↪ Time consuming and tricky technique
- Loss of ultrastructural details
  - ↪ Particularly membranes, due to the use of detergents.

2.3   FROZEN TISSUE TECHNIQUE

2.3.1   Principles

In situ hybridization is carried out on unembedded tissue, usually frozen tissue or cell fractions directly mounted on an electron microscope grid.

- ↪ This technique is used on ultrathin sections of frozen tissue.

After fixation, the preparation of ultrathin cryosections is as follows:

- Cryoprotection
  - ↪ Hyperosmolar and/or hypo-osmolar
- Freezing
  - ↪ In a gradient of liquid nitrogen vapors
- Cryo-ultramicrotomy
  - ↪ Cutting cryosections

Then *in situ* hybridization is performed:

- Pretreatments
- Hybridization
- Post-hybridization treatments

Visualization is carried out depending on the type of label used (radioactive or antigenic). The final stages before observation are coating and staining.

1. Cryoprotection:
   The aim is to limit the consequences of freezing on the cell morphology.
   The water in the cell is replaced by an aqueous mixture to prevent the formation of ice crystals during freezing.

   ⮑ Indispensable

   ⮑ Cryoprotectors must not effect nucleic acids or cell or tissue structures.

2. Freezing:
   The aim is to solidify the tissue to allow ultrathin sections to be cut.

   ⮑ Adequate preservation of the ultrastructure depends on this complex step.

3. Cryo-ultramicrotomy:
   The thickness of ultrathin sections of frozen tissue is around 100 nm.

   ⮑ Sections are cut with a cryo-ultramicrotome

4. Pretreatments:
   The thickness of the section is not limiting, so this stage can be cut to a minimum. The proteins associated with the nucleic acids may be limiting.

   ⮑ Light deproteinization is sometimes necessary.

5. Hybridization:
   Carried out directly on ultrathin frozen sections by contact with the reaction medium. Post-hybridization treatment is short.

   ⮑ Accessibility is not a limiting factor.

6. Post-hybridization treatments:
   Limited due to good accessibility

7. Visualization:

   - Immunocytochemistry

     ⮑ Antigenic labels (usually colloidal gold)
     ⮑ The best technique

   - Autoradiography

     ⮑ Radioactive labels (usually ^{35}S)
     ⮑ Not often used

8. Coating and staining:
   After simple staining, ultrathin frozen sections are usually coated to prevent desiccation.

   ⮑ Methylcellulose is mostly used (*see* Chapter 6, Section 6.11).
   ⮑ Artifacts will result from the loss of water during drying.

9. Observation:
   After staining and coating quantification can be carried out.

   ⮑ The conservation of water and soluble molecules in the cell gives a distinctive appearance to frozen sections.

2.3.2 Summary of Different Stages

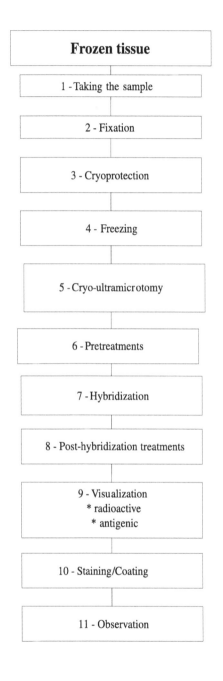

Frozen tissue

1 - Taking the sample

2 - Fixation

3 - Cryoprotection

4 - Freezing

5 - Cryo-ultramicrotomy

6 - Pretreatments

7 - Hybridization

8 - Post-hybridization treatments

9 - Visualization
* radioactive
* antigenic

10 - Staining/Coating

11 - Observation

☞ *See* Chapter 6.

1: *See* Chapter 3, Section 3.2.1.

2: *See* Chapter 3, Section 3.2.4.

3: *See* Chapter 6, Section 6.4.

4: *See* Chapter 6, Section 6.6.

5: *See* Chapter 6, Section 6.7.
- Ultrathin cryosections (80–100 nm)
- Semithin cryosections (0.5–1 μm)

6: *See* Chapter 6, Section 6.8.

7: *See* Chapter 6, Section 6.9.

8: *See* Chapter 6, Section 6.9.5.

9: *See* Chapter 6, Section 6.10.
- Radioactive hybrids (*see* Chapter 6, Section 6.10.1)
- Antigenic hybrids (*see* Chapter 6, Section 6.10.2)

10: *See* Chapter 6, Section 6.11.

11: *See* Examples of Results.

Figure 2.4 Principles of the frozen tissue technique.

2.3.3 Advantages/Disadvantages

2.3.3.1 Advantages

• Highly sensitive	➪ The absence of an embedding medium allows the penetration of reagents throughout the thickness of the section, increasing the availability of targets and the possibility of detection.
	➪ Maximum preservation of nucleic acids due to the presence of water in the tissue, which ensures that the structure of the molecules is maintained
• High resolution	➪ Colloidal gold can be used.
• Rapid	➪ Freezing avoids the need for embedding. ➪ Pretreatment is unnecessary.
• Multiple labeling	➪ Preservation of antigens is comparable to that of nucleic acids. ➪ The use of colloidal gold labeling makes detection of different substances easy.
• Storage	➪ Ultrathin sections at 4°C (*see* Chapter 6, Section 6.7.2.3)
	➪ As a tissue block in liquid nitrogen (*see* Chapter 6, Secton 6.6.3.1.3)
• Multiple labeling is relatively easy	➪ Nucleic acids and antigens are equally well preserved.

2.3.3.2 Disadvantages

• Need specific equipment	➪ Freezing system ➪ Cryo-ultramicrotome
• Preservation of ultrastructure	➪ Inferior to embedded tissue
• Sample storage	➪ Liquid nitrogen is necessary for storing frozen tissue and has to be replenished regularly.

2.4   SUMMARY

Table 2.1   Summary Table of the Different Methods Used for Ultrastructural *In Situ* Hybridization

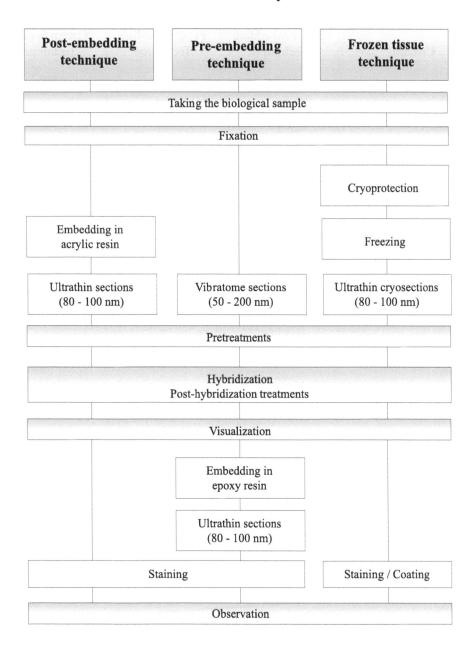

2.5   SELECTION CRITERIA

Table 2.2   Criteria Used for Choosing One of the Three Techniques for Ultrastructural *In Situ* Hybridization

Criteria	Post-Embedding Technique	Pre-Embedding Technique	Frozen Section Technique
Materials	+	++	+++
Morphology	++	+++	+
Sensitivity	+	++	+++
Resolution	+++		+++
Concentration of probe	+++	++	+
Detection system:			
• Autoradiography	++	+++	++
• Colloidal gold	+++		+++
• Enzymatic		+++	+
Storage	+++	+	++
Quantification	+++	+	+++
Toxicity	+++	+	
Semithin	0	++	+++

Chapter 3

Sample Preparation

CONTENTS

The techniques used to ensure that both morphology and nucleic acids are preserved for
in situ hybridization are similar to those used for
classical electron microscopy. This includes the
sampling and fixation of tissue under near sterile
conditions.

Fixation of biological material, including tissue (organs, biopsies, etc.) or cells in a culture
(monolayers, suspensions), must be carried out
using the appropriate technique:

⇝ For example, immunocytochemistry
⇝ Avoid contact with RNase

- Post-embedding technique:
  The sample is embedded in acrylic resin
  and thin sections are cut (ultrathin and/or
  semithin) prior to *in situ* hybridization.
- Pre-embedding technique:
  In situ hybridization is carried out on thick
  vibratome sections. Visualization of the
  probe can then be carried out before or after
  embedding in resin.
- Frozen sections:
  Samples are cryoprotected frozen sections
  by cryo-ultramicrotomy (ultrathin and/or
  semithin) prior to *in situ* hybridization.

3.1   SAMPLE ORIGIN

All types of biological material may be fixed:
1. Tissues

   - Organs
   - Biopsies

2. Cells

   - Suspensions
   - Monolayers

3.2 SUMMARY OF DIFFERENT STEPS

↪ Depending on the sample type

3.2.1 Tissue

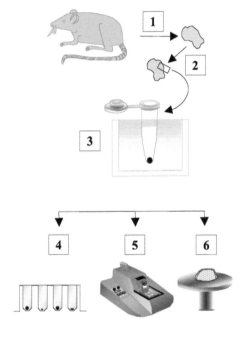

1 = **Dissection**
   With or without perfusion fixation

2 = **Sample size**
   Biological samples ≈ 1 mm³ or ≈ 0.5 cm³.

3 = **Immersion fixation** (4°C)
   Washes

❑ *Next stages:*
4 = **Embedding in resin** (*see* Chapter 4)
5 = **Vibratome sections** (*see* Chapter 5)
6 = **Cryoprotection/freezing** (*see* Chapter 6)
   Coating with sucrose

Figure 3.1 Fixation of tissue.

3.2.2 Cell Suspensions

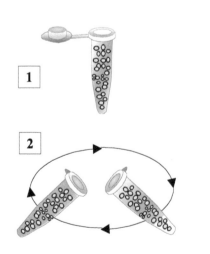

1 = **Cell suspension**

2 = **Centrifugation**

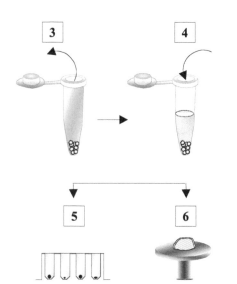

3 = **Aspirate supernatant**
4 = **Fix the pellet.**
  Washes
  Optional: the cell pellet may be embedded in agarose (*see* Figure 3.12).

❏ *Next stages:*
5 = **Embedding in resin** (*see* Chapter 4)
6 = **Cryoprotection/Freezing** (*see* Chapter 6)
  Coating in sucrose

Figure 3.2 Fixation of a cell fraction.

3.2.3 Cell Monolayers

1 = **Treat cells with trypsin.**
2 = **Fixation before or after removal of cell layer**
  F = fixative

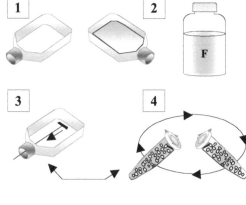

3 = **Scrape off cells.**
4 = **Centrifugation**
  Optional: the cell pellet may be embedded in agarose (*see* Figure 3.12).

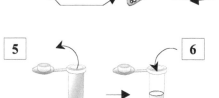

5 = **Aspirate supernatant**
6 = **Fixation** (after removal of cell layer)
  • Centrifugation (optional)
  • Washes

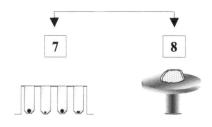

❏ *Next stages:*
7 = **Embedding in resin** (*see* Chapter 4)
8 = **Cryoprotection/freezing** (*see* Chapter 6)
  Coating in sucrose

Figure 3.3 Fixation of a cell pellet from the cell culture.

3.3   PRINCIPLES OF FIXATION

Chemical fixatives enter the aqueous parts of tissues and stop cell metabolism by inactivating endogenous enzymes, such as lysosomal enzymes and RNase, while stabilizing the cell morphology and the integrity of nucleic acids.

⇝ This should be in a medium which is close to physiological to maintain normal morphology.

⇝ Preservation of cell morphology is achieved by chemical bridges being formed between cellular components.

⇝ Conservation of RNA is more difficult than conservation of DNA.

3.3.1   Conditions for Fixation

To obtain the best results, fixation must be carried out without delay:

- *In vivo,* or
- On samples

Conditions used are similar to sterile conditions.

⇝ Fixation by perfusion (*see* Figure 3.10)

⇝ Fixation by immersion (*see* Figure 3.11)

⇝ Contamination by RNases or DNases must be avoided:
- Wear gloves
- All instruments must be sterilized (*see* Appendix A2).

3.3.2   Parameters for Fixation

To achieve good fixation it is necessary to consider the following conditions:

- Speed of penetration

⇝ Chemicals in solution penetrate the cell at different speeds, depending on the chemical and the tissue. Thus, the size of the sample should be gauged to ensure rapid fixation.

- Temperature

⇝ At low temperatures, autolytic enzyme activities are reduced as well as the rate of fixative penetration.

- pH

⇝ The cellular ionic equilibrium and proteins are at a neutral pH.

⇝ Partial destruction of nucleic acids occurs at an acid pH.

⇝ Partial breakdown of ribosomes and cellular membranes occurs at a basic pH.

- Concentration of fixative

⇝ Too strong a concentration may over-fix surface structures and prevent penetration to the nucleic acids.

⇝ Too high a dilution results in a long fixation and risks diffusion and loss of cellular constituents.

- Osmolarity

↪ Osmolarity of the fixative (buffer + fixative) should equal physiological osmotic pressure. Osmolarity can be adjusted by adding sodium chloride or larger molecules such as sucrose.
↪ Inappropriate osmolarity can cause plasmolysis or cell swelling.

- Sample size

↪ ≈ 1 mm^3 is considered optimal to ensure good fixation for electron microscopy.

- Duration

↪ This varies depending on the origin, size, and permeability of the sample.

3.4   TYPES OF FIXATIVE

↪ The best fixative has to be found for each method.

Fixatives are classified into two groups:

- Crosslinker fixatives
- Chemical fixatives

↪ Used prior to *in situ* hybridization
↪ Used after *in situ* hybridization

The success of an *in situ* hybridization depends greatly on the choice of fixative.

↪ A polyvalent fixative that gives a good compromise between preservation of morphology and preservation of the nucleic acids should be used.

A wide range of fixatives and fixation protocols appropriate for each type of biological material are available.

↪ Each fixative has unique properties that result in both advantages and disadvantages. A fixative that fulfills all requirements does not exist.

3.4.1   Crosslinker Fixatives

3.4.1.1   Nature
These fixatives are most often used for *in situ* hybridization.
  The least crosslinker aldehyde fixatives are generally used:

↪ Use under a fume hood that vents to the outside.

- Paraformaldehyde

↪ Polymer of formaldehyde. The advantage of paraformaldehyde over prepared formol is that there are no stabilizing additives present, but this means that the solution must be made just prior to use (*see* Appendix B4.3).

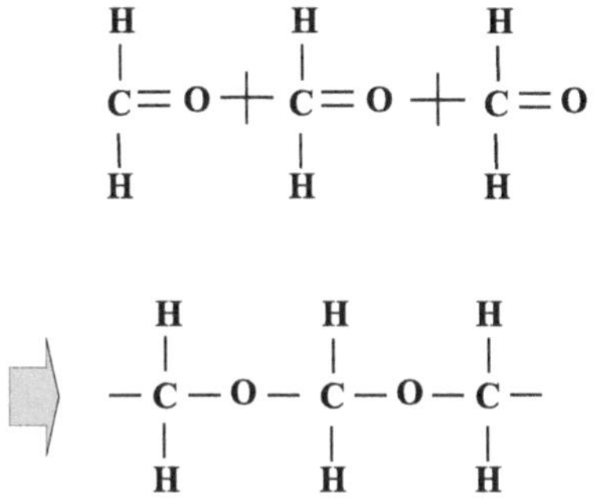

Figure 3.4   Polymerization of paraformaldehyde.

73

• Glutaraldehyde

$$CHO - (CH_2)_3 - CHO$$

⇨ Using small amounts of this crosslinker fixative results in a vast improvement in tissue morphology. It is used at concentrations of 0.02 to 0.5%.

⇨ **Toxic by inhalation**

Figure 3.5 Formula of glutaraldehyde.

3.4.1.2 Properties

Aldehyde fixatives easily penetrate tissue.

Fixation takes place by the formation of methylenic bridges favored at a pH of 7.5.

⇨ It is still best to use small pieces of tissue to ensure good fixation.

⇨ These fixatives give the best compromise between morphology and the *in situ* hybridization signal.

⇨ A sufficient quantity of fixative must be used (minimum of 10 × the volume of the sample) in a buffered solution.

3.4.1.3 Mechanisms of Action

Fixation forms bridges between different structures at the cellular and tissue levels (e.g., formaldehyde and glutaraldehyde) (*see* Figures 3.6 and 3.7).

• Formaldehyde

$$\boxed{1}$$

$$R - NH_2 + HCHO \longrightarrow R - NH - CH_2OH$$

1 = Formation of complex via NH_2 groups

$$\boxed{2}$$

$$R-NH-CH_2OH + NH_2-R$$

$$\longrightarrow R-NH-CH_2-NH-R + H_2O$$

2 = Formation of bridges between neighboring molecules

Figure 3.6 Formation of methylenic bridges.

• Glutaraldehyde

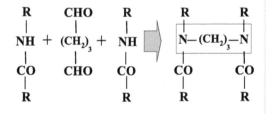

⇨ The methylenic bridges are formed even if the NH_2 groups are not free (as in the case of peptides).

Figure 3.7 Formation of bridges by glutaraldehyde.

3.4.1.4 Advantages/Disadvantages

❏ *Advantages*

- Rapid penetration
- The molecular linkage gives good accessibility for the probe to reach the target nucleic acid sequences.
- Good fixative for proteins

↬ Concentration between 2 and 4%
↬ Good for large probes.

↬ Paraformaldehyde is the aldehyde fixative most often used for *in situ* hybridization.

❏ *Disadvantage*

- Degrades rapidly after dilution

↬ 4% paraformaldehyde should be prepared just before use.

3.4.1.5 Preparation

- 2.5% glutaraldehyde
- 4% paraformaldehyde (PF)
- 4% paraformaldehyde/0.05% glutaraldehyde

↬ Ready to use (*see* Appendix B4.2)
↬ Ready to use (*see* Appendix B4.3.1)
↬ Ready to use (*see* Appendix B4.3.2)

3.4.2 Chemical Fixative

↬ Osmium tetroxide

3.4.2.1 Nature

↬ Sometimes incorrectly called osmic acid
↬ MW = 245.2

Figure 3.8 Formula of osmium tetroxide.

3.4.2.2 Properties

Osmium tetroxide is a volatile hydrophilic and very dangerous.

It penetrates tissue very slowly.

The speed of diffusion is slowed down by the cold but less so than autolysis.

A buffered solution must be used to counter-act the development of an acid zone next to the fixed areas during fixative penetration.

↬ The vapors are very irritating.
↬ **Use under a fume hood and store in a double container.**

↬ Samples must be small.
↬ Used on sections after hybridization.
↬ Fixation must be carried out at a low temperature.
↬ A slightly alkaline pH gives good results.
↬ Do not use salt buffers.

3.4.2.3 Mechanism of Action

Osmium tetroxide is a powerful reducing agent that creates bridges between ethylenic bonds in neighboring molecules (*see* Figure 3.9).

↬ The most reactive amino acids are those that contain ethylenic bonds (e.g., tryptophan, cysteine, histidine).

It works equally well with SH groups.

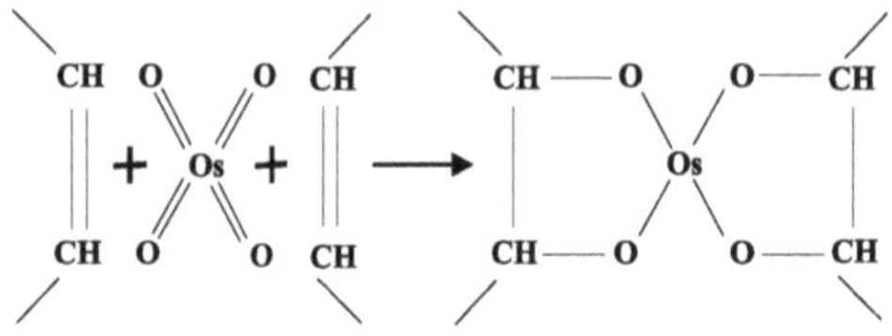

Figure 3.9 Mode of action of osmium tetroxide.

3.4.2.4   Advantages/Disadvantages

❑ *Advantages*

- There is good preservation of structures, particularly membranes.
- Stains tissue for electron microscopy

↬ Osmium atoms stop or strongly diffuse electrons.

❑ *Disadvantages*

- Dangerous and volatile

↬ **Use under a fume hood and store in a double container in dry conditions** (*see* Appendix A1.2).

- Destroys nucleic acids
- Disposal of osmium tetroxide

↬ Cannot be used prior to hybridization

3.4.2.5   Preparation

Osmium tetroxide 1 or 2%

↬ Ready to use (*see* Appendix B4.1)

3.5   MATERIALS/SOLUTIONS

3.5.1 Materials

↬ **All material that touches the tissue must be sterile** (*see* Appendix A2).

3.5.1.1   Tissue

- Artery clamp/forceps

↬ For small animals, it is sometimes necessary to wrap the arms of the forceps in soft material to protect the delicate vascular tissue.

- Glass tubes with plastic tops
- Pasteur pipettes and bulbs
- Peristaltic pump with a steady pressure

↬ For sample fixation
↬ Sterile
↬ May be replaced by an inverted bottle suspended at an appropriate height to give the correct pressure

- Propipette
- Refrigerator
- Scalpels, dissection forceps
- Scissors for dissection, metal cannula

↬ To aspirate dangerous chemicals
↬ 4°C
↬ Sterile

3.5.1.2 Cell Suspensions/Cell Monolayers

• Centrifuge	↝ Slow
• Centrifuge tubes	↝ Homogenization of samples
• Embedding molds and support	↝ Sterile
• Eppendorf tubes	↝ Sterile
• Pasteur pipettes and bulbs	↝ Sterile
• Refrigerator	↝ 4°C
• Vortex	↝ Homogenization of samples

3.5.2 Solutions

• 2% agarose in same buffer as fixative	↝ Molecular biology quality (*see* Appendix B2.4) ↝ PBS buffer, phosphate, or cacodylate (*see* Appendix B3.1/3.4)
• Buffers	↝ *See* Appendix B3.
– 100 m*M* phosphate buffer; pH 7.4	↝ *See* Appendix B3.4.1.
– 100 m*M* cacodylate buffer; pH 7.4	↝ *See* Appendix B3.1. ↝ This buffer is recommended for plant tissue and avoids precipitates, which can be a problem with a phosphate buffer.
– PBS buffer	↝ *See* Appendix B3.4.3.
• 0.1% collagenase in PBS	↝ *See* Appendix B2.10.
• 4% paraformaldehyde (PF)	↝ *See* Appendix B4.3.1.
• Trypsin 0.25%, ± EDTA 0.05%, PBS	↝ *See* Appendix B2.23.

3.6 PROTOCOLS

Fixation is generally carried out using 4% paraformaldehyde in PBS.	↝ This is considered the basic technique.

3.6.1 Tissue

3.6.1.1 Fixation by Perfusion

↝ Fixation by perfusion is carried out by placing a cannula into the aorta via the left ventricle.

3.6.1.1.1   SUMMARY OF DIFFERENT STEPS

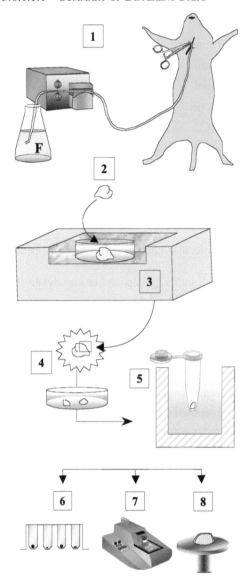

1 = Opening the thoracic cavity and exposing the pericardium:
  - A small incision is made into the left ventricle close to the point of the heart.
  - A cannula is gently inserted up to the aorta.
  - Fix the cannula in place with an artery clamp.
  - Cut the right auricle.
  - Perfuse with fixative (F) using a peristaltic pump.

2 = Dissection of tissue

3 = Immersion of the tissue sample in fixative (4°C)

4 = Size of tissue sample (4°C)
  Biological samples ≈ 1 mm³ or ≈ 0.5 cm³

5 = Fixation (4°C)
  Washes

❏ *Next stages:*
6 = Embedding in resin (*see* Chapter 4)

7 = Vibratome sections (*see* Chapter 5)

8 = Cryoprotection (*see* Chapter 6)
  Coat with sucrose.

Figure 3.10 Method for fixation of tissue.

3.6.1.1.2   PROTOCOL
Fixation by perfusion is followed by immersion fixation:

1. Wash
   - Buffer   **3 min**
2. After the wash, perfuse with fixative at a speed lower than 30 mL per min.   **30 min**

↪ When the tissue to be fixed is fragile, it is best to fix by perfusion (*see* Figure 3.10).

↪ Use ≈ 100 mL of warm buffer for each rat.
↪ Make up ≈ 400 mL of fixative to perfuse each rat. The efficiency of the perfusion may be observed by the stiffening of the neck and blanching of the liver.

3. After perfusion, dissect the tissue rapidly and immerse in the same fixative.   **1–2 h at 4°C**

↝ The time of fixation varies depending on the size of the samples, which must not be > 1 mm³ for frozen or resin embedding.
↝ Temperature should be 4°C to inhibit enzyme activity.

4. Wash
   • Buffer   **2 × 15 min**

❏ *Next stages:*

   • Embedding in acrylic resin
   • Cutting vibratome sections
   • Cryoprotection and freezing

↝ *See* Chapter 4, Section 4.3.
↝ *See* Chapter 5, Section 5.3.4.
↝ *See* Chapter 6, Section 6.4.

3.6.1.2   Fixation by Immersion

3.6.1.2.1   SUMMARY OF DIFFERENT STAGES

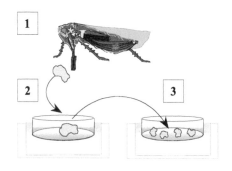

1 = **Dissection**

2 = **Immersion in fixative** (4°C)

3 = **Chopping samples** (if necessary)
   Biological samples ≈ 1 mm³ or ≈ 0.5 cm³

4 = **Fixation** (4°C)
   Washes

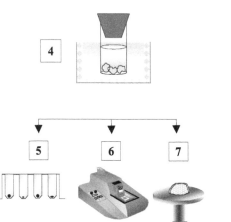

❏ *Next stages:*
5 = **Embedding in resin** (*see* Chapter 4)
6 = **Vibratome sections** (*see* Chapter 5)
7 = **Cryoprotection/freezing** (*see* Chapter 6)

Figure 3.11   Method of fixation by immersion.

3.6.1.2.2   PROTOCOL

1. After sacrificing the animal, dissect the samples.

↝ ≈ 1 mm³ for good fixation
↝ ≈ 5 to 10 mm in thickness if thick sections are to be cut (a vibratome section is between 50 and 100 µm) (*see* Chapter 5, Section 5.3.4).

2. Immerse immediately in a buffered fixative at 4°C.

⮑ The fixative may be used only once.
⮑ The sample should sink to the bottom of the solution. If this doesn't happen, a vacuum should be applied for a few minutes to get rid of any air bubbles stopping the penetration of fixative into the tissue (e.g., plant tissue, lung, etc.).
⮑ The sample can be trimmed after it has hardened for several minutes in the fixative.

- Buffered fixative **2 h at 4°C**

⮑ *See* Appendix B4.3.1. The time can be prolonged for larger samples of tissue that will be thick-sectioned.
⮑ Overfixation does not occur with aldehydes.

3. Wash:
   - Buffer **4 × 15 min at 4°C**

⮑ Several changes of buffer are needed to eliminate any traces of free aldehyde (there is a risk of nonspecific binding with immunocytochemistry) (*see* Chapter 7, Section 7.6.2.4).

4. May be stored **1–5 days at 4°C**

⮑ Add 0.01% sodium azide if necessary.

❏ *Next stages:*

- Embedding in acrylic resin
- Cutting thick vibratome sections followed by possible storage in 70% alcohol
- Cryoprotection and freezing

⮑ *See* Chapter 4, Section 4.3.
⮑ *See* Chapter 5, Section 5.3.4.

⮑ *See* Chapter 6, Section 6.4.

3.6.2   Cell Suspensions

1. Wash the cells in a serum-free medium

⮑ The presence of serum proteins may give rise to artifacts.

2. After centrifugation of the cell suspension, aspirate the supernatant.

⮑ It may be necessary to warm the fixative solution if the cells are cultured at 37°C to preserve the cell morphology.

3. Fix immediately in a buffered fixative

⮑ *See* Figure 3.2.
⮑ *See* Appendix B4.3.1.

4. Resuspend

⮑ With fragile or sparse cells, the pellet should be embedded in agarose prior to fixation. This is then treated in the same way as tissue (*see* Section 3.6.1.2).

- Buffered fixative **10 min at 4°C**

⮑ Fixation is short because the diameter of the cells is no more than a few μm.

5. Aspirate the fixative after centrifugation and replace immediately with the buffer.
6. Resuspend

⮑ Wash several times, with centrifugation between each change.

- PBS buffer **3 × 5 min at 4°C**

⮑ If necessary, centrifuge after each wash.

7. Keep the fixed tissue in a pellet and treat as a tissue.

    ↝ Embedding in 1% agarose makes it easier to handle the pellet (*see* Figure 3.11).

8. Storage is possible.   **1–5 days at 4°C**

    ↝ Add 0.01% sodium azide if necessary.

❏ *Next stages:*

• Embedding in acrylic resin   ↝ *See* Chapter 4, Section 4.3.
• Cryoprotection and freezing   ↝ *See* Chapter 6, Section 6.4.

3.6.3   Cell Monolayers

1. Loosen the cell monolayer from the plastic culture with enzyme treatment prior to or after fixation:

    ↝ This treatment may cause ultrastructural modification.
    ↝ *See* Figure 3.3. Monolayer cell cultures after enzyme treatment are treated as cell suspensions.

    • 0.25% trypsin in PBS, or   **5–9 min at 37°C**

    • 0.1% collagenase in PBS   **15 min at 4°C**

    ↝ This protocol should be tested in advance.

2. Centrifuge the cell suspension.   **600–9000 g 5 min**

    ↝ Centrifuge between each wash if necessary.

3. Wash cells in serum-free medium:

    ↝ The presence of serum proteins may give rise to artifacts.
    ↝ After centrifugation, the cell pellet is treated as a cell suspension (*see* Figure 3.8).

    • PBS buffer   **3 × 5 min at 4°C**

    ↝ Centrifuge between each wash if necessary.

4. Embed in agarose.

    ↝ **Optional.** If the cells are fragile or sparse, the pellet should be embedded in agarose to avoid centrifugation (*see* Figure 3.10). This can then be treated as a tissue (*see* Section 3.6.1.2).

5. Fixation
    • Buffered fixative   **10 min at 4°C**

    ↝ This stage is identical for cell suspensions.

6. Wash   **3 × 5 min**
    • PBS buffer   **at 4°C**

    ↝ Centrifuge between each wash if necessary.

7. Storage is possible.   **1–5 days at 4°C**

    ↝ Add 0.01% sodium azide if necessary.

❏ *Next stages:*

• Embedding in acrylic resin   ↝ *See* Chapter 4, Section 4.3.
• Cryoprotection and freezing   ↝ *See* Chapter 6, Section 6.4.

3.7   EMBEDDING CELLS IN AGAROSE

3.7.1   Aims

If the sample to be fixed is fragile (pellets, cells in culture), it is best to embed in agarose prior to fixation.

⮞ Embedding a pellet in agarose avoids the need for centrifugation (*see* Figure 3.10). This is then treated in the same way as tissue (*see* Section 3.6.1.2).

3.7.2   Summary of Different Stages

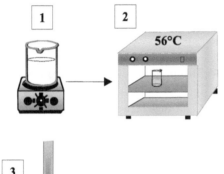

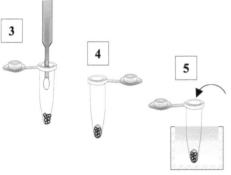

1 = **Dissolving the agarose**
2 = **Keeping the agarose warm** (56°C)

3 = **Embedding the cell pellet in agarose**
     Homogenization
4 = **Chilling** (4°C)
5 = **Fixation** (4°C)
     Washes

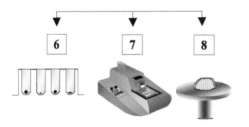

❑ *Next stages:*
6 = **Embedding in resin** (*see* Chapter 4)
7 = **Vibratome sections** (*see* Chapter 5)
8 = **Cryoprotection** (*see* Chapter 6)

Figure 3.12   Embedding a pellet in agarose.

3.7.3   Protocol

PREPARATION OF THE AGAROSE SOLUTION
1. Dissolve the agarose in the fixation buffer.

↬ *See* Appendix B2.4.
↬ Use a water bath. Cover the solution to avoid evaporation.

- Agarose **2 g**
- Fixation buffer **100 mL**

↬ Salt buffer (i.e., phosphate or cacodylate)

2. After dissolving, keep the solution in an oven
- Prepare in advance **at 56°C**

↬ Several hours

PROTOCOL
1. Centrifuge the cell suspension. **800–1200 g** **20 min**

↬ Too high a speed will break the cells.

2. Pour the agarose onto the pellet
- 2% agarose **2–3 drops** **at 56°C**

3. Mix rapidly with the tip of a pipette

↬ The cells should be evenly distributed throughout the solution.

4. Chill.

↬ Centrifuge if necessary (400 to 600 *g*) to concentrate the cells and eliminate bubbles.

5. Turn out of the tube and wash:
- Buffer **3 × 5 min** **at 4°C**

↬ *See* Appendix B3.4.1 and 3.4.3.

6. Fix
- Buffered fixative **1 h** **at 4°C**

↬ *See* Appendix B4.3.1.
↬ The pellet and agarose may be suspended in the fixative.

7. Wash:
- Buffer **4 × 5 min** **at 4°C**

❑ *Next stages:*
- Embedding in acrylic resin
- Cutting thick vibratome sections and storing in 70% alcohol
- Cryoprotection and freezing

↬ *See* Chapter 4, Section 4.3.
↬ *See* Chapter 5, Section 5.3.4.

↬ *See* Chapter 6, Section 6.4.

3.8   PROTOCOL TYPES

3.8.1   Tissue

FIXATION
4% PF in 100 m*M* phosphate buffer **2 h** **at 4°C**

↬ *See* Appendix B4.3.1.

WASHES
Buffer   **4 × 15 min**   ⇨ *See* Appendix B3.4.1.

❑ *Next stages:*

- Embedding in acrylic resin   ⇨ *See* Chapter 4, Section 4.3.
- Cutting thick vibratome sections and stor-   ⇨ *See* Chapter 5, Section 5.3.4.
  ing in 70% alcohol
- Cryoprotection and freezing   ⇨ *See* Chapter 6, Section 6.4.

3.8.2   Cell Suspensions

⇨ A cell fraction pellet is treated in the same way as a cell suspension.

1. Washes:
   - Serum-free medium
2. Centrifugation   **800–1200 g**   ⇨ After centrifugation, aspirate the supernatant.
             10 min
3. Fixation:
   - 4% PF in 100 m*M* phosphate   **10 min**   ⇨ *See* Appendix B4.3.1.
     buffer   **at 4°C**
4. Washes:
   - Phosphate buffer   **3 × 5 min**   ⇨ *See* Appendix B3.4.1.
             at 4°C

❑ *Next stages:*

- Embedding in acrylic resin   ⇨ *See* Chapter 4, Secton 4.4.3.
- Cryoprotection and freezing   ⇨ *See* Chapter 6, Section 6.6.4.

3.8.3   Cell Monolayers

1. Removing the cell layer

   - Mechanical, or   **5–15 min**
   - Enzymatic
     - 0.25% trypsin in PBS   **5–9 min**   ⇨ If necessary, centrifuge after each wash.
               at 37°C

2. Centrifugation   **600–900 g**   ⇨ After centrifugation, aspirate the supernatant.
             for 5 min
3. Washes:
   - Buffer   **3 × 5 min**   ⇨ Or serum-free medium
             at 4°C
4. Fixation:
   - 4% PF in 100 m*M* phosphate   **10 min**   ⇨ *See* Appendix B4.3.1
     buffer   **at 4°C**
5. Washes:
   - Buffer   **3 × 5 min**   ⇨ *See* Appendix B3.4.1.
             at 4°C

❏ *Next stages:*

- Embedding in acrylic resin   ↝ *See* Chapter 4, Section 4.3.
- Cryoprotection and freezing   ↝ *See* Chapter 6, Section 6.4.

3.9 SUMMARY TABLE

Table 3.1   Criteria Determining the Technique to be Used

Types and Methods of Sample Fixation	Post-Embedding Technique	Pre-Embedding Technique	Frozen Sections
Tissue: organ, biopsy, etc.			
Perfusion + immersion			
Immersion			
Cell suspensions			
Pellet by immersion			
Pellet in agarose by immersion			
Monolayer cultures			
Monolayers			
Pellet by immersion			
Pellet in agarose by immersion			

Chapter 4

Post-Embedding Technique

CONTENTS

Embedding in hydrophilic resin allows a number of tissue samples to be tested at the same time. *In situ* hybridization can be carried out under conditions similar to those used for immunocytochemistry.

↬ To validate *in situ* hybridization or immunocytochemistry

Limitations are:
• Resin embedding
• Sensitivity

↬ Partial dehydration is necessary.
↬ Some tissue components will be lost in preparation.

• Quality of the morphology

Treatment of tissue for post-embedding *in situ* is similar to that for immunocytochemistry.

4.1 PRINCIPLES

For electron microscopy, tissue is embedded in hydrophilic resin, and ultrathin sections between 80 and 100 nanometers are cut. Nucleic acids are exposed at the section surface where they may be detected.

↬ Sections correspond to a single cell thickness

Only the part of the nucleic acid sequence that is exposed at the surface of the section may be detected (*see* Figure 4.1); nucleic acids present in the interior of the section cannot be detected.

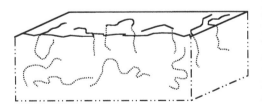

Figure 4.1 Localization of nucleic acid sequences in cell sections.

The section surface is irregular and is consequently larger than sections from epoxy embedded tissues.

↬ Cutting the surface of the resin and the cellular structures embedded gives the probes access to the nucleic acids.

4.2 SUMMARY OF DIFFERENT STAGES

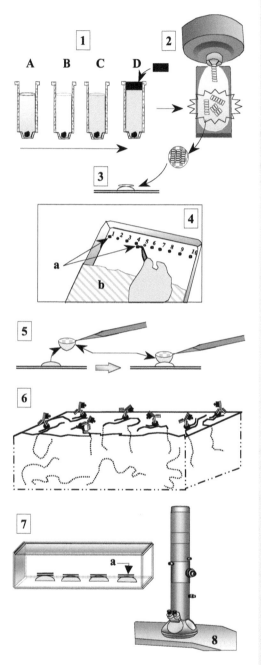

1 = Embedding
  (A) Dehydration in ethanol
  (B) Infiltration of solvent/resin
  (C) Impregnation of resin
  (D) Closing capsule

2 = Ultramicrotomy
  Mounting ultrathin sections on grids (collodion coated, carbon coated)

3 = Pretreatments (optional)
  • Denaturation of the probe
  • Denaturation of the target molecule
  • Alkaline treatment

4 = Hybridization
  (a) Place the grids in solution
  (b) Humid chamber

5 = Washes

6 = Immunocytochemical visualization
  Direct immunocytochemistry reaction

7 = Staining
  (a) Grid

8 = Observation with the electron microscope

Figure 4.2 Post-embedding technique.

4.3   EMBEDDING

4.3.1   Principles

Tissue has to be hardened to facilitate the cutting of thin sections. The water in the tissue is replaced with fluid resin, which is then polymerized. The final hardness of the tissue depends on the type of resin used.

Epoxy or methacrylate resin allows ultrathin sectioning of tissue.

After progressive dehydration, the tissue is infiltrated with a mixture of solvent and resin, with the concentration of resin increasing successively.

The final state of hydration of the tissue depends on the type of resin used.
The time taken for resin infiltration depends on the viscosity of the resin (*see* Section 4.3.2.1).

The final stage of embedding is impregnation with pure resin, eliminating all traces of solvent. To ensure impregnation, the tissue is transferred to several changes of pure, fresh resin before polymerization. The length of this stage varies according to the size of the sample and may be prolonged without any loss of morphology.

The presence of solvent in the resin inhibits polymerization.
Good embedding results from tissue water being gradually replaced by resin.
The quality of the sections obtained is dependent on the complete replacement of solvent by resin.

After embedding in resin the tissue is sectioned.

The thickness of the sections obtained depends on the hardness of the resin.

4.3.2   Types of Acrylic Resin

For embedding biological material, acrylic resins are hydrophilic and compatible with *in situ* hybridization.

↪ Different protocols are required for different resins.
↪ Lowicryls HM20 and HM23 are the only hydrophobic resins.

This was developed to preserve the structures of antigens, but has been shown to preserve the structure of nucleic acids equally well.

↪ The structures of proteins are preserved.

Different resins may be used for *in situ* hybridization. These are:

• GMA or glycol methacrylate

↪ 2-hydroethyl methacrylate
↪ Hydrophilic resin

• Different types of Lowicryl resin

↪ Acrylic methacrylate resin
↪ Hydrophilic and/or hydrophobic resins

1

$$CH_2 = C\ (CH_3) - CO - O - R$$

1 = General formula of methacrylate

2

$$CH_2 = CH - CO - O - R$$

2 = General formula of acrylate

3

$$CH_2 = C\ (CH_3) - CO - O - R$$
$$CH_2 = C\ (CH_3) - CO - O^{\diagup}$$

3 = General formula of dimethylacrylate cross linker (accelerator)

Figure 4.3 General formulae of Lowicryl resins.

• LR White (London Resin Gold)

⇒ Methacrylate resin/hardener
⇒ Hydrophilic resin

• Unicryl

⇒ Methacrylate resin
⇒ Hydrophilic resin

4.3.2.1 Characteristics

These are resins made up of a mixture of acrylates–methacrylates with low viscosity before polymerization, allowing rapid tissue infiltration.

⇒ The viscosity at 25°C is low (0.7 centipoises) for methacrylate resins (very rapid infiltration). The viscosity is higher (≈ 150–1650 centipoises) for epoxy resins due to the large size of the monomers present (very slow infiltration) (*see* Glossary).

Shrinkage is an important factor.

⇒ 10 to 20%, only 4% for epoxy resins.

Ultrathin surface sectioning of a tissue embedded in acrylic resin reveals profiles of the different structural components.

⇒ Irregular surface (*see* Figure 4.1).

The polymerization of methacrylate is catalyzed by the addition of benzoyl peroxide. This chain reaction maybe brought about:

⇒ 1,2-dichlorobenzoylperoxide
⇒ Three-dimensional polymerization for epoxy resins.

1. At 4°C
   • LR White, GMA

⇒ In the presence of an amine accelerator.

2. By heat
   • LR White

⇒ Without an accelerator.

3. At low temperatures by ultraviolet light

⇒ At 360 nm for all resins, in the presence of a light-sensitive activator to start the chain reaction.

   • Glycol methacrylate (GMA)

⇒ Polymerization at 4°C

- Lowicryls
  - K4M
  - HM20
  - K11M
  - HM23
- London Resin Gold
- Unicryl

↬ Polymerization at –20°C to –80°C
↬ Polymerization at –20°C to –35°C
↬ Polymerization at –40°C
↬ Polymerization at –60°C
↬ Polymerization at –80°C
↬ Polymerization at –25°C to –35°C
↬ Polymerization at –20°C to –35°C

The stability of ultrathin sections in the electron beam varies, but is never good.

↬ A support film is necessary (*see* Appendix A4).

4.3.2.1.1 LOWICRYLS

These resins are a mixture of acrylate and methacrylate monomers and have low viscosities at low temperatures.

A catalyst is necessary to start the polymerization chain reaction.

All the dehydration and infiltration stages are carried out at low temperatures to limit the loss of nucleic acids and preserve ultrastructural morphology.

Lowicryls have low viscosities and polymerize at low temperatures (K4M: –35°C; HM20: –70°C) under UV light (360 nm).

↬ Generally this is benzoyl peroxide (e.g., O-methylbenzoine for Lowicryl K4M).
↬ Lowicryls do not sublime under the electron beam.

↬ Heat induced polymerization may be used.
↬ Polymerization is started by a catalyst sensitive to UV light. The reaction is strongly exothermic and controlled by low temperatures (–20 to –60°C).
↬ They both have the same viscosity

There are 2 types of Lowicryl:

1. Hydrophilic
   - Lowicryl K4M
   - Lowicryl K11M
   These two resins are fluid between –31°C and –40°C but have different hydrophobicity.

↬ Polar

2. Hydrophobic
   - Lowicryl HM20
   - Lowicryl HM23
   Both resins are specialized for embedding at very low temperatures (–70°C), a necessary property for rapid freezing and cryosubstitution techniques.

↬ Nonpolar

↬ May be used for *in situ* hybridization
↬ These characteristics permit the conservation of antigens. The protocols are complex and it is necessary to adapt both tissues and techniques for their use.
↬ A mask and gloves **must** be worn throughout the embedding procedure.

Lowicryl resins are very toxic and can provoke eczema and/or other allergic reactions.

Lowicryl resins are sold in kits comprising three components:

1. Crosslinker

↬ Triethylene glycol dimethacrylate. Solution A (accelerator) (*see* Figure 4.4)

$$-(CH_2-CH_2-O_3)-$$

2. Monomer
   • Hydroxypropyl methacrylate

$$- CH_2 - CH_2 - CH_2 - OH$$

   • Hydroxyethyl acrylate

$$- CH_2 - CH_2 - OH$$

   • n Hexyl methacrylate

$$- CH_2 - (CH_2)_4 - CH_3$$

3. Initiator
   1,2-dichlorobenzoylperoxide

Figure 4.4 Formula of triethylene glycol dimethacrylate.

↝ *See* Figures 4.5, 4.6, and 4.7

Figure 4.5 Formula of hydroxypropyl methacrylate.

Figure 4.6 Formula of hydroxyethyl acrylate.

Figure 4.7 Formula of n Hexyl methacrylate.

↝ Benzoyl peroxide, catalyst for methacrylates

4.3.2.1.2 UNICRYL

1. This is a methacrylate resin of low viscosity, very hydrophilic, permitting rapid infiltration into tissue.
2. It polymerizes at low temperatures under UV light or heat.
3. Unicryl resin is composed of glycol methacrylate and hydroxypropyl methacrylate. The composition of the mixture is as follows:
   • 2-hydroxyethyl methacrylate
   • 2-hydroxypropyl methacrylate
   • *n*-butyl methacrylate
   • Styrene
4. The commercially available solution is ready to use and has a long shelf life if stored at 4°C.
5. This product is very **toxic**.
6. The embedding protocol is rapid.
7. The hydrophilic nature of the resin preserves subcellular structures without chemical reactions or the formation of intermolecular bridges.

↝ It is possible to speed up the infiltration.

↝ The embedding protocol will be different.

↝ 40%
↝ 30%
↝ 26%
↝ 4%

↝ No preparation is necessary.

↝ **Mask and gloves must be worn.**
↝ There is no substitution stage.

4.3.2.1.3 LONDON RESIN GOLD

1. This acrylic resin was specifically designed to be very hydrophilic. It is a mixture of methacrylate and hardener.

↪ More widely known as LR White

↪ Polyhydroxyaromatic acrylic resin

2. It is sold in solution and must be kept at 4°C.

↪ It has a short shelf life.

3. LR White may be polymerized by two procedures, by either a chemical catalyst, or heat treatment:

↪ According to the degree of hardness required

• 4°C

↪ Polymerization at 4°C is carried out by the addition of the accelerator provided with the resin.

• 50°C

↪ Polymerization takes place in the absence of the accelerator.

4. The stability of ultrathin sections under the electron beam is good.

↪ Superior to those in Lowicryl K4M

5. The low viscosity permits embedding of larger specimens.

6. This product is not very toxic.

7. Many varieties of this resin with different levels of hardness are available:

↪ Medium and hard are used for electron microscopy.

• LR White medium

↪ Medium is used most for *in situ* hybridization.

• LR White hard

↪ The hardest

• LR White soft

↪ The softest

4.3.2.1.4 Glycol Methacrylate (GMA)

1. This acrylic resin is hydrophilic with very low viscosity and is a mixture of several methacrylates and Luperco; it polymerizes under UV light at 4°C.

↪ One of the first resins used for electron microscopy

2. It has become less popular because the preparation of the mixture is tricky and **the components are very toxic.**

4.3.2.2 Advantages/Disadvantages

❏ *Advantages*

• Good preservation of ultrastructure

↪ Generally greater than that obtained with frozen tissue

↪ This approach is a good compromise between the necessity to preserve the structure of the nucleic acids and the necessity for good morphology.

• Weak sensitivity

↪ Less than that obtained using other methods

• Polymerization at low temperatures

↝ Numerous cellular constituents may be lost during dehydration and impregnation, or denatured during the polymerization of the resin. These problems are considerably reduced at lower temperatures.

• Easily stored
• Material embedded in resin blocks may be stored for unlimited periods.

❑ *Disadvantages*
• **This product is very toxic and is a skin irritant.**

↝ For Lowicryl resins it is particularly important to take precautions during handling. **Avoid breathing the fumes (all manipulations should take place in a fume hood while wearing a mask and gloves).**
↝ LR White resin is less toxic.

• Its use is not compatible with certain tissues.
• The hardness of the resin is variable.
• The resin does not bond well with the biological material.
• The resin is unstable under the electron beam.

↝ Some tissues are almost never embedded in acrylic resin (e.g., plant tissue).
↝ Problems occur during sectioning.
↝ Problems occur during sectioning and mounting the sections on grids.
↝ It is necessary to coat the grids before mounting the sections.

4.3.2.3 Choice of Resin

The choice of resin depends on the type of material to be studied and the nucleic acid to be detected (DNA vs. RNA). In practice, if the use of high temperatures does not prevent the recognition of nucleic acids with *in situ* hybridization, then resins which polymerize at 4°C may be used.

↝ For example, LR White medium

Conversely, if problems occur with the preservation of nucleic acids, then the material must be embedded at lower temperatures.

↝ For example, Lowicryl, Unicryl

Acrylic resins have different viscosities and degrees of hydrophobicity. In the case of large samples, a resin with very low viscosity should be used.

↝ For example, Unicryl

The hydrophobicity of a resin influences the degree of wetting that occurs at the section surface. This property greatly influences the binding of the probe to the target and the amount of nonspecific binding.

↝ Very hydrophilic resin (e.g., LR White medium)
↝ To improve the immunocytochemical signal when nonradioactive probes are used (*see* Chapter 8)

Table 4.1   Summary of the Characteristics of Acrylic Resins

Resin	Type	Viscosity	Polymerization				
			Heat > 19°C	Room Temp.	4°C	Low temperature + UV	
						−20°C −35°C	−40°C −80°C
GMA	Hydrophilic	Very low					
K4M	Hydrophilic	Low					
HM20	Hydrophobic						−40°C
K11 M	Hydrophilic	Very low					−60°C
HM23	Hydrophobic						−80°C
LR White medium	Hydrophilic	Low					
Unicryl	Hydrophilic	Very low					

4.3.3   Materials/Products/Solutions

↪ For use with all acrylic resins

4.3.3.1   Materials

• Cold store (0°C to 4°C)

↪ For storage of the embedding products required for Epon

• Freezer (−30°C)

↪ For storage of the embedding mixture Lowicryl K4M during impregnation

• Magnetic stirrer

↪ To mix the resin

• Oven

↪ 60°C for heat polymerization of resin

• Polymerization chamber

↪ For embedding at low temperatures
↪ Commercially available; improves the ease of manipulation of the resin and helps avoid allergies

• UV lamps

↪ 365 nm
↪ 2 × 15 watts and/or 2 × 8 watts, for polymerization of resin.

4.3.3.2   Small Items

• Aluminium foil
• Bulbs, Pasteur pipettes
• Disposable plastic pipettes
• Embedding tubes with rubber stoppers

↪ Sterile
↪ Propipette
↪ Sterile

• Gelatin capsules (dry)	⇨ The use of small gelatin capsules (size 00) allows rapid trimming of the embedded sample. ⇨ Embedding in a very dry gelatin capsule (≥ 30 min at 37°C). ⇨ Polyethylene and polypropylene capsules are relatively transparent to UV light at 365 nm. Beem capsules are recommended.
• Glass vials	⇨ ≈ 10 mL capacity for keeping the different solutions and mixing in the automatic cryo-substitution chamber (AFS) (*see* Figure 4.11).
• Gloves, mask	⇨ **Indispensable (embedding products are very toxic)**
• Micro-pipettes	
• Plexiglas support with holes	⇨ To support the gelatin capsules under the UV light
• Scalpels, dissection forceps	⇨ Sterile
• Setup for keeping solutions cold	⇨ Ice bucket

4.3.3.3 Products

Mixture embedding	⇨ *See* Appendix B5.
• LR White medium	⇨ *See* Appendix B5.3.2.
• Lowicryl K4M resin	⇨ *See* Appendix B5.3.1.
• Unicryl	⇨ Commercially available resin solution, ready for use (*see* Appendix B5.3.3)

4.3.3.4 Solutions

• Dehydration	
• Increasing concentrations between 35% and 100%	⇨ Ethanol may be replaced by methanol.

4.3.4 Embedding in Lowicryl K4M Resin

4.3.4.1 Principles

After fixation and washing, specimens undergo several successive treatments:

1. Dehydration is carried out according to PLT (progressive lowering of temperature). Fixed biological samples are dehydrated by increasing concentrations of methanol or ethanol while the temperature of the solvent is gradually lowered.	⇨ The water contained in the tissue is replaced at low temperatures by an organic solvent (cryosubstitution), either methanol or ethanol. ⇨ AFS (automatic freeze-substitution-system) is a system for cryodehydration in ethanol and cryo-infiltration in acrylic resin (*see* Section 4.3.7).
Dehydration starts at 0°C, then the temperature is progressively lowered to the temperature at which polymerization of the resin takes place (e.g., −31°C for Lowicryl K4M).	⇨ 0°C/−20°C/−30 to −35°C ⇨ The temperature of dehydration decreases with each change of solvent.

2. Cryo-infiltration is carried out with a solvent miscible with Lowicryl K4M.

⮑ K4M resin is miscible with ethanol, acetone, and methanol.

⮑ The dehydrated samples are first infiltrated with a mixture of solvent–resin with increasing concentrations of resin, then with pure Lowicryl K4M resin.

3. Cryo-embedding at low temperature:
   This replaces solidified gelatin with resin.

⮑ Final embedding

4. Cryopolymerization:
   This is carried out using UV light at low temperatures and takes several days.

⮑ Hardens the resin

❑ *Parameters of polymerization*

Many factors are important in polymerization:

- Distance between the block and the UV source

⮑ Greater than or equal to 9 cm

- Wavelength of the UV light

⮑ 360 nm

- Power of the UV

⮑ 2×15 watts and/or 2×8 watts

- Temperature

⮑ $-20°C$ to $-31°C$

- Time

⮑ Greater than or equal to 5 days

4.3.4.2   Summary of Different Stages

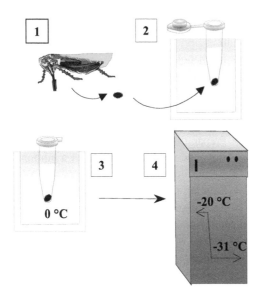

1 = **Dissection**

2 = **Fixation by immersion**
   Washes

3 = **Dehydration in ethanol**
   Increasing concentrations of ethanol temperatures from 0°C

4 = **Cryo-infiltration** ($-20°C$ down to $-31°C$)
   Mixture of ethanol–Lowicryl K4M in increasing concentrations

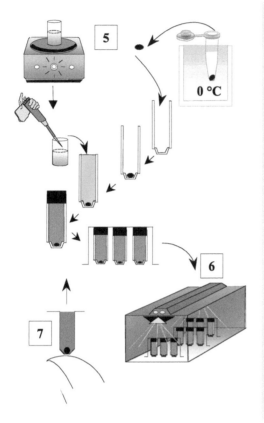

5 = Cryo-embedding
Mixture of monomeric Lowicryl K4M

6 = Cryopolymerization
Under UV light at –31°C for 5 days

7 = Removal of molds

Figure 4.8 Preparation of samples embedded in Lowicryl K4M.

4.3.4.3 Products/Solutions

• 35%, 55%, 70%, 95%, 100% ethanol

↪ Prepare 1 mL of mixture in separate glass vials.
↪ Methanol may be used.
↪ The different solutions may be kept at the appropriate temperature before use.

• Lowicryl K4M resin

↪ **The resin is very toxic and must be used under a fume hood** (*see* Appendix A1.2).
↪ Preparation (*see* Appendix B5.3.1)
↪ Keep the mixture at –31°C for a minimum of 30 min before use.

– Monomer
– Crosslinker

– Initiator

↪ Solution supplied by the manufacturer
↪ Accelerator solution supplied by the manufacturer
↪ To be dissolved in the monomer and crosslinker

• Infiltration solution

↪ Prepare the different infiltration solutions and transfer to the cryosubstitution chamber (–31°C) a minimum of 30 min before each change of solution.

4.3.4.4   Protocol

1. Dehydration

• 35% ethanol	**30 min at 0°C**
• 55% ethanol	**1–2 h at 0°C**
• 70% ethanol	**2 × 1 h at –20°C**
• 95% ethanol	**2–3 h at –31°C**

2. Cryo-infiltration
   • Cryo-infiltrator

– 95% ethanol + resin (2 vol/1 vol)	**2–5 h at –31°C**
– 95% ethanol + resin (vol/vol)	**24 h at –31°C**
– 95% ethanol + resin (1 vol/2 vol)	**6–24 h at –31°C**

   • Replace the solution with the embedding mixture of Lowicryl K4M

– 1st change/resin K4M	**overnight at –31°C**
– 2nd change/resin K4M	**≥ 6 h at –31°C**

3. Cryo-embedding
   • Prepare the mixture of Lowicryl K4M for the final stage.

   • Place the molds or the gelatin capsules in the AFS chamber on crushed ice.
   • Fill the gelatin capsules with the freshly prepared resin.
   • Carefully place the sample at the bottom of the capsule.

↪ **Under a fume hood**

↪ The times indicated may be modified according to the size of the sample. The times are determined by the volume of the sample in mm³.

↪ **Optional**

↪ For each change of solution, aspirate the liquid with a disposable pipette and replace with the next solution already chilled.

↪ This resin allows the presence of 5% water:
   • Partial dehydration of samples is possible (95% ethanol) for animal tissues.
   • Complete dehydration is necessary for plant tissue to extract any pigments which may absorb UV light.

↪ Do not prolong this stage more than several hours.
↪ These times may be increased.

↪ Use freshly prepared resin.

↪ Two changes of pure resin, ranging from several hours to 24 prior to polymerization

↪ *See* Appendix B5.3.1.
↪ Prepare ≈ 1 mL of resin to fill each gelatin capsule.
↪ Oxygen inhibits polymerization. During the preparation of the resin, it is necessary to limit contact with air (i.e., keep in an airtight container; agitate very gently). Keep the mixture at –31°C for ≥ 30 minutes before use.
↪ *See* Figure 4.16.

↪ Each embedded sample must be clearly labeled.
↪ If the sample has to be oriented, a flat support may be used (*see* Figure 4.14).

- Completely fill the capsule to limit the volume of air trapped at the surface of the resin.

⇨ *See* Figure 4.8.

- Close the capsules tightly.

⇨ Plastic molds or gelatin capsules. These must be airtight to avoid contact with air.

- Place the capsules on a thin plastic support.

⇨ UV light must pass through the support (e.g., Plexiglas).

4. Cryopolymerization
  - Position the UV lamp.   **2 × 15 watts**
                                               360 nm

⇨ UV lamps lose their power as they age.

  – Polymerization   **5 days**
                                               at –31°C

⇨ After this time, it is preferable to let the samples warm to room temperature for several hours before trimming.

  - Check the polymerization.

⇨ If polymerization is not complete, replace the samples under UV light at room temperature for 24 hours.

❏ *Next stage:*
  - Cutting ultrathin sections

⇨ Ultramicrotomy (*see* Section 4.4).

4.3.5   Embedding in LR White Resin

⇨ London Resin white

4.3.5.1   Principles

This resin was specially developed to be very hydrophilic and to have good stability under the electron beam. It also has low toxicity.

This resin has a low viscosity, allowing the embedding of larger samples. Dehydration may be partial to increase sensitivity of detection.

⇨ 8 centipoises
⇨ Greater than or equal to several mm^3
⇨ 5% water

This resin is available in three levels of hardness (medium, hard, soft). The choice is determined by the type of biological material.

⇨ Fatty tissue vs. bone, cartilage, etc.

After fixation and washes, biological samples are submitted to the following treatments:

1. Dehydration is carried out at room temperature in a series of increasing concentrations of ethanol (55% to 95% and/or 100%).

⇨ Water contained in the tissue is replaced by organic solvents (i.e., ethanol).
⇨ Do not use acetone; it inhibits polymerization.

2. Infiltration is carried out at room temperature:
  - Ethanol–resin mixture
  - Monomer mixture

⇨ LR White is miscible with ethanol.

⇨ Increasing concentrations of resin
⇨ Methacrylate

3. Embedding:
  This is carried out by placing the specimen at the bottom of a gelatin capsule filled with:

⇨ Final embedding
⇨ These must be airtight since oxygen prevents the hardening of the resin.

  - Mixture (LR White, pure + accelerator)
  - LR White, pure (without accelerator)

⇨ If polymerization takes place at 4°C
⇨ If polymerization takes place at 60°C

4. Polymerization
   - Low temperature
   - High temperature

⮡ With an accelerator
⮡ Without an accelerator
⮡ An accelerator may be added to start the chain reaction more rapidly.

❑ *Parameters of polymerization*
Several parameters play a role in polymerization:

   - Duration
   - Homogeneity of the resin–accelerator mixture
   - Temperature

⮡ More than 24 h
⮡ The mixture must be perfectly homogenous for correct polymerization.
⮡ 4°C or 60°C

4.3.5.2 Summary of Different Stages

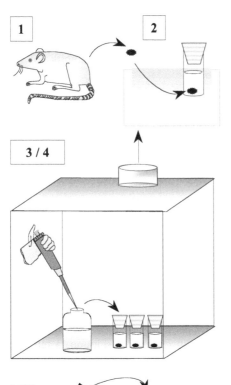

1 = **Dissection**
(With or without perfusion fixation)

2 = **Fixation by immersion (4°C)**
Washes

3 = **Dehydration in ethanol**
Changes of ethanol

4 = **Infiltration**
   - Mixture of ethanol and LR White
   - Pure LR White

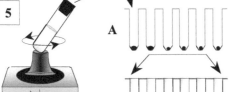

5 = **Embedding in LR White**
   - LR White mixture
     – With an accelerator if polymerization is to take place at a low temperature
     – Without an accelerator if polymerization is to take place at a high temperature
     (A) Positioning of the sample
     (B) Filling and closing the molds

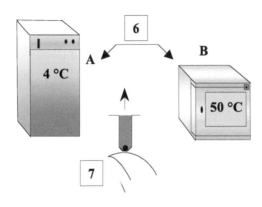

6 = Polymerization
(A) Low temperature
(B) High temperature

7 = Removing the resin block from the mold.

Figure 4.9 Preparation of samples embedded in LR White.

4.3.5.3 Materials/Products/Solutions

- 55%, 70%, 95%, 100% ethanol

↪ Prepare 1 mL of solvent in a glass vial.
↪ Do not use acetone; it inhibits polymerization.

- LR White medium

↪ Preparation (*see* Appendix B5.3.2)
↪ **Important: keep the two components at 4°C until use**.

  – Methacrylate monomer

↪ Monomer of polyhydroxylated aromatic acrylic resin

  – Accelerator

↪ Viscous liquid, difficult to pipette. Measure with a precise propipette (5 µL/5 mL of resin).

- Vortex

↪ The mixture is difficult to homogenize (LR White + accelerator).

4.3.5.4 Protocol

↪ **Under fume hood**

1. Dehydration

↪ At room temperature
↪ The times indicated may be modified according to the size of the samples (e.g., 1 mm³).

• 55% ethanol	**30–40 min**

↪ At each change of solution, aspirate the liquid with a disposable pipette and refill with the next solution.

• 70% ethanol	**3 × 30–40 min**
• 100% ethanol	**3 × 30–40 min**

↪ A partial dehydration (95% ethanol) is possible.

2. Infiltration

↪ At room temperature

• 100% ethanol + resin (2 vol/1 vol)	**1–2 h**

↪ Do not prolong the first solution for more than a few hours.

• 100% ethanol + resin (vol/vol)	**2–4 h**
• 100% ethanol + resin (1 vol/2 vol)	**2–4 h**

↪ May be prolonged

Replace the resin–ethanol mixture with pure resin

⇁ Resin without accelerator

- 1st change/pure LR White resin   **1 h at 4°C**
- 2nd change/pure LR White resin   **overnight at 4°C**

⇁ The impregnation stage may be prolonged for several hours at 4°C.

3. Embedding

⇁ Each embedding must be well labeled.

- Place the sample at the bottom of the capsule or mold filled with pure resin or a mixture (resin + accelerator)

⇁ Use the smallest possible molds since it is easier to position the specimen and makes trimming the block easier.

⇁ The temperature of polymerization depends on the presence or absence of an accelerator.

  - Pure LR White
  - Mixture (LR White + accelerator)

⇁ Polymerization at 60°C

⇁ Polymerization at 4°C

⇁ Preparation (*see* Appendix B.5.3.2)

⇁ Polymerization of mixture for less than 15 min at room temperature:
  - Take out the products just before use
  - Prepare small quantities (5 mL) of mixture (resin–hardener)
  - Homogenization of the mixture must be complete. The very small quantity of accelerator (μL) in the viscous resin (5 mL) makes homogenization difficult. Vortex thoroughly.

- Completely fill the capsules or molds with the resin mixture maintained at 4°C.
- Close the gelatin capsules or molds tightly.

⇁ **Important:** this limits the air– resin contact; oxygen inhibits polymerization.

4. Polymerization

⇁ Good polymerization results in a very hard resin.

- Without accelerator:
  - Temperature   **60°C**
  - Time of polymerization   **24–48 h**

⇁ During polymerization, localized heating may deform the ultrastructure. It is preferable to carry out polymerization at 4°C.

- With accelerator:
  - Temperature   **4°C**
  - Time of polymerization   **48 h**

⇁ Transfer to an oven at 37°C for several hours after polymerization; this makes the mold easier to remove.

⇁ Let the blocks equilibrate at room temperature for several hours before trimming them.

❑ *Next stage:*
- Cutting ultrathin sections

⇁ Ultramicrotomy (*see* Section 4.4)

4.3.6　Embedding in Unicryl Resin

4.3.6.1　Principles

This resin is a single type of monomer that penetrates cells easily.

This hydrophilic resin is easy to use:

- The embedding procedure is rapid.
- Resin is ready to use.

After fixation and washes, the samples are treated as follows:

1. Dehydration is carried out at progressively low temperatures (from 0°C to –20°C) in a graded series of ethanol (55% to 100%).

2. Infiltration — After the last change in ethanol, the samples are infiltrated with pure resin at a low temperature.

3. Embedding — Place the sample at the bottom of a capsule and fill with resin.

4. Polymerization is carried out:

   - By UV light at –20°C for 5 days

   - By UV light at 4°C for 3 days

   - At 50°C for 3 days

❏ *Factors which affect polymerization*
Several factors play a role in the rate of polymerization:

- Power of UV lamps

- Wavelength
- Temperature
- Distance between the block and the UV light
- Time

➥ Embedding tissue in Unicryl is relatively easy.
➥ There is no substitution stage.
➥ This limits the risks associated with handling.

➥ The water contained in the tissues must be replaced by an organic solvent (substitution by ethanol).
➥ Unicryl is miscible with ethanol.

➥ Final embedding
➥ The capsule must be airtight to prevent contact with oxygen which prevents hardening of the resin.
➥ Hardening of the resin

➥ Distance of ≈ 10 cm between the sample and the lamps (2 × 8 watts)
➥ Distance of ≈ 15 cm between the sample and the lamps (2 × 8 watts)
➥ For 1 mL of Unicryl

➥ 2 × 8 watts
➥ 2 × 15 watts
➥ 360 nm
➥ –20 to 50°C
➥ Greater than or equal to 10 to 15 cm

➥ 3–5 days

4.3.6.2   Summary of Different Stages

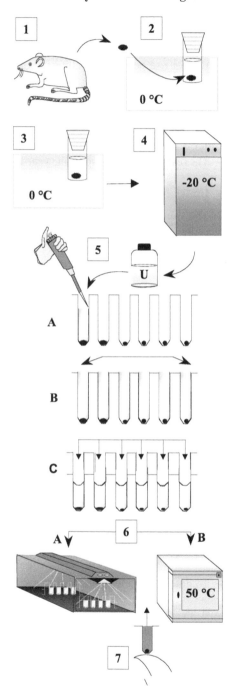

1 = Dissection
(With or without fixation by perfusion)

2 = Fixation by immersion (4°C)
Washes

3 = Dehydration in ethanol
Increasing concentrations of ethanol at temperatures from 0°C to –20°C

4 = Cryo-infiltration
Mixture of Unicryl monomer

5 = Embedding
Mixture of Unicryl monomer (U)
(A) Positioning of samples
(B) Filling capsules
(C) Closing capsules

6 = Polymerization
(A) Low temperature — Under UV light at –20°C for 5 days
(B) High temperature

7 = Removing the capsule

Figure 4. 10 Preparation of biological samples embedded in Unicryl.

4.3.6.3 Products/Solutions

- 35%, 55%, 70%, 95%, 100% ethanol

↝ Prepare 1 mL of mixture in a glass vial.
↝ The different solutions are kept at the appropriate temperature before use.
↝ Do not use acetone; it inhibits polymerization.

- Unicryl

↝ Ready to use resin is commercially available.
↝ **The resin is very toxic and must be used under a fume hood** (*see* Appendix A1.2).
↝ The resin is stable for up to one year at 4°C.

4.3.6.4 Protocol

1. Dehydration

↝ **Under fume hood**

↝ The times indicated may be modified according to the size of the samples. Those indicated are for samples ≤ 1 mm³.
↝ For each change of solution, aspirate the liquid with a disposable pipette and replace with the next solution.

• 35% ethanol	**30 min at 0°C**
• 55% ethanol	**60 min at 0°C**
• 70% ethanol	**2 h at –20°C**
• 100% ethanol	**2–3 h at –20°C**

↝ **Optional**

↝ Partial dehydration is possible (95% ethanol)
↝ The samples do not go through a substitution.

2. Cryo-infiltration

↝ Unicryl must be kept, at a minimum, for 30 min at –30°C before use.
↝ Do not prolong this time for more than a few hours.

• 1st change of Unicryl	**1 h at –20°C**
• 2nd change of Unicryl	**overnight at –20°C**

3. Cryo-embedding
   - Place the molds or gelatin capsules in the AFS chamber on crushed ice.
   - Fill the capsules with resin kept at –20°C.
   - Position the samples carefully at the bottom of the capsules.

   - Completely fill the capsules to minimize the volume of air enclosed.
   - Close the capsules.

   - Place the capsules on a thin plastic support.

↝ *See* Figure 4.16.

↝ To limit contact with air trapped above the resin
↝ If the sample must be oriented, a flat embedding mold may be used with the AFS system (*see* Figure 4.14).

↝ Or plastic molds, which must be airtight to avoid contact between the resin and oxygen, which inhibits polymerization.
↝ The UV light must not be stopped by the support.

4. Cryopolymerization

- Position the UV lamps. **2 × 15 watts 360 nm**

  – Length of polymerization **5 days at –20°C**
- Check the polymerization.

↪ Cryopolymerization takes place under UV light; the length of the polymerization depends on the power of the UV lamps used.
↪ UV lamps lose power as they get older.
↪ Distance of the UV to the specimen is ≈ 10 cm
↪ Let the samples warm to room temperature slowly before trimming.
↪ If polymerization is not complete, replace the samples under UV light for 24 hours at room temperature.

❏ *Final stage*
Cutting ultrathin sections

↪ Ultramicrotomy (*see* Section 4.4).

4.3.7 PLT Procedure

↪ Progressive Lowering of Temperature

4.3.7.1 Aims

PLT consists of the dehydration and inclusion in acrylic resin of biological samples after chemical fixation at low temperatures.

The tissue is fixed in paraformaldehyde with a light fixation and is thus fragile and sensitive to the dehydration and embedding conditions. The effect of both is considerably reduced at low temperatures.

The acrylic resins used are very toxic. There is a venting system incorporated into the apparatus that prevents any solvent or monomer vapors from entering the room.

↪ Progressive dehydration by chilling with liquid nitrogen vapors.
↪ For example, automatic system for cryo-substitution at low temperatures
↪ Limits the risks of oxidation
↪ Cell constituents may be lost in the course of dehydration and impregnation, or denatured during polymerization of the resin.

↪ Take precautions at low temperatures.
↪ **Always wear gloves throughout the embedding procedure.**

4.3.7.2 Advantages/Disadvantages

❏ *Advantages*

- Good reproducible results
- Built-in venting system
- Embedding may be carried out under specific conditions maximized for a particular sample.

↪ Embedding parameters are programmed.
↪ Good security for the user
↪ Eppendorf tubes, perforated capsules, flat embedding

❏ *Disadvantages*

- A single program for embedding for each run
- A single type of embedding for each sample
- Limited number of samples

↪ Only one type of embedding may be carried out each week.
↪ Either flat embedding perforated capsules or Eppendorf tubes may be used.
↪ Maximum 24 blocks

4.3.7.3 Equipment

The apparatus has nine programs that can be saved. Each has two or three possible variations of temperature and time (T1, T2, T3) (*see* Figure 4.13).

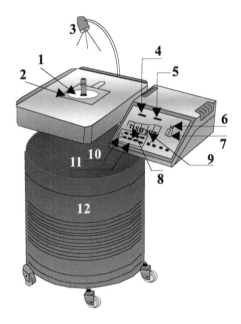

Temperature is lowered by liquid nitrogen vapor.

The apparatus contains a maximum of 50 L of liquid nitrogen.

The apparatus has a venting system that prevents solvents and monomer vapors from entering the room.

4.3.7.4 Accessories

1. Embedding in perforated capsules
   • Chamber 1

   • Gelatin capsules, size 1
   • Glass beakers

   • Hooked cover

   • Plastic capsules perforated at the base

⮑ Cryosubstitution apparatus (AFS)
⮑ From 65°C to –40°C
⮑ There is a semi-automatic system that programs the time of the rate of temperature descent (S1 and S2).

1 = Sliding cover with removable glass plate
2 = Substitution chamber (venting system)
3 = Light
4 = Temperature display
5 = Time display
6 = UV lamp control
7 = Nitrogen level control
8 = Time and temperature program control
9 = Rate of temperature descent control
10 = Key for storing program in memory
11 = Choice of program (0 to 9)
12 = Liquid nitrogen reservoir

Figure 4.11 Cryosubstitution apparatus.

⮑ Nitrogen is liquid at –196°C. The vapors cool the sample and also ensure the exclusion of oxygen.
⮑ ≈ 40 L are needed for a single embedding procedure in a resin like Lowicryl K4M; polymerization at –31°C (*see* Section 4.3.4).
⮑ Safety precaution for the user

⮑ Choice of embedding system

⮑ *See* Figure 4.12. To position the plastic capsules
⮑ For the final embedding
⮑ ≈ 10 mL volume for two changes of solution in chamber 1
⮑ To lift and position eight capsules around the central cavity of chamber 3
⮑ *See* Figure 4.14 (1). For embedding biopsies or cell pellets embedded in agarose

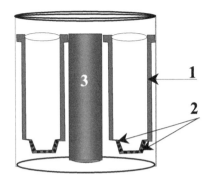

1 = **Plastic capsules**
2 = **Perforated base**
3 = **Central cavity**

Figure 4.12 Perforated capsules positioned in chamber 1.

2. Embedding in Eppendorf tubes
   • Chamber for Eppendorf tubes

   • Eppendorf microtubes
   • Glass beaker
   • Screen ring for polymerization

3. Flat embedding — Molds

4. Polymerization with UV light — Switch on the UV lamp

⮑ *See* Figure 4.14 (2).
⮑ For embedding cell pellets
⮑ There is a risk of losing some of the sample during changes of solution.
⮑ Capacity of 0.5–1.5 mL
⮑ 10 mL capacity
⮑ Small diameter for 0.5 mL Eppendorf tubes
⮑ Large diameter for 1.5 mL Eppendorf tubes
⮑ *See* Figure 4.14 (3).
⮑ All the equipment for flat embedding must be thoroughly dried at 50°C before use.
⮑ Closed and numbered
⮑ Seven samples/molds may be used.
⮑ With light control

4.3.7.5   Protocol

1. Filling the liquid nitrogen
   • Start the apparatus under pressure.
   • Fill the liquid nitrogen reservoir.

   • Cool the substitution chamber.
2. Setting the program
   • Select the program.
     – T1   **1 h 30 min at 0°C**
     – S1   **30 min**
     – T2   **2 h at –20°C**
     – S2   **30 min**
     – T3   **8 h at –31°C**

⮑ Lowicryl K4M was never meant to be used with plastic molds with perforated bases.

⮑ ≈ 40 L of liquid nitrogen are needed for a protocol lasting from 7 to 8 days.

⮑ *See* Figure 4.13.

⮑ 35%, 55% ethanol

⮑ 70% ethanol

⮑ 95% ethanol; mixture of ethanol and increasing concentration of Lowicryl K4M, then pure Lowicryl
⮑ Time of polymerization ≈ 5 days

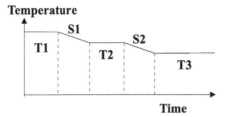

(T1) Temperature 0°C
(S1) Time of the rate of temperature descent
(T2) Temperature –20°C
(S2) Time of the rate of temperature descent
(T3) Temperature –31°C

Figure 4.13 Programming the rate of cooling for embedding in Lowicryl K4M with the automatic system.

3. Dehydration
   • Place chamber 1 in the cryosubstitution chamber.
   • Place the fixed and washed samples in the perforated capsules and place these in chamber 1 (*see* Figure 4.12).

↪ Up to four chambers can be accommodated.

↪ Identify the samples by placing a small piece of numbered card in the capsules. The position of the samples in the support should be noted from 1 to 8 starting at the arrow.

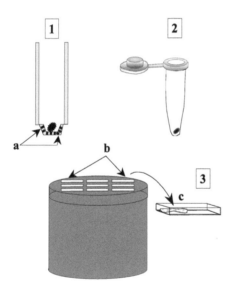

1 = **Plastic capsule with perforated base**
   (a) Samples

2 = **Eppendorf tubes**
   Cell suspensions

3 = **Mold for flat embedding**
   (b) cavity
   (c) oriented samples

Figure 4.14 System for embedding biological samples.

   • Chill the different solutions of ethanol to the appropriate temperature before use.
   • Add (or remove) 5 to 10 mL of solvent to the central opening of the cylinder of chamber 1 (*see* Figure 4.15).

↪ Using a disposable pipette
↪ Dehydration starts at 0°C (T1); at the selected time, the temperature is lowered progressively (rate of cooling, S1) to reach the 2nd temperature chosen (T2). On reaching the 2nd temperature (T2), the second stage of cooling (S2) is started automatically, down to temperature T3 (*see* Figure 4.13).

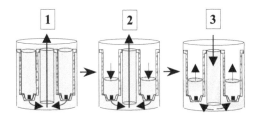

1 = Aspiration of solutions from the central opening of the cylinder
2 = Complete elimination
3 = Filling the chamber with the next solution

Figure 4.15 Changing solutions using capsules with perforated bases (chamber 1).

The samples must never freeze.

↝ For each stage, the temperature in the chamber must be higher than the freezing point of the dehydration mixture.

4. Cryo-infiltration
   • Prepare the different infiltration solutions (with increasing concentrations of resin), chilled before use.

↝ *See* Appendix B5.3.
↝ At temperature T3 (–31°C), the samples are first in a solvent–resin mixture and then in pure resin.

   • Remove the used liquid from the central opening and replace with the next solution (*see* Figure 4.15).

↝ With a disposable pipette

5. Cryo-embedding
   • Prepare 20 mL of resin and chill in the cryo-substitution chamber (–31°C).

↝ Use a glass beaker.
↝ Chill the resin for a minimum of 30 min before use.

   • Place chambers 2 and 3 in the cryo-substitution chamber at the polymerization temperature (T3):
     – Chamber 2

↝ T3 = –31°C for Lowicryl K4M

↝ *See* Figure 4.16 (2). Place the gelatin capsules in the chamber and fill with freshly prepared resin.
↝ There is only a single type of embedding protocol.

     – Chamber 3

↝ *See* Figure 4.16 (3).
↝ Pour 15 mL of 100% ethanol in the chamber.
↝ Ethanol improves the polymerization (improved transduction of UV light).
↝ Do not use water; it will dissolve the gelatin.

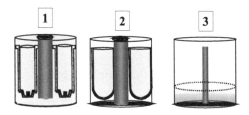

1 = **Chamber 1**
Samples which have been cryosubstituted in capsules with perforated bases
2 = **Chamber 2**
Gelatin capsules + Lowicryl K4M
3 = **Chamber 3**
15 mL of 100% ethanol

Figure 4.16 Transfer support for embedding samples in gelatin capsules.

- Fill the capsules with perforated bases containing the specimens with freshly prepared resin and chill (chamber 1).
- Fix the removable cover to a cryomanipulator and close all the perforated capsules tightly.
- Transfer all the perforated capsules at the same time into the gelatin capsules filled with freshly prepared resin.

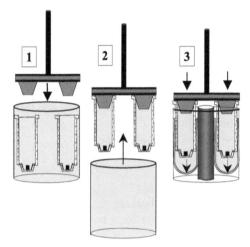

- Place the reference mark (o) of the support in line with the removable cover.

- Press very hard so that the removable cover is forced down on the gelatin capsules.
- Remove the removable cover with the perforated capsules and push into chamber 3, which contains the chilled ethanol.
- Repeat if necessary with a 2nd support into chamber 2.

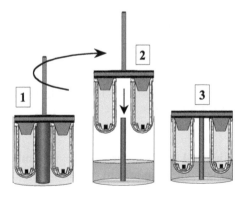

↝ *See* Figure 4.16 (1).
↝ Possible to accommodate four chambers.

↝ *See* Figure 4.17 (1).

↝ *See* Figure 4.17 (2).
↝ A large part of the resin is displaced by the capsules and remains in chamber 2. This may be used for the next stage.

1 = Fixing the removable cover to the plastic capsules
Chamber 1

2 = Lifting the capsules

3 = Positioning in the gelatin capsules
Chamber 2

Figure 4.17 Transferring the capsules with perforated bases into the gelatin capsules.

↝ *See* Figure 4.17 (3).
↝ Match the arrow on the chamber with that of the removable cover.
↝ The sample n°1 is facing the point of the removable cover.
↝ *See* Figure 4.18 (1).

↝ *See* Figure 4.18 (2).

↝ *See* Figure 4.18 (3).

1 = Fixing the removable cover
2 = Transferring the samples on the support
Chamber (3)
3 = Positioning the sample molds
Chamber (3)
100% ethanol

Figure 4.18 Positioning the samples in the chamber (3) for polymerization.

6. Cryopolymerization

- Place the polymerization screen ring on the different supports.
- Place the UV lamp over the opening of the cryosubstitution chamber.
- Polymerize.   **5 days at –31°C**
- Check the polymerization.

➯ *See* Figure 4.19.
➯ Cryopolymerization takes place in direct UV light.
➯ For a better spread of the UV light

➯ If the polymerization is incomplete, replace under UV light for 24 h at room temperature.

1 = **UV light**

2 = **Polymerization screen**

3 = **100% ethanol**

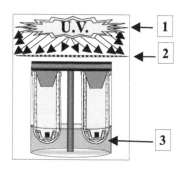

Figure 4.19 Polymerization in gelatin capsules.

❏ *Final stage:*
- Cutting ultrathin sections

➯ Ultramicrotomy (*see* Section 4.4).

4.4   ULTRAMICROTOMY

4.4.1   Summary of Different Stages

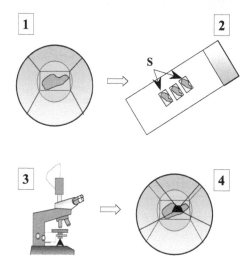

1 = **Microtomy**
Original size of block face

2 = **Semithin sections** (1 µm)
Mounting on a glass slide
s = section

3 = **Light microscopy**
After staining the semithin section, the region of interest is located

4 = **Ultramicrotomy**
The block is trimmed around the area of interest prior to cutting ultrathin sections.

Figure 4.20 Trimming the block for ultramicrotomy.

4.4.2 Materials/Products/Solutions

4.4.2.1 Large Equipment

- Carbon evaporating system
- Hot plate
- Knife maker

- Oven
- Ultramicrotome

⇝ For making carbon films
⇝ For drying semithin sections onto slides
⇝ For making glass knives. Diamond knives may also be used for ultramicrotomy.
⇝ *See* Appendix A5.1.
⇝ 37°C for drying support films
⇝ For cutting ultrathin sections of embedded tissue

4.4.2.2 Small Items

- Carbon electrodes
- Dental wax or nail polish
- Diamond knife

- Filter paper

- Glass
- Glass Petri dishes or watch glass

- Glass slides
- Laboratory consumables (e.g., pipettes, tubes, forceps)
- Nickel or gold grids, mesh ≥ 200 with collodion and carbon films (*see* Appendix A4.3)

- Plastic boats
- Probe with fine point

⇝ For carbon evaporation
⇝ For mounting boats on the glass knives
⇝ Precludes making glass knives (*see* Appendix A5.2).
⇝ The type used for rapid filtration is best for collodion-coated grids.
⇝ For making glass knives
⇝ A drop of collodion should be placed in the receptacle without touching the sides.
⇝ For mounting semithin sections
⇝ New or sterile
⇝ Do not use copper grids as they may cause chemical reactions at a later stage.
⇝ Diameter 3.05 mm
⇝ Ultrathin sections of acrylic resin are unstable under the electron beam and must be supported on coated grids.
⇝ LKB
⇝ *See* Figure 4.21. For positioning sections without damage

Figure 4.21 Probe with fine point.

- Single-edged razor blades

- Titanium forceps, ultrafine and antimagnetic
- Wire loop

⇝ Fine blades for trimming blocks, thoroughly cleaned in ethanol
⇝ Indispensable for manipulating nickel grids (static electricity is a problem)
⇝ *See* Figure 4.22; or comet for lifting semithin sections (*see* Appendix A6.1, making a comet).

1 = **Making a comet by punching a hole in plastic**
2 = **Wire loop**

Figure 4.22 Equipment for manipulating semithin sections.

4.4.2.3   Products

- Acetone
- Formvar powder

⇝ For removing any grease from the grids

4.4.2.4   Solutions

- 1% aqueous toluidine blue

  ⇝ For staining semithin sections
  ⇝ *See* Appendix B7.1.1.
- 2% collodion in isoamyl acetate

  ⇝ *See* Appendix A4.3.
  ⇝ Used at 0.5%
- Distilled water
- 0.15–0.25% formvar

  ⇝ *See* Appendix A4.4.
  ⇝ Used at these concentrations
- Isoamyl acetate

  ⇝ AR quality

4.4.3   Cutting Sections

Sectioning is an important element in the quality of the structure and the signal.

⇝ Special materials are necessary to cut good sections.

Semithin sections are important:

- To check the quality of the embedding and the tissue morphology

  ⇝ Before commencing ultrastructural studies
- To find the area of interest in a homogeneous tissue

  ⇝ This must be small enough for cutting good quality ultrathin sections.
- To provide a smooth block surface

  ⇝ To ensure that the chosen zone appears in the first semithin sections

4.4.3.1   Materials

Sections of embedded tissue obtained by ultramicrotomy may be of different thicknesses and are:

- 0.5 to 2 μm in thickness (semithin sections)

  ⇝ *See* Chapter 7, Section 7.4.
  ⇝ Mounted on glass slides for light microscopy. These sections allow the localization of the area of interest, ensuring that the block face is small enough for cutting good ultrathin sections.
- ≤ 100 nm in thickness (ultrathin sections)

  ⇝ Mounted on nickel grids for subcellular observations

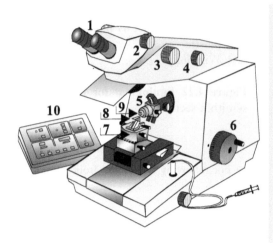

1 = Eyepieces
2 = Microscope with adjustable magnification
3 = Focus
4 = Lateral movement
5 = Orientation head in the specimen
6 = Hand-wheel with macro/micro advance
7 = Knife holder
8 = Knife with boat
9 = Specimen holder
10 = Control unit

Figure 4.23 Ultramicrotome.

4.4.3.2 Sectioning Parameters

A number of fine adjustments are indispensable for cutting good sections:

- The meniscus of the water in the boat

- The angle of the knife

- The speed of cutting

- The thickness of the sections:

  - Semithin

  - Ultrathin

↪ These are specific for cutting sections embedded in acrylic resin.

↪ For cutting acrylic resin, it must be slightly lower than that used when cutting epoxy resin.

↪ This depends on the hardness of the block:
  - Larger than 5° for harder blocks (Lowicryl K4M, Unicryl)
  - Smaller than 5° for softer blocks (LR White medium)

↪ At least 1 to 2 mm/s. A slower speed of ≈ 1 mm/s is advised for harder blocks.

↪ Judged from the interference color of the section

↪ A thickness of ≤ 2 μm is obtained with macro advance (protocol for cutting semithin sections, *see* Chapter 7, Section 7.2.2).

↪ A thickness of 80–100 nm is obtained with automatic advance. Variations in the hardness of the block can pose problems with variations in the thickness of sections.

4.4.3.3 Protocol for Ultrathin Sections

Preparation of the block prior to cutting ultrathin sections:

1. Trim the block (*see* Figure 4.24).

↪ The upper and lower edges of the block must be parallel with the cutting edge of the knife.

The block face must be small enough to give good quality ultrathin sections. A block face that is too large causes vibrations during cutting, which reduces the quality of the sections.

In contrast to semithin sections, no resin should remain around the sample (*see* Chapter 7, Section 7.4.2.4).

A well-trimmed block allows a ribbon of sections to be cut.

Figure 4.24 Trimming the block.

2. Use the left side of the razor to cut the sides of the block that run parallel to the edge of the knife.

The best part of the cutting edge (*see* Figure 4.25)

Ensure that the knife and the block are tightened securely.

1 = **Part of the knife most chipped**
2 = **Part used to smooth the block**
3 = **Best part of the knife**
4 = **This part of the knife is used from right to left to cut ultrathin sections.**

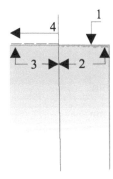

Figure 4.25 The glass knife.

3. Choose the angle of the knife according to the hardness of the block.
4. Adjust the level of the water, lowering it if necessary.

For Lowicryl K4M and Unicryl: 6–7°

For LR White medium: 3–4°

Resin is very hydrophilic and the block may get wet and drag sections behind the knife, in which case the level of the water in the boat should be lowered.

5. Ultrathin sections are cut with the left side of the glass knife (*see* Figure 4.25).

The edge of the block and the cutting edge must be oriented so that the area of interest will be wholly contained within the section.

Section speed: slow (1 mm/s)

Level of water in the boat: very low

Figure 4.26 Cutting ultrathin sections of tissue embedded in acrylic resin.

6. Cut at a slow speed.

→ 1 mm/s

→ During sectioning, the difference in the hardnesses of materials may cause tearing, and cellular components may be lost, leaving holes in the sections; this phenomenon is reduced at slower cutting speeds.

7. Evaluate the thickness of the sections by the interference color (*see* Table 4.2).

→ Ultrathin sections display interference colors, which vary according to the section thickness (between 70 and 110 nm). Only sections that are between gold and grey should be used for electron microscopy ($\approx$ 90 nm in thickness).

Table 4.2   Newton's Colors

Newton's Colors	Section Thickness
Grey	Less than 60 nm
Silver	60 to 90 nm
Gold	90 to 150 nm
Purple	150 to190 nm
Blue	190 to 240 nm
Green	240 to 280 nm
Yellow	280 to 320 nm

8. Pick up ultrathin sections on a nickel grid coated in collodion and carbon (*see* Appendix A4).

→ The coating is **indispensable** as acrylic resin is unstable under electron beams.

→ Hold the edge of the grid with a fine forceps, keeping the grid horizontal.

→ The grid is submerged in the boat and brought parallel to the surface of the water (*see* Figure 4.27); the water will cover the surface of the grid.

→ To be able to pick up ultrathin sections on a nickel grid which has been coated in collodion and carbon, it may be necessary to ionize the grid with an electronic discharge under a vacuum (glow-discharge).

9. The grid is dried on filter paper at room temperature and should be kept covered in a dry place away from dust.

→ Use the grids quickly after treatment.

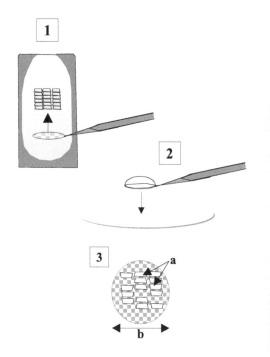

1 = **Mounting ultrathin section on grids**

2 = **Dry on filter paper**

3 = **Ultrathin sections on a grid**
    (a) ultrathin sections
    (b) 3.05 mm

Figure 4.27 Ultrathin sections on a grid (nickel or gold) collodion and carbon coated.

❏ *Next stages:*
  • Pretreatments
  • Hybridization

➯ *See* Section 4.5.
➯ *See* Section 4.6.

4.5   PRETREATMENTS

4.5.1   Aims

Pretreatment allows the probe access to the nucleic acids.

➯ At the same time preserving the ultrastructure

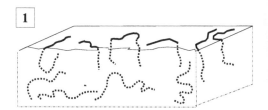

1 = **Before pretreatment:**
    Very little nucleic acid accessible

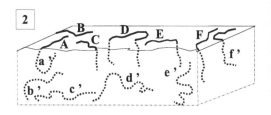

2 = **After pretreatment:**
    The zones A, B, C, D, E, and F are accessible to the probe (solid lines). The zones a', b', c', d', e', and f' remain inaccessible to the probe (dotted lines).

Figure 4.28 Principles of pretreatment.

4.5.2   Summary of Different Stages

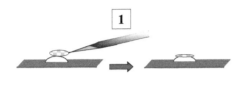

1 = Place the grid on the pretreatment solution and incubate.

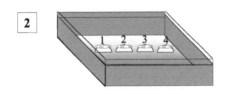

2 = Grids are labeled.

Figure 4.29 Incubation of grids.

4.5.3   Materials/Products/Solutions

4.5.3.1   Materials

- Agitator
- Centrifuge
- Gloves
- Microwave 245 mHz, ≈ 750 W
- Vortex

↩ Weak horizontal agitation
↩ More than 14,000 *g*

↩ The wavelength is important.

4.5.3.2   Products

- Calcium chloride
- Disodium phosphate
- DNase
- Hydrochloric acid
- Paraformaldehyde
- Pepsin
- Potassium phosphate
- Pronase
- Proteinase K
- RNase A

- Sodium chloride
- Tris-hydroxymethyl-aminomethane
- Triton X-100

↩ **Use only for *in situ* hybridization**
↩ Molecular biology quality
↩ AR grade
↩ Powder
↩ Destruction of all DNA
↩ AR grade
↩ Powder
↩ Powder or solution
↩ Powder
↩ Powder or solution
↩ Solution is better.
↩ With no DNase contamination. Available in solution
↩ Keep at −20°C
↩ Destruction of all RNA
↩ AR grade
↩ Powder, keep dry
↩ Keep away from the light.

4.5.3.3   Solutions

➭ **All solutions must be sterile for _in situ_ hybridization (_see_ Appendix A2).**

- Buffers
  - 1X or 10X PBS

    ➭ _See_ Appendix B3.

    ➭ _See_ Appendix B3.4.3. Keep at room temperature after autoclaving.

  - 10X Phosphate; pH 7.4

    ➭ _See_ Appendix B3.4.1. Keep at room temperature for 1 month after autoclaving.

  - 20X SSC (standard saline citrate); pH 7.0

    ➭ _See_ Appendix B3.5. Keep at 4°C or at room temperature.

  - Hybridization buffer

    ➭ Without the probe

  - TE (100 mM Tris–HCl/10 mM EDTA) 10X; pH 7.6

    ➭ _See_ Appendix B3.6.1. Keep at 4°C or at room temperature.

  - TE (20 mM Tris–HCl/300 mM NaCl); pH 7.6

    ➭ _See_ Appendix B3.7.5. Keep at 4°C or at room temperature.

  - TE/NaCl (100 mM Tris–HCl/10 mM EDTA/5 M NaCl) 10X; pH 7.6

    ➭ _See_ Appendix B3.6.2. Keep at 4°C or at room temperature.

  - Tris–HCl/CaCl$_2$ (20 mM Tris/2 mM CaCl$_2$); pH 7.5

    ➭ _See_ Appendix B3.7.2. Keep at 4°C or at room temperature.

  - Tris–HCl/glycine (50 mM Tris/50 mM glycine); pH 7.4

    ➭ _See_ Appendix B3.7.3. Solution 1X

  - Tris–HCl/MgCl$_2$ (10 mM Tris/5 mM MgCl$_2$); pH 7.3

    ➭ _See_ Appendix B3.7.4.

- 100 mg/mL DNase in sterile water

  ➭ Keep at –20°C in 10 μL aliquots (_see_ Appendix B3.3).

- 0.2 N Hydrochloric acid
- 1 mg/mL pronase in a TE buffer (10 mM Tris–HCl/1 mM EDTA; pH 7.6)

  ➭ Keep at –20°C in 50 μL aliquots.

- 1 mg/mL proteinase K in sterile water

  ➭ _See_ Appendix B2.15. Keep at –20°C in 50 μL aliquots.

- 4% paraformaldehyde (PF) in 100 mM of phosphate buffer; pH 7.4

  ➭ _See_ Appendix B4.3. Keep at –20°C.

- Pepsin in 0.2 N HCl; pH 5.0
- 100 mg/mL RNase in sterile water

  ➭ _See_ Appendix B2.16. Keep at –20°C in 10 μL aliquots.

- Sterile water

  ➭ _See_ Appendix B1.1. Use only once when the bottle is opened or use DEPC water (_see_ Appendix B1.2).

- 0.1% Triton X-100

  ➭ Diluted in sterile water

4.5.4   Deproteinization

The treatment partially eliminates proteins, particularly those structurally associated with nucleic acid to improve accessibility for probes targets.

➭ Enzyme treatment (e.g., proteinase K, pronase, pepsin, etc.).

4.5.4.1 Proteinase K

Proteinase K is a protease selective for proteins associated with nucleic acids.

↪ This pretreatment provides a compromise between signal intensity and tissue preservation or staining.

4.5.4.1.1 USE

The enzyme action is modulated by the following conditions:

↪ This is a key stage for the success of the reaction.

1. Concentration:
   The enzyme concentration may be varied according to the type of sample embedded (tissue, cells) or the nature of the nucleic acid target.

   ↪ A moderate degree of digestion (1–10 µg/mL) gives good results. A higher concentration can affect the morphology.

2. Buffer:
   Proteinase K (stock solution) is diluted in a Tris buffer (20 mM)/CaCl$_2$ (2 mM); pH 7.5.

   ↪ This buffer contains calcium, which is a cofactor for the enzyme. It is possible to use other buffers (e.g., TE) to limit the action of proteinase K.

3. Temperature:
   Optimal temperature   **37°C**

   ↪ The enzyme is less active at lower temperatures.

4. Duration   **5–30 min**

   ↪ Regulates the final effect

4.5.4.1.2 PROTOCOL

1. Warm the buffer (Tris/CaCl$_2$) to 37°C and float the grid, section side down, on a drop of buffer.

   ↪ The choice of buffer modifies the activity of proteinase K.
   ↪ Add proteinase K just before use.

2. Enzyme treatment:
   Proteinase K   **3 µg/mL**
                                         15 min
                                         at 37°C

   ↪ The concentration, duration, and temperature must be controlled.

3. Stopping the reaction:
   • Tris/CaCl$_2$ **2 min**
   • Buffer: phosphate 100 mM **5 min**

   ↪ Rapid washing
   ↪ To stop the action of proteinase K, by removing NH$_2$ and calcium chloride by **changing the buffer.**

4. Post-fixation

   ↪ Restabilizes the tissue. **Indispensable before *in situ* hybridization**.

   • 4% PF in a phosphate buffer   **5 min**
     100 mM; pH 7.4

   ↪ This fixative provides the best compromise between the efficiency of *in situ* hybridization and the preservation of morphology.

5. Washes:
   • Buffer: phosphate   **3 × 5 min**
   • 150 mM NaCl   **3 min**

   ↪ Removes traces of fixative
   ↪ **Useful**

❑ *Next stage:*
Hybridization

↪ *See* Section 4.6.

4.5.4.2   **Pronase**

4.5.4.2.1   USE

1. Concentration	**1–10 mg/mL**

↬ Variable activity between batches

↬ The concentration depends on the thickness of the sections and the fixative used.

2. Buffer
TE buffer (10 mM **pH 7.6**
Tris–HCl/1 mM EDTA)

↬ Other buffers may be used (e.g., 10 mM Tris–HCl).

3. Temperature   **RT or 37°C**

↬ The effect of this protease is very dependent on the temperature.

↬ **Use with care.**

4. Duration   **1–10 min**

4.5.4.2.2   PROTOCOL

1. Wash:
10 mM Tris–HCl buffer/1 mM **5 min**
EDTA

↬ *See* Appendix B3.6.1.

2. Incubate in the pronase   **1–10 min**
solution.

↬ Prepare just before use.

3. Wash in Tris–glycine buffer   **5 min**
(50 mM Tris/50 mM glycine;
pH 7.4)

↬ *See* Appendix B3.7.3.

↬ Stop the action of pronase in a Tris–glycine buffer.

4. Wash in 150 mM NaCl.   **5 min**

❏ *Next stage:*
• Hybridization

↬ *See* Section 4.6.

4.5.4.3   **Pepsin**

↬ This enzyme is pH dependent.

4.5.4.3.1   USE

1. Concentration   **1–10 mg/mL**

↬ The concentration depends on the thickness of the sections and the fixative used.

2. Buffer
Pepsin/HCl   **pH 5.0**

↬ Pepsin is active at an acid pH. Its activity is reduced by increasing the pH.

3. Temperature   **RT or 37°C**

↬ The optimal temperature is 37°C, but it remains active at room temperature.

4. Duration   **1–10 min**

↬ Depends on the temperature and the pH

4.5.4.3.2   PROTOCOL

1. Wash in Tris–NaCl buffer
(50 mM Tris–HCl/150 mM NaCl); pH 7.6.

↬ *See* Appendix B3.7.5.

2. Treat sections with pepsin diluted in HCl
0.2 N; pH 5.0.

↬ The concentration, duration, and temperature must be determined for each batch.

3. Wash in Tris–NaCl buffer   **5 min**
50 mM Tris–HCl/150 mM NaCl).

↬ Inactivates the enzyme

❏ *Next stage:*
• Hybridization

↬ *See* Section 4.6.

4.5.5   Destruction of Nucleic Acids

4.5.5.1   RNase

4.5.5.1.1   USE

⇨ **Optional**
⇨ RNase A degrades single-stranded sequences associated with proteins or other macromolecules. It is used to reduce background labeling.

1. Concentration	**1–10 mg/mL**	⇨ Solution in sterile water or 2X SSC
2. Buffer		
2X SSC; pH 7.4		⇨ *See* Appendix B3.5.
3. Temperature	**RT or 37°C**	⇨ Optimal temperature is 37°C, but the enzyme remains active at room temperature.
4. Duration	**60 min**	⇨ It is not possible to overdigest the sample.

4.5.5.1.2   PROTOCOL

1. Wash in 2X SSC; pH 7.4 — **5 min**
2. Treat the sections with RNase **1 h** diluted to 10 mg/mL in 2X SSC; **37°C** pH 7.4
   ⇨ *See* Appendix B2.16.
3. Wash in 2X SSC. — **2 × 5 min**
   ⇨ **Indispensable** for the next stages

❑ *Next stage*
 • Hybridization
   ⇨ *See* Section 4.6.

4.5.5.2   DNase

4.5.5.2.1   USE

⇨ Optional
⇨ DNase degrades double-stranded sequences.

1. Concentration	**1 mg/mL**	⇨ The concentration is not important.
2. Buffer		
10 mM Tris–HCl/5 mM MgCl$_2$; pH 7.3		⇨ *See* Appendix B3.7.4. ⇨ It is recommended to add 2% RNasine and 2 mM DTT to inhibit any RNase activity.
3. Temperature	**37°C**	⇨ Optimal temperature is 37°C
4. Duration	**60 min**	⇨ The time may be reduced.

4.5.5.2.2   PROTOCOL

1. Wash in 10 mM Tris–HCl **10 min** buffer/5 mM MgCl2; pH 7.3
2. Treat the sections with DNase **1 h** diluted to 1 mg/mL in 10 mM **37°C** Tris–HCl buffer/5 mM MgCl$_2$; pH 7.3
   ⇨ *See* Appendix B2.12.
3. Wash in sterile distilled water. **3 × 2 min**
   ⇨ Under a stream of water from a wash bottle

❑ *Next stage:*
 • Hybridization
   ⇨ *See* Section 4.6.

4.5.6 Prehybridization

☞ **Optional**

4.5.6.1 Aims

The prehybridization stage is carried out just before hybridization. Sections are incubated in the hybridization mixture without the probe.

☞ Prehybridization is used to reduce the non-specific signal.

❑ *Advantage*
- Reduction of the nonspecific signal

☞ Saturation of nonspecific binding sites by the macromolecules in the buffer (dextran sulfate, Denhardt's solution, RNA, DNA)

❑ *Disadvantage*
- Reduction of the sensitivity

☞ There is a risk of diluting the hybridization buffer and thus the probe. Saturation of target sites will also occur.

4.5.6.2 Protocol

1. Float the grid, section side down, on a drop of hybridization mixture without a probe.
   - Hybridization buffer   **≥ 40 µL/grid**

☞ *See* Section 4.6.4.1. It is possible to increase the concentrations of DNA, RNA, and Denhardt's solution, leaving out the formamide.

   - Duration of incubation   **1–2 h at RT**

☞ This stage may be delayed (several hours).

2. Eliminate most of the buffer before hybridization.

☞ Do not dry the sections. The sections should remain damp to help hybridization.

❑ *Next stage:*
- Hybridization

☞ *See* Section 4.6.

4.5.7 Denaturation of Nucleic Acid Targets

☞ For double-stranded sequences

4.5.7.1 Aims

Some nucleic acid targets are double-stranded or have a secondary structure. Thus it is necessary to denature that structure for hybridization with a single-stranded complementary sequence (the probe) to take place.

☞ It is always useful to check the structure of a nucleic acid target.

Simultaneous denaturation of the nucleic acid target and the probe is not possible.

Double-stranded intracellular structures may be denatured by chemical treatment:

☞ It is not possible to heat the probe and sections to 100°C to denature them.
☞ Find the best compromise for the signal between background ratio and morphology.

- Alkaline treatment (0.5 N NaOH)

☞ This treatment must be carefully controlled.

❑ *Advantages*

- Simple
- Rapid

❑ *Disadvantage*
  • Alters the morphology

4.5.7.2 Protocol

1. Float the grid on the denaturing solution:
   • 0.5 N NaOH   **4 min**
2. Wash in sterile distilled water   **3 × 2 min**
3. Air dry.   **10 min**

❑ *Next stage:*
  • Hybridization

⮥ **Chemical treatment**

⮥ Risk of destroying tissue structure
⮥ At 4°C
⮥ To avoid dilution

⮥ *See* Section 4.6.

4.6 HYBRIDIZATION

4.6.1 Principles

To expose suitably pretreated sections to a labeled probe in hybridization solution under conditions that allow the formation of hybrids

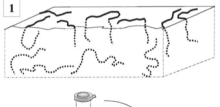

1 = **Pretreated section**

2 = **Labeled probe**

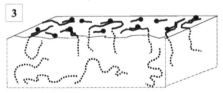

3 = **Formation of hybrids**

Figure 4.30 Principles of hybrid formation.

4.6.2 Summary of Different Stages

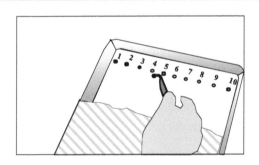

⮥ The temperature is chosen to maximize the formation of specific hybrids and to minimize the formation of nonspecific hybrids.
⮥ The duration of the protocol is unimportant (16 h). An incubation of several hours will give the same results.

Figure 4.31 Incubation of ultrathin sections with the reaction mixture (humid chamber).

4.6.3 Materials/Products/Solutions

4.6.3.1 Materials

- Centrifuge   ↝ More than 14,000 *g*
- Eppendorf tubes   ↝ Sterile
- Filter paper   ↝ For providing the humidity in the incubation chamber and for drying grids
- Gloves
- Humid chamber   ↝ A Petri dish (≈ 24 × 24 cm) with a hydrophobic film in the base for drops of solutions and a source of humidity (e.g., filter paper soaked in 5X SSC) (*see* Figure 4.31)
- Magnetic stirrer   ↝ For agitating the wash solution for the grids; the speed must be slow and stable.
- Spotting tile   ↝ Sterile for the hybridization and for the washes with as gentle an agitation as possible and placed on a magnetic stirrer
- Vortex   ↝ Slow

4.6.3.2 Products

- DNA   ↝ Molecular biology quality; preferably in ready-to-use solutions
- RNA
- Dextran sulfate   ↝ Available in solution
- Poly A   ↝ Molecular biology quality. Preferably in ready-to-use solutions
- RNase A   ↝ Without DNase
                                  ↝ Available in solution
- Sodium chloride   ↝ AR grade, anhydrous
- Sodium citrate   ↝ AR grade, anhydrous

4.6.3.3 Solutions

- 50% dextran sulfate   ↝ *See* Appendix B2.21. Keep at –20°C.
- Deionized formamide   ↝ *See* Appendix B2.14. Keep at –20°C in 500 µL aliquots. If it thaws at –20°C, do not use.
- DNA (10 mg/mL)   ↝ *See* Appendix B2.3. Sonicated. Keep at –20°C. Freezing and thawing causes breakdown.
- Poly A (10 mg/mL)   ↝ Equivalent to RNA.
- RNA (10 mg/mL)   ↝ *See* Appendix B2.5. Keep at –20°C. Freezing and thawing causes breakdown.
- RNase A [100 µg/mL] (TE/NaCl)   ↝ *See* Appendix B2.16.
- 20X SSC; pH 7.0   ↝ *See* Appendix B3.5. Keep at 4°C or at room temperature.
- Sterile water   ↝ *See* Appendix B1. Use only once after the bottle is opened or use DEPC water.
- TE Buffer/NaCl   ↝ *See* Appendix B3.6.2.

4.6.4 Protocol

⇨ **Wear gloves**
⇨ cDNA, single-stranded DNA, RNA, and oligonucleotide probes.

4.6.4.1 Hybridization Buffer

1. Make up the solution in the following order in a sterile Eppendorf tube:

Solutions	Final concentration
• SSC buffer	**4X**
• Deionized formamide	**30–50%**
• 50% dextran sulfate	**10%**
• RNA (10 mg/mL)	**100–250 µg/mL**
• DNA (10 mg/mL)	**500 µg/mL**

2. Vortex.
3. Centrifuge.

⇨ The hybridization buffer may be kept at –20°C; with or without the probe.

⇨ **Indispensable**
⇨ **Useful**
⇨ The concentration will vary according to the specificity chosen.
⇨ **Useful**
⇨ **Useful**
⇨ **Useful**
⇨ Gently
⇨ Eliminates bubbles and ensures that everything is at the bottom of the tube.

4.6.4.2 Reaction Mixture

1. Add the probe to the hybridization buffer:

• cDNA probe	**10–20 µg/mL with hybridization buffer**
• Single-stranded DNA probe	**5–10 µg/mL with hybridization buffer**
• cRNA probe	**1–20 µg/mL with hybridization buffer**
• Oligonucleotide probe	**30–200 pmoles/ mL with hybridization buffer**

2. Vortex.
3. Centrifuge.

⇨ The probes are labeled with an antigenic tag (*see* Chapter 1, Section 1.2.2).
⇨ **Indispensable:** Double-stranded probes must be denatured (*see* Chapter 1, Section 1.6.2).
⇨ After centrifugation and drying, the labeled probe is dissolved in the hybridization buffer.
⇨ These probes do not have to be denatured.

⇨ These probes may be denatured.

⇨ These probes do not have to be denatured, but should be hydrolyzed to produce sequences of around 300 nucleotides for the best results.
⇨ Gently
⇨ Gently

4.6.4.3 Hybridization

Float grids on a drop of reaction mixture on a hydrophobic film in a humid chamber (*see* Figure 4.31):

⇨ **Wear gloves**

⇨ It is possible to use a porcelain plate. 100 µL of solution is enough for 5 grids.
⇨ The humidity in the chamber must be high enough to prevent any evaporation of the hybridization buffer: place several layers of filter paper soaked in 5X SSC in the base of the box (*see* Figure 4.31).

• Drop size	**≥ 30 μL**	↝ For one grid.
• Incubation	**1 h–overnight** **at RT or 37°C** **(humid chamber)**	

❑ *Next stage:*
  • Post-hybridization treatments   ↝ *See* Section 4.7.

4.7   POST-HYBRIDIZATION TREATMENTS

4.7.1   Principles

To eliminate the nonspecific hybrids while keeping specific binding

↝ It is difficult to find the right conditions.

4.7.2   Summary of Different Stages

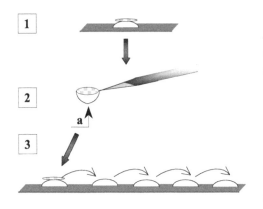

1 = Grid on drop of buffer

2 = Transfer grid without removing excess liquid.
(a) excess liquid

3 = Place grid on the next solution.

Figure 4.32   Moving the grid.

4.7.3   Solutions

↝ After hybridization, sterility is no longer important.

• 20X SSC; pH 7.0

↝ *See* Appendix B3.5. Keep at 4°C or at room temperature.

4.7.4   Protocols

4.7.4.1   Parameters

• Duration of washes	**10 min/bath**	↝ 30 min–1 hour
• Quantity of buffer	**≥ 40 μL/grid**	↝ For each grid

↝ The number and volume of washes are more important than the duration.

- Gloves do not need to be worn

  ⇒ The presence of RNase is no longer a danger except for riboprobes, which can still be broken down by this enzyme. Hybridization products, which are double-stranded cannot be broken down.

- Washing in SSC at room temperature

  ⇒ A basic pH may cause a partial degradation of hybrids.
  ⇒ It is possible to gently agitate using a spot-test porcelain tile on a magnetic stirrer.

- Room temperature

  ⇒ Washing at higher temperatures is usually too stringent.

- All solutions are passed through a 0.22 μm filter.

  ⇒ Or sterilized.

- All the changes of solution involve transferring the grid without removing the excess liquid (*see* Figure 4.32).

  ⇒ The grid should not sink to the bottom of the drop.

4.7.4.2   cDNA Probes

- 4X SSC + 30% formamide	**5 min**	⇒ The first wash is to remove the reaction mixture.
- 4X SSC	**2 × 10 min**	⇒ **Necessary**
- 2X SSC	**2 × 5 min**	⇒ **Necessary**
- 1X SSC	**2 × 5 min**	⇒ **Optional stage**
- 0.5X SSC	**2 min**	⇒ **Extra stage**

❑ *Next stage:*
- Visualization

  ⇒ *See* Section 4.8.

4.7.4.3   Single-Stranded DNA Probes

- 4X SSC	**5 min**	⇒ The first wash is to remove the reaction mixture.
- 2X SSC	**2 × 5 min**	⇒ **Necessary**
- 1X SSC	**2 × 5 min**	⇒ **Optional stage**
- 0.5X SSC	**2 min**	⇒ **Extra stage**

❑ *Next stage:*
- Visualization

  ⇒ *See* Section 4.8.

4.7.4.4   cRNA/Riboprobes

- 4X SSC + 30% formamide	**5 min**	⇒ The first wash is to remove the reaction mixture.
- 4X SSC	**2 × 10 min**	⇒ **Necessary**
- 2X SSC	**2 × 5 min**	⇒ **Necessary**
- 1X SSC	**2 × 5 min**	⇒ **Optional stage**
- 0.5X SSC	**2 min**	⇒ **Optional stage**

❑ *Next stage:*
- Visualization

  ⇒ *See* Section 4.8.

4.7.4.5   Oligonucleotides Probes

• 4X SSC	**5 min**	↝ The first wash is to remove the reaction mixture.
• 2X SSC	**2 × 10 min**	↝ **Necessary**
• 1X SSC	**2 × 5 min**	↝ **Optional stage**
• 0.5X SSC	**2 min**	↝ **Extra stage**

❑ *Next stage:*
• Visualization   ↝ *See* Section 4.8.

4.8   IMMUNOCYTOCHEMICAL VISUALIZATION

4.8.1   General Case

4.8.1.1   Principles

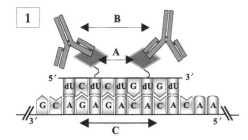

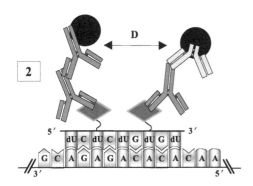

↝ The label (biotin, digoxigenin, or fluorescein) attached to the probe (*see* Chapter 1) is detected by indirect immunocytochemistry using an antibody (in general, IgG) or an immunoglobulin fragment labeled with colloidal gold.

1 = **The first stage** is the formation of an antigen–antibody complex with the hapten (biotin, digoxigenin, or fluorescein)
  (A) Hapten
  (B) Anti-hapten (raised in species X) [IgG]
  (C) Target
2 = **The next stage** is to visualize the antigen–antibody complex with a secondary antibody labeled with colloidal gold (*see* Chapter 1).
  (D) Anti-species X (IgG, Fab, F(ab′)₂); conjugated with colloidal gold

Figure 4.33   Principles of hybrid detection by indirect immunocytochemistry.

❑ *Advantages*
• Sensitivity is greater than that of direct immunocytochemistry

↝ The antigen–antibody reaction is not limited by colloidal gold.
↝ Several secondary antibody molecules may react with one primary antibody molecule amplifying the reaction.

• Availability of reagents

↝ IgG, Fab, and F(ab′)₂ conjugated with colloidal gold are commercially available.

• Multiple markers

↝ Different sizes of colloidal gold are available.

❏ *Disadvantages*
- Time consuming
- There is only one type of marker— colloidal gold.

↪ Successive incubations with each antibody
↪ Nonspecific absorption onto the resin may occur.

4.8.1.2 Solutions

- Antibodies
  1. IgG anti-hapten unconjugated
  2. IgG, F(ab′)$_2$, Fab
     Anti-species X conjugated

↪ IgG monoclonal or polyclonal
↪ All the markers may be used
↪ Colloidal gold: 5, 10, or 15 nm

- Buffers
  1. Blocking buffer
     100 mM phosphate buffer/300 mM NaCl/1% serum albumin; pH 7.4
     ± 0.1–0.01% Triton X-100

↪ *See* Appendix B6.1.
↪ It is possible to use other agents in the blocking buffer (*see* Appendix B6.1) such as:
  - Goat serum
  - Fish gelatin
  - Ovalbumin
  - Nonfat milk powder

  2. 100 mM phosphate buffer; pH 7.4
  3. 100 mM phosphate buffer/300 mM NaCl; pH 7.4
  4. 20X SSC; pH 7.0

↪ *See* Appendix B3.4.1.
↪ *See* Appendix B3.4.2.

↪ *See* Appendix B3.5. Keep at 4°C or at room temperature.

  5. 20 mM Tris–HCl buffer/300 mM NaCl; pH 7.6

↪ *See* Appendix B3.7.5.

- Fixatives
  1. 2.5% glutaraldehyde in 2X SSC; pH 7.0

↪ *See* Appendix B4.2.
↪ Maintains the salt concentration

  2. 4% paraformaldehyde (PF) in 2X SSC; pH 7.0

↪ *See* Appendix B4.3.
↪ Maintains the salt concentration

- Triton X-100

↪ Useful but not indispensable

4.8.1.3 Protocol

↪ The next stages are carried out at room temperature to avoid evaporation of the solutions.
↪ Immunodetection is carried out on drops of 40 to 50 µL on a hydrophobic film.

1. To stabilize the structures after the last wash in SSC:
   4% PF in 2X SSC   **5 min**
2. Wash:
   2X SSC   **3 × 5 min**

↪ The grids are transferred from one drop to the next without removing the excess solution (*see* Figure 4.32).

3. To block nonspecific sites
   100 mM phosphate   **15–30 min**
   buffer/300 mM NaCl/1%
   serum albumin; pH 7.4

↪ This stage is **indispensable** to reduce the nonspecific binding.

4. Primary antibody raised in species X
   - Diluted 1:50 in 100 mM **≥ 40 µL/grid**
     phosphate buffer/
     300 mM NaCl; pH 7.4

↪ Formation of an antigen–antibody complex
↪ The dilution is between 1:20 and 1:10.

• Incubation	**60 min**	⇝ Humid chamber (*see* Figure 4.31)

5. Wash:
   • Phosphate buffer/NaCl **2 × 5 min**
   • 20 m*M* Tris–HCl **2 × 5 min**   ⇝ Change of buffer
   buffer/300 m*M* NaCl; pH 7.6
6. Secondary antibody directed against spe-   ⇝ Detection of complex
   cies X and conjugated
   • Diluted 1:50 in 20 m*M*   **≥ 20 μL/grid**   ⇝ Secondary antibody: IgG, Fab fragments,
   Tris–HCl buffer/300 m*M*   conjugated and diluted 1:25–1:50 (according
   NaCl; pH 7.6   to supplier's instructions)
   • Incubation **60 min**   ⇝ Humid chamber
7. Wash:
   • Tris–HCl/NaCl buffer **2 × 5 min**
   • 2X SSC **2 × 5 min**   ⇝ Change of buffer
8. Fix: ⇝ Useful but not indispensable
   2.5% glutaraldehyde in 2X SSC **5 min**   ⇝ Paraformaldehyde may be used
9. Wash:
   • 2X SSC **5 min**
   • Sterile water **5 min**   ⇝ After fixation, tissue may be washed in
   water without damaging the hybrids.

❏ *Next stage:*
   • Staining   ⇝ *See* Section 4.9.

4.8.2   Biotin

4.8.2.1   Principles

Two possibilities:

⇝ Biotin is incorporated into the probe (*see* Chapter 1, Section 1.2.2).

1. Direct reaction

⇝ The hybrids containing one or more molecules of biotin are detected by:

1 = Streptavidin conjugated to colloidal gold
2 = Streptavidin–biotin complex conjugated to colloidal gold (commercially available)
3 = Anti-biotin immunoglobulin conjugated to colloidal gold
   (S) streptavidin
   (B) biotin
   (C) target

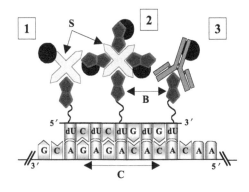

Figure 4.34 Principles of direct detection of a biotin-labeled probe.

❏ *Advantages*
   • Rapidity   ⇝ One step
   • Different diameters of colloidal gold may   ⇝ The smaller the gold particle, the more sen-
   be used   sitive the technique (e.g., 5 nm vs.15 nm = × 3 sensitivity).

- Amplification is possible.

❏ *Disadvantage*
- Weak sensitivity

2. Indirect reaction

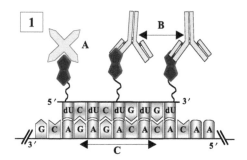

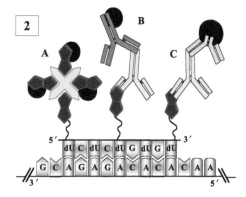

❏ *Advantages*

- Detection of the hapten by a molecule distinct from the immunoglobulins
- Specific
- Sensitive

- Multiple labeling

❏ *Disadvantages*

- Endogenous biotin

- Time consuming

⇨ But is limited

⇨ Essentially due to steric hindrance and the weak labeling of the probe

1 = First stage
  The reaction uses:
  (A) unlabeled streptavidin
  (B) anti-biotin IgG (primary antibody) raised in species X
  (C) target

2 = Second stage
  (A) Incubation with labeled biotin. Streptavidin binding sites are saturated with biotin conjugated with colloidal gold.
  (B) Incubation with anti-species X IgG conjugated with colloidal gold.
  (C) Incubation with anti-species X F(ab′)$_2$ fragments conjugated with colloidal gold

Figure 4.35 Principles of indirect detection of biotinylated hybrids.

⇨ Multiple labeling

⇨ High affinity of streptavidin for biotin
⇨ Due to the high affinity between streptavidin and biotin
⇨ Several diameters of colloidal gold are available.

⇨ Biotin is present in certain animal tissues (kidney, heart, muscle, and liver). Endogenous biotin must be inhibited by a treatment with proteases and after checking the absence of labeling from endogenous biotins.
⇨ Two successive incubations

4.8.2.2 Solutions

- Antibodies
  - IgG anti-biotin
    ➥ Conjugated (direct immunocytochemical reaction) (*see* Figure 4.34)
  - Goat serum
    ➥ Nonspecific antibody may be replaced by a blocking solution.

- Buffers
  - Blocking buffer:
    20 mM Tris–HCl buffer/300 mM NaCl/ 1% goat serum/± Triton X-100; pH 7.6
    ➥ *See* Appendix B6.1.
    ➥ The high concentration of Na$^+$ ions serves to preserve the hybrids.
    ➥ Other agents may also be used:
      - Serum albumin
      - Fish gelatin
      - Ovalbumin
      - Nonfat milk powder
  - 20X SSC; pH 7.0
    ➥ *See* Appendix B3.5. Keep at 4°C or at room temperature.
  - 20 mM Tris–HCl buffer/300 mM NaCl; pH 7.6
    ➥ *See* Appendix B3.7.5.
- Fixatives
  - 2.5% glutaraldehyde in 2X SSC; pH 7.0
    ➥ *See* Appendix B4.2.
    ➥ To preserve the salt concentration
  - 4% paraformaldehyde (PF) in 2X SSC; pH 7.0
    ➥ *See* Appendix B4.3.
    ➥ To preserve the salt concentration
- Inhibition of endogenous biotin
- Streptavidin
  ➥ Conjugated (direct immunocytochemical reaction)
- Streptavidin–biotin complex conjugated
  ➥ Direct immunocytochemical reaction (*see* Figure 4.34).

4.8.2.3 Protocol for the Direct Reaction

➥ Short protocol: 90–120 min
➥ All of the following steps are carried out **at room temperature in a humid atmosphere.**
➥ Immunodetection is carried out on drops of solution 30 to 50 μL on a hydrophobic film (*see* Figure 4.31).

1. After the final wash in SSC, the sections are fixed:
   4% PF in 2X SSC   **5 min**
   ➥ High concentration of NaCl
2. Wash:
   ➥ The grids are transferred without removing excess solution (*see* Figure 4.32).
   2X SSC   **3 × 5 min**
   ➥ Constant concentration of NaCl
3. Block nonspecific sites:
   Blocking buffer   **10–30 min**
   ➥ **Indispensable:** to eliminate any reaction on nonspecific sites, sections are preincubated with serum.

4. Inhibition of endogenous biotin

5. Formation of the antigen–antibody complex (i.e., hapten–IgG)

- Streptavidin conjugate diluted between 1:20–1:50 in 20 mM Tris–HCl buffer/300 mM NaCl **≥ 20 μL/ grid 60–90 min**
- IgG, Fab, F(ab′) anti-biotin conjugated and diluted between 1:20 and 1:50 in 20 mM Tris–HCl buffer/300 mM NaCl

6. Wash:
- 100 mM Tris–HCl buffer/300 mM NaCl **3 × 10 min**
- 2X SSC **2 × 5 min**

7. Fixation
2.5% glutaraldehyde in 2X SSC **5 min**

8. Wash:
- 2X SSC **5 min**
- Sterile water **5 min**

❏ *Next stage:*
Staining

↬ The concentration of Triton X-100 may be varied between 0.01 and 0.5%. **Optional.** Too high a concentration of Triton may cause excessive background or inhibit the reaction.

↬ **Optional**

↬ **Immunodetection.** Formation of hybrid–antibody complex

↬ Streptavidin may be replaced by an IgG anti-biotin conjugate.

↬ The dilution of the antibody is always weak to compensate for the weak sensitivity (between 1:10 and 1:100, according to the density of labeling).

↬ The signal/background ratio is generally low.

↬ Change of buffer, as Tris reacts with glutaraldehyde

↬ Necessary for staining without losing the signal

↬ After glutaraldehyde fixation, labeling is stable. Without this stage, NaCl crystals may form on the section.

↬ *See* Section 4.9

4.9 STAINING

↬ Positive staining

4.9.1 Principles

The direct observation of cellular and subcellular structures at the electron microscope is not possible without staining with heavy metals. Unless the electron beam is modified as it passes through the section there will be no signal. Electrons that are strongly deviated from their course or stopped completely will result in an area of high contrast appearing on the screen. Staining with heavy metals allows the discrimination of subcellular structures.

↬ Uranyl and lead salts

4.9.2 Summary of Different Stages

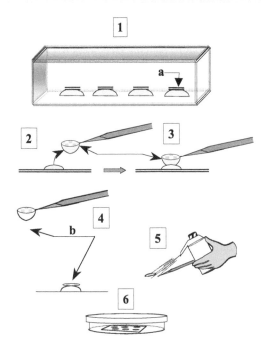

1 = **Incubation with heavy metal salts**
In darkness, float grids on uranyl acetate solution
(a) grid

2 = **Wash in distilled water.**
3 = **Transfer the grids.**

4 = **Last wash on a drop of distilled water**
(b) excess liquid
5 = **Wash freely in a jet of water.**
6 = **Dry on filter paper.**
• At room temperature in a dust-free atmosphere
• 37°C oven

Figure 4.36 Staining ultrathin sections of embedded tissue.

4.9.3 Materials/Products/Solutions

4.9.3.1 Materials

• Filter paper
• Glass Petri dish

• Grid box
• Hydrophobic film like Parafilm
• Light-tight bottle

• Plastic wash bottle
• Ultrafine, anti-magnetic, titanium forceps

⇝ For drying grids
⇝ For drying grids
⇝ For staining ultrathin sections
⇝ For storing grids
⇝ For drops of solution during staining
⇝ For storing the uranyl acetate solution (light sensitive)
⇝ Do not store the water, or filter before use.
⇝ Limits the problems of magnetism

4.9.3.2 Products

• Lead nitrate crystals
• Ruthenium red
• Sodium citrate
• Sodium hydroxide
• Uranyl acetate

⇝ Electron microscope quality (EM)
⇝ Crystals, $(PbNO_3)_2$
⇝ Mixture of certain stains
⇝ $Na_3C_6H_5O_7, 2H_2O$
⇝ NaOH (pellets)
⇝ $(CH_3COO)_2UO_2, 2H_2O$
⇝ Crystals; **radioactive (wear a mask when weighing)**

4.9.3.3 Solutions

- Aqueous uranyl acetate
  - ⇨ Light sensitive
  - ⇨ Filter before use.
  - ⇨ *See* Appendix B7.2.1.2.

  - – 2 to 5% uranyl acetate
  - – 0.5% uranyl acetate
  - ⇨ For routine staining (*see* Section 4.9.4.1)
  - ⇨ For staining with ruthenium red (*see* Section 4.9.4.2)

- Lead citrate
  - ⇨ Reynolds lead citrate (*see* Appendix B7.2.2)

- 0.1% aqueous ruthenium red
  - ⇨ Very sensitive to light

4.9.4 Protocols

4.9.4.1 Routine Staining

- Staining with uranyl acetate
  - ⇨ **In the dark**
  1. Float the grids on a drop of solution with the sections face down
     - ⇨ Place the drops of stain on hydrophobic film like Parafilm.
     - 5% aqueous uranyl acetate   **In the dark**
       - ⇨ *See* Appendix B7.2.1.2.
       - ⇨ Uranyl is sensitive to light.
       - ⇨ Ethanol degrades Lowicryl K4M.
     - Drop size   **40–50 µL/grid**
     - Time   **20–30 min**
       - ⇨ The time varies according to the fixation and the resin used.
  2. Wash the grids one at a time:
     - Distilled water   **5 × 4 min**
       - ⇨ Transfer between different drops (*see* Figure 4.32)
     - Distilled water   **Wash bottle**
       - ⇨ Wash freely in a jet of distilled water from a wash bottle (*see* Figure 4.36).
  3. Dry on filter paper.
     - ⇨ At room temperature in a dust-free atmosphere

- Staining with lead citrate
  - ⇨ According to Reynolds
  1. Place pellets of sodium hydroxide in a small container in a Petri dish with a hydrophobic film in the bottom.
     - ⇨ At room temperature in a dust-free atmosphere
  2. Keep the Petri dish closed.   **5–10 min**
     - ⇨ Keep carbon dioxide away from the solution.
  3. Rapidly place the drops of lead citrate on the film.
     - ⇨ Place the grids on the drops of stain.
     - ⇨ It is better to limit the number of grids to be stained (8 maximum).
  4. Rapidly place the grid on the lead citrate solution:
     - ⇨ Place the sections face down on the stain.
     - Reynolds lead citrate solution   **40–50 µL/grid**
       - ⇨ *See* Appendix B7.2.2.
       - ⇨ Prevent carbon dioxide from contacting the solution.
     - Time   **2–5 min**
  5. Wash the grids one after the other:
     - ⇨ Place the sections face down on the water.
     - Distilled water   **5 × 2 min**
       - ⇨ Several washes on drops of distilled water
     - Distilled water   **Wash bottle**
       - ⇨ Wash freely in a jet of distilled water from a wash bottle (*see* Figure 4.36).

6. Dry on filter paper.

➯ At room temperature in a dust-free atmosphere or in an oven at ≤ 37°C

❏ *Next stage:*
Observation

➯ The sections should be observed soon after staining.
➯ *See* Examples of Results (Figure 3).

4.9.4.2   Staining with Ruthenium Red

1. Place the grids on the drops of stain with the sections face down on the stain:
   • 0.1% aqueous solution of **3 min**
     ruthenium red **In darkness**
2. Wash freely:
   • Distilled water **5 × 4 min**

3. Place on stain:
   • 5% aqueous uranyl acetate **5–10 min**
     in darkness
4. Wash several times in drops of distilled water:
   • Distilled water **5 × 4 min**

❏ *Next stage:*
   • Observation

➯ Place several drops of stain on Parafilm.

➯ The solution is sensitive to light.

➯ Transfer the grids to several drops (*see* Figure 4.32).

➯ Uranyl is sensitive to light.
➯ *See* Appendix B7.2.1.2.

➯ The sections should be observed soon after staining.

4.10   PROTOCOL TYPE

➯ Antigenic probes

• Fixation
• Embedding in Lowicryl K4M resin
• Ultramicrotomy
• Pretreatments
• Hybridization
• Post-hybridization treatment
• Visualization of hybrids

➯ General case (indirect immunocytochemical reaction)

• Staining
• Observation

1. Fixation
   • 4% paraformaldehyde in **3–24 h**
     100 m*M* phosphate buffer **at 4°C**
     • Wash buffer **4 × 15 min**
2. Embedding in Lowicryl K4M resin
   • Dehydration
     – 35% ethanol **30 min**
       at 0°C
     – 55% ethanol **1–2 h**
       at 0°C

➯ *See* Chapter 3, Section 3.6.
➯ *See* Appendix B4.3.

➯ *See* Appendix B3.4.
➯ *See* Section 4.3.4.

– 70% ethanol	**2 × 1 h** **at −20°C**	
– 95% ethanol	**2–3 h** **at −31°C**	
• Cryo-infiltration		
– 95% ethanol + resin (2 vol/1 vol)	**2–5 h** **at −31°C**	
– 95% ethanol + resin (vol/vol)	**24 h** **at −31°C**	
– 95% ethanol + resin (1 vol/2 vol)	**6–24 h** **at −31°C**	
– 1st change/resin K4M	**overnight** **at −31°C**	⇨ *See* Appendix B5.3.1. ⇨ Use freshly prepared resin.
– 2nd change/resin K4M	**6 h** **at −31°C**	
• Cryo-embedding		⇨ Each embedding must be referenced.
• Cryopolymerization		
– UV	**2 × 15 watts** **360 nm**	
– Duration of polymerization	**5 days** **at −31°C**	

3. Ultramicrotomy — ⇨ *See* Section 4.4.
   - Ultrathin sections on grids **90–110 nm** — ⇨ Color: grey-gold
4. Pretreatments — ⇨ *See* Section 4.5.
   - Deproteinization — ⇨ *See* Section 4.5.4.
     - Proteinase K — ⇨ *See* Section 4.5.4.1.

· 20 mM Tris buffer/ 2 mM $CaCl_2$; pH 7.5	**2 min**	⇨ *See* Appendix B3.7.2.
· Proteinase K in Tris/$CaCl_2$ buffer	**3 μg/mL** **15 min** **at 37°C**	⇨ *See* Appendix B2.15.
· 20 mM Tris buffer/ 2 mM $CaCl_2$; pH 7.5	**2 min**	
· Phosphate buffer	**5 min**	
· 4% PF in phosphate buffer	**5 min**	⇨ *See* Appendix B4.3.
· Phosphate buffer	**3 × 5 min**	⇨ Removes traces of fixative

   - Pronase — ⇨ *See* Section 4.5.4.2.

· Pronase in TE buffer (10 mM Tris–HCl/ 1 mM EDTA)	**1–10 min** **pH 7.6** **at RT** **or 37°C**	⇨ Prepare just before use.
· Tris–HCl/glycine buffer	**5 min**	⇨ Stops the action of pronase
· 150 mM NaCl	**5 min**	

   - Pepsin — ⇨ *See* Section 4.5.4.3.

· Tris–HCl/NaCl buffer		
· Pepsin/0.2N HCl; pH 5.0		⇨ The concentration, the duration, and the temperature must be controlled for each batch.
· Tris/NaCl buffer; pH 7.6	**5 min**	⇨ The pH must be basic.

• Prehybridization   ↪ *See* Section 4.5.6.
  – Hybridization buffer   **≥ 40 µL/grid**
  – Length of incubation   **1–2 h**
                                       at RT
• Denaturation of nucleic acid targets   ↪ *See* Section 4.5.7.
                                               ↪ **Optional;** solely in the case of double-
                                               stranded DNA (*see* Chapter 2)
  – 0.5 N NaOH   **4 min** ↪ Controlled duration
  – Ice-cold distilled water   **2 × 3 min**
    Drying

5. Hybridization   ↪ *See* Section 4.5.6
  • Add the labeled probe in the hybridiza-
    tion buffer:
    – cDNA probe   **10–20 µg/mL** ↪ **Indispensable.** These probes are double-
                        of hybridization buffer stranded and must be denatured.

    – Single-stranded   **5–10 µg/mL** ↪ These probes are not denatured.
      DNA probe   **of hybridization buffer**

    – cRNA probe   **1–20 µg/mL** ↪ These probes may be denatured.
                        of hybridization buffer

    – Oligonucleotide **30–200 pmoles/mL**   ↪ Do not denature.
      probe   **of hybridization buffer**
  • Vortex.
  • Centrifuge.
  • Denature .   **3 min** ↪ Only for DNA probes
                                      at 100°C

  • Chill rapidly on ice
  • Hybridize
    – Place on solution   **≥ 30 µL/grid** ↪ Volume could be reduced to a few µL.
    – Incubation   **1 h–overnight**   ↪ Humid chamber
                                       at RT
                                     or 37°C

6. Post-hybridization treatments   ↪ *See* Section 4.7.
  • cDNA probe   ↪ *See* Section 4.7.4.2.
    – 4X SSC + 50% formamide   **5 min**
    – 4X SSC   **2 × 10 min**   ↪ **Necessary**
    – 2X SSC   **2 × 5 min**   ↪ **Necessary**
    – 1X SSC   **2 × 5 min**   ↪ **Optional**
  • Single-stranded DNA probe   ↪ *See* Section 4.7.4.3.
    – 4X SSC   **5 min**
    – 2X SSC   **2 × 5 min**   ↪ **Necessary**
    – 1X SSC   **2 × 5 min**   ↪ **Optional**
  • cRNA probe   ↪ *See* Section 4.7.4.4.
    – 4X SSC + 30% formamide   **5 min**
    – 4X SSC   **2 × 10 min**   ↪ **Necessary**
    – 2X SSC   **2 × 5 min**   ↪ **Necessary**
    – 1X SSC   **2 × 5 min**   ↪ **Optional**
    – 0.5X SSC   **2 × 5 min**   ↪ **Optional**

- Oligonucleotide probe   ⇨ *See* Section 4.7.4.5.
  - 4X SSC   **5 min**
  - 2X SSC   **2 × 10 min**   ⇨ **Necessary**
  - 1X SSC   **2 × 5 min**   ⇨ **Optional**
7. Immunocytochemical visualization   ⇨ *See* Section 4.8.1.3.
                                ⇨ Indirect reaction
                                ⇨ Immunodetection is carried out on drops of 40 to 50 µL on a hydrophobic film.

- Stabilizes the structures
  - 4% PF in 2X SSC   **5 min**
- Wash
  - 2X SSC   **3 × 5 min**
- Block nonspecific sites
  - 100 mM phosphate   **15–30 min**   ⇨ **Indispensable**
    buffer/300 mM NaCl/
    1% serum albumin;
    pH 7.4
- Incubate the 1st antibody anti-hapten   ⇨ Raised in species X
  - Diluted 1:50 in   **≥ 40 µL/grid**
    100 mM phosphate
    buffer/300 mM NaCl;
    pH 7.4
  - Duration   **60 min**
                 Humid chamber
- Wash
  - Phosphate buffer/NaCl   **2 × 5 min**
  - 20 mM Tris–HCl   **2 × 5 min**   ⇨ Change of buffer
    buffer/300 mM NaCl;
    pH 7.6
- Conjugated antibody (anti-species X)   ⇨ Detection of complex
  - Diluted 1:50 in   **≥ 20 µL/grid**   ⇨ Secondary antibody: IgG, Fab fragments,
    20 mM Tris–HCl   conjugated
    buffer/300 mM NaCl;
    pH 7.6
  - Incubation   **60 min**   ⇨ Humid chamber
- Wash
  - Tris–HCl buffer/NaCl   **2 × 5 min**
  - 2X SSC   **2 × 5 min**   ⇨ Change of buffer
- Fix
  - 2.5% glutaraldehyde 2X SSC   **5 min**
- Wash
  - 2X SSC   **5 min**
  - Sterile water   **5 min**   ⇨ To remove salt
8. Staining   ⇨ *See* Section 4.9.
  - Staining in uranyl acetate
    - 5% aqueous uranyl   **40–50 µL/grid**   ⇨ *See* Appendix B7.2.1.2.
      acetate   **20–30 min**
                    In darkness

– Distilled water	**5 × 4 min at RT**	⇨ Finish washing in a jet of distilled water from a wash bottle (*see* Figure 4.35).
– Drying		
• Staining in lead citrate		
– Lead citrate solution	**40–50 µL/grid 2–5 min (+ sodium hydroxide pellets)**	⇨ *See* Appendix B7.2.2. ⇨ Keep the dish closed.
– Distilled water	**5 × 4 min**	⇨ Finish wash in a jet of distilled water from a wash bottle.
9. Observation		⇨ *See* Examples of Results (Figures 3 and 23).

Chapter 5

Pre-Embedding Technique

CONTENTS

5.1   PRINCIPLES

Hybridization is carried out on tissue before it is embedded in resin. Semithin and ultrathin sections can then be observed with light and electron microscopes, respectively.

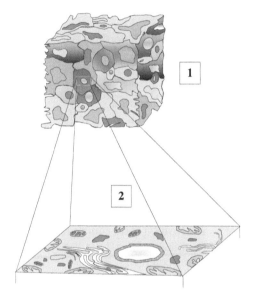

⇨ Thick sections of unembedded tissue (pre-embedding technique)

The thickness of the tissue should not be greater than 100 μm.

1 = Zone for ultrathin sections

2 = Ultrathin section

Figure 5.1 Diagram of tissue sample.

❑ *Advantages*

• Preservation of nucleic acid sequences

• Sensitivity

• Ultrastructure

⇨ Nucleic acids do not suffer any deterioration due to the embedding procedure.
⇨ Hybridization is carried out with only the tissue fixation and the penetration of the probes into the tissue to limit the reaction.
⇨ The tissue appears as it would after normal fixation and embedding in epoxy resin.

❑ *Disadvantages*

• Storage of samples

• Quantification is difficult

• Resolution

• Can only be used for antigenic probes

⇨ The technique should be carried out directly after the samples are taken.
⇨ The penetration of the probe and revelation products into the tissue is variable.
⇨ If colloidal gold cannot penetrate into the tissue, the peroxidase system is used.
⇨ The use of antibodies means that some loss of ultrastructure occurs due to the use of detergents to allow penetration into the tissue.

5.2   SUMMARY OF DIFFERENT STAGES

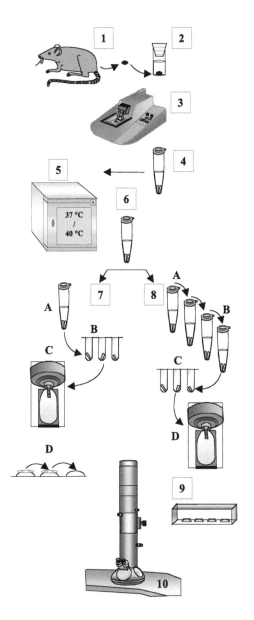

1 = **Tissue sampling**
2 = **Fixation**
     Washes

3 = **Cutting thick sections** (vibratome)
     50–200 μm

4 = **Prehybridization**
5 = **Hybridization**
6 = **Washes**

7 = **Developing the reaction on ultrathin sections** (after embedding thick sections):
     (A) Fixation
     (B) Embedding
     (C) Cutting ultrathin sections
     (D) Developing the reaction on ultrathin sections.
        • Autoradiography
        • Immunocytochemistry
8 = **Developing the reaction on thick sections (before embedding):**
     (A) Immunocytochemistry on thick sections
     (B) Post-fixation
     (C) Embedding in epoxy resin
     (D) Cutting ultrathin sections

9 = **Staining**
10 = **Observation**

Figure 5.2 Pre-embedding technique.

5.3   CUTTING VIBRATOME SECTIONS

5.3.1   Material

The vibratome is a microtome that allows the cutting of thick sections (50–200 μm) of fixed tissue. Sections are cut in PBS.

➥ Sections are called floating sections.
➥ Can be kept in 70% ethanol (*see* Section 5.3.5)

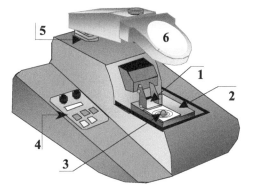

1 = Blade holder
2 = Bath filled with buffer
3 = Tissue is glued to a flat surface which cannot be oriented.
4 = Controls
5 = Support for magnifying lens and lamp
6 = Magnifying lens

Figure 5.3 Vibratome.

5.3.2   Parameters for Cutting Vibratome Sections

The tissue is in a fixed position in the bath containing buffer. A vibrating blade advances toward the tissue.
Three parameters control sectioning:

- Vertical displacement
- Speed of advance
- Oscillation of the vibration

5.3.2.1   Vertical Displacement
The displacement determines the thickness of the sections. This cannot be less than 20 μm or greater than 200 μm.

➥ The structure and the hardness of the tissue limits the thickness that can be cut.
➥ The thickness of the sections is limited by this method.

5.3.2.2   Speed of Advance
This is the speed that the blade cuts through the tissue.
   The speed should be adjusted according to the heterogeneity and hardness of the tissue.

➥ The harder the tissue, the faster the speed.
➥ The softer and the more heterogeneous the tissue, the slower the speed.

5.3.2.3 Oscillation

This allows the tissue to be sectioned.

⇨ The frequency used depends on the tissue.

5.3.2.4 Determination of Parameters

The choice of parameters depends on the size of the tissue, its hardness, and heterogeneity:

- Size
  The thickness of the section increases with the size of the tissue.

  ⇨ Vertical displacement: ↗
  ⇨ Speed of advance: ↘
  ⇨ Oscillation: →

- Hardness
  The thickness of the section decreases with the hardness of the tissue.

  ⇨ Vertical displacement: ↘
  ⇨ Speed of advance: ↘
  ⇨ Oscillation: ↗

- Heterogeneity
  Slow oscillations help to section these tissues.

  ⇨ Vertical displacement: →
  ⇨ Speed of advance: →
  ⇨ Oscillation: ↘

5.3.3 Materials/Products/Solutions

1. Material
   - Freezer

     ⇨ –20°C, –80°C

2. Minor material
   - Eppendorf tubes

     ⇨ Sterile
   - Superglue

     ⇨ Glue that is unaffected by water
   - Tool for picking up sections

     ⇨ Sterile (e.g., dissecting needle or Pasteur pipette)

3. Products

     ⇨ Electron microscope quality
     ⇨ **To be used only for *in situ* hybridization**

   - Dimethylsulfoxyde (DMSO)
   - Glycerol
   - Potassium phosphate

     ⇨ Powder
   - Sodium phosphate

     ⇨ Powder
   - Sucrose

4. Solutions
   - Cryoprotective agents
     - 10% dimethylsulfoxyde (DMSO) in 100 mM phosphate buffer

       ⇨ Good cryoprotectant, but is an organic solvent and may alter membranes
     - 10% glycerol in 100 mM phosphate buffer

       ⇨ Toxic product but has good penetration
     - 30% sucrose in 100 mM phosphate buffer

       ⇨ The most chemically neutral
   - 70% ethanol
   - 100 mM phosphate buffer

     ⇨ *See* Appendix B3.4.1.

5.3.4 Protocol

1. Orient the tissue sample.

⇨ The surface on which the tissue is mounted cannot be oriented.

2. Glue the sample to the base.

⇨ Use Superglue (e.g., cyanolite) which is unaffected by water.

3. Fill the vibratome bath with buffer.

⇨ Usually a phosphate buffer is used (*see* Appendix B3.4.1).

4. Adjust the height of the blade to just above the sample.

5. Use several advances of the blade to cut superficial and incomplete sections of the tissue.

⇨ This permits adjustments to be made (speed of advance and oscillation).

6. Lower the blade to the **50–100 μm** thickness of section required.

7. Remove the tissue section with the aid of a sterile implement (e.g., dissecting needle).

⇨ Detach the section which will often remain attached to the main body of the tissue.

8. Reverse the blade.

⇨ Start again at step 6.

❑ *Next stages:*
- Storage
- Pretreatments

⇨ *See* Section 5.3.5.
⇨ *See* Section 5.4.

5.3.5 Storage of Vibratome Sections

The only way to delay this technique is to freeze sections before pretreatment and hybridization.

⇨ This is a time-consuming step.

❑ *Freezing protocol:*
1. Place a number of sections in an Eppendorf tube containing cryoprotectant

⇨ 30% sucrose in a phosphate buffer
⇨ 10% glycerol in a phosphate buffer
⇨ 10% DMSO in a phosphate buffer

2. Freeze.   **−20°C**
                              −80°C
                            −196°C

⇨ Very slow freezing
⇨ Slow freezing
⇨ Rapid freezing
⇨ According to the method chosen, the ultrastructure of the tissue will be altered. The faster the freezing, the less damage will occur.

3. Store.   **−20°C**
                      Several days
                              −80°C
                    Several months
                            −196°C
                      No time limit

⇨ The sections are in liquid and there is less chance of dehydration.
⇨ It is better to store the sections in cryoprotectant.
⇨ The lower the temperature at which the tissue is stored, the slower any recrystallization may occur (*see* Section 5.3.5).

❑ *Next stage:*
- Pretreatments

⇨ *See* Section 5.4.

5.4 PRETREATMENTS

5.4.1 Aims

Pretreating allows the labeled probe to reach the target nucleic acid present within the cells. In sections, several layers of intact cells may be present.

↪ The thickness of the sections makes this stage indispensable.

↪ Permeabilization and deproteinization are the principal components.

1 = Before pretreatment
Nucleic acid is one of the following types:
- Double-stranded DNA
- Single-stranded mRNA

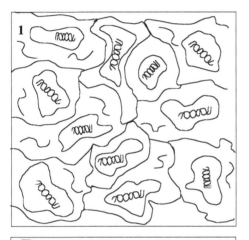

2 = After pretreatment
- Deproteinization
- Permeabilization

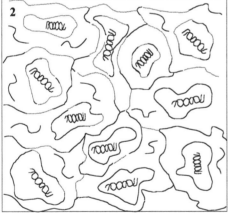

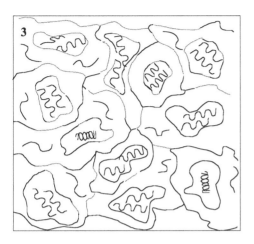

3 = Denaturation of the target nucleic acid
(double-stranded DNA)

Figure 5.4 Pretreatments.

5.4.2   Materials/Products/Solutions

5.4.2.1   Materials

- Centrifuge.   ⇌ More than 14,000 g
- Oven at 37°C
- Vortex.   ⇌ Slowly

5.4.2.2   Products   ⇌ **To be used solely for *in situ* hybridization.**

- Calcium chloride
- Dithiothreitol (DTT)   ⇌ Molecular biology quality
                                       ⇌ $C_4H_{10}O_2S_2$
- DNA   ⇌ Salmon or herring sperm, in solution/ sonicated
                                       ⇌ Preferably in a ready-to-use solution
- Hydrochloric acid   ⇌ AR quality
- Potassium phosphate   ⇌ Powder
- Proteinase K   ⇌ Molecular biology quality
- RNA   ⇌ Transfer RNA in solution
- Sarcosyl or saponin   ⇌ Molecular biology quality
- Sodium chloride   ⇌ AR quality, anhydrous
- Sodium citrate   ⇌ AR quality
- Sodium phosphate   ⇌ Powder
- Tris-hydroxymethyl-aminomethane (Tris) ⇌ Powder, keep dry.
- Triton X-100   ⇌ Keep away from light.

5.4.2.3   Solutions

- Buffers:
  - 10X phosphate buffer; pH 7.4

  - 20X SSC; pH 7.0

  - Tris–HCl buffer/CaCl$_2$ (20 mM Tris/2mM CaCl$_2$); pH 7.6
- Deionized formamide

- 50X Denhardt's solution

- Detergents:
  - 0.2% Saponin in phosphate buffer
  - 0.2% Sarcosyl in phosphate buffer
  - 0.05% Triton X-100 in phosphate buffer
- DNA (10 mg/mL)

- 70%, 95%, 100% ethanol
- 4% paraformaldehyde (PF) in 100 mM phosphate buffer; pH 7.4
- Pronase (0.1 mg/mL) in phosphate buffer; pH 7.4
- Proteinase K 1 mg/mL in Tris–HCl buffer/ CaCl$_2$ (20 mM Tris/2mM CaCl$_2$); pH 7.6
- RNA (10 mg/mL)

- Sterile water

- 10 N Sodium hydroxide

⮑ **Sterility of the solutions is very important for hybridization** (*see* Appendix A2).

⮑ Store at room temperature for 1 month after sterilization (*see* Appendix B3.4.1).
⮑ Store at 4°C or at room temperature (*see* Appendix B3.5).
⮑ Store at 4°C or at room temperature (*see* Appendix B3.7.2).
⮑ Store at –20°C in 500 μL aliquots. If the solution thaws at –20°C, do not use (*see* Appendix B2.14).
⮑ Store at –20°C in aliquots. It can be thawed several times (*see* Appendix B2.19).

⮑ *See* Appendix B2.18.

⮑ Sonicated; store at –20°C. Repeated freezing and thawing brings about breakage of the strand (*see* Appendix B2.3).

⮑ *See* Appendix B4.3. Store at –20°C.

⮑ Storage at –20°C in aliquots of 50 μL.

⮑ *See* Appendix B2.15. Storage at –20°C in aliquots of 50 μL.
⮑ Store at –20°C. Repeated freezing and thawing brings about breakage of the strand (*see* Appendix B2.5).
⮑ Use only once or use DEPC water (*see* Appendix B1.1).
⮑ Do not keep if a precipitate appears (*see* Appendix B2.20).

5.4.3   Permeabilization

5.4.3.1   Principles

Allows access of the labeled probe to the nucleic acid target by partial destruction of the cell membranes (*see* Figure 5.4).

Cell membranes are made up of phospholipids that may be solubilized:

- Detergents
- Alcohol
- Enzymes

⮑ This stage must be carefully controlled.

⮑ Triton X-100, saponin, or sarcosyl
⮑ Ethanol
⮑ Lipase

❏ *Advantages*
- Increased sensitivity
- Produces a partial deproteinization

➥ Allowing penetration of the labeled probe
➥ Limits the deproteinization stage

❏ *Disadvantages*
- Alteration of the ultrastructure

- Increasing reaction time

➥ Breakage of the cell membranes results in loss of some cell contents.
➥ The longer the reaction, the greater the destruction of cell contents. It is necessary to fix the tissue to stabilize it.

5.4.3.2   Protocol

1. Incubate sections:
   - 0.05% Triton X-100 in phosphate buffer, or   **30 min**
   - 0.2% saponin in phosphate buffer, or   **30 min**
   - 0.2% sarcosyl in phosphate buffer   **30 min**
2. Wash sections in the same buffer
   - Phosphate buffer   **3 × 10 min**

➥ Each new batch should be tested.

❏ *Next stages:*
- Deproteinization
- Prehybridization

➥ *See* Section 5.4.4.
➥ *See* Section 5.4.5.

5.4.4   Deproteinization

5.4.4.1   Principles
To eliminate proteins associated with nucleic acids before hybridization

This obstacle to hybridization may be eliminated by:

- Enzymes
- Detergents

➥ The best conditions should be determined in advance.

➥ Specific and rapid
➥ Simultaneously permeabilizes

5.4.4.2   Protocol

5.4.4.2.1   PROTEINASE K

1. Incubate sections in the following solutions:
   - 20 mM Tris buffer/ CaCl$_2$ 2 mM; pH 7.6   **15 min at RT**
   - Proteinase K in Tris–HCl buffer/CaCl$_2$ **1 µg/mL 1 h at RT**

➥ Proteolytic enzyme

➥ The concentrations and the time of incubation must be assessed in advance.
➥ The choice of buffer can modify the activity of proteinase K (e.g., absence of CaCl$_2$ diminishes the activity of the enzyme).

- 100 mM phosphate buffer; pH 7.4   **2 × 15 min**
    ⮑ This stops the action of proteinase K by eliminating calcium chloride and NH$_2$ groups in the Tris buffer.
    ⮑ Indispensable before hybridization

2. Post-fixation
   - 4% PF in 100 mM phosphate buffer; pH 7.4   **5 min**
       ⮑ *See* Appendix B4.3.1.
3. Wash sections
   - Phosphate buffer   **3 × 5 min**
       ⮑ Removal of the enzyme
       ⮑ Removal of traces of fixative

❏ *Next stage:*
- Prehybridization
    ⮑ *See* Section 5.4.5.

5.4.4.2.2 PRONASE

1. Wash:
   - 100 mM phosphate buffer   **10 min**
2. Incubate sections with proteolytic enzyme
    ⮑ Varies with each new batch of enzyme
   - Pronase B in 100 mM phosphate buffer; pH 7.4   **25 μmoles/mL**   **15 min**   **at RT**
3. Wash sections in the same buffer
   - Phosphate buffer   **3 × 10 min**
       ⮑ Eliminates the enzyme
   - PF 4% in phosphate buffer   **5 min**
       ⮑ *See* Appendix B4.3.1.
   - Phosphate buffer   **3 × 5 min**
       ⮑ Eliminates traces of fixative

❏ *Next stage:*
- Prehybridization
    ⮑ *See* Section 5.4.5.

5.4.5 Prehybridization

5.4.5.1 Principles

The aim of this stage is to saturate all nonspecific sites.

This is carried out before hybridization and consists of incubating the thick sections in the hybridization mixture in the absence of the probe.

⮑ The labeled probe may adhere to nonspecific sites on proteins and nucleic acids.
⮑ The presence of macromolecules and double-stranded nucleic acids in the mixture saturates nucleic acids and proteins present in the tissue section, which may bind the probe.

❏ *Advantage*
- Diminishes the nonspecific signal

❏ *Disadvantages*
- Diminishes the sensitivity
    ⮑ Sometimes causes the reassociation of proteins with nucleic acids
- Lengthens the reaction time
    ⮑ This stage is generally useful.

5.4.5.2 Prehybridization Buffer

⮑ The buffer may be kept at –20°C

1. Place the following in a sterile Eppendorf tube:

Solutions	Final concentration
• 20X SSC	**4X**
• 50X Denhardt's solution	**1X**
• tRNA 10 mg/mL	**100 µg/mL**
• DNA 10 mg /mL	**100 µg/mL**
• Sterile water	**to x mL**

2. Vortex to mix.
3. Centrifuge.

↪ Eliminates bubbles

5.4.5.3 Protocol

1. Incubate floating sections in the hybridization mixture without the probe:
   - Prehybridization buffer ≈ **100 µL**

   ↪ The sections must be immersed in the buffer.
   - Duration of incubation **1 h**
     at 37°C

   ↪ May be delayed (several hours at room temperature)
2. Remove as much of the buffer as possible.

   ↪ The thick sections must not be allowed to dry out.

❏ *Next stage:*
   • Hybridization

↪ *See* Section 5.5.

5.5 HYBRIDIZATION

5.5.1 Principle

Incubate floating sections with the labeled probe in a hybridization buffer at a temperature that allows specific hybridization.

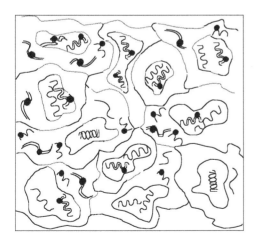

Figure 5.5 Hybridization.

5.5.2   Materials/Products/Solutions

5.5.2.1   Materials

- Centrifuge
- Gloves
- Oven
- Vortex

⇝ **Sterilization of material is important.**

⇝ More than 14,000 *g*

⇝ 37°C–40°C
⇝ Slow

5.5.2.2   Products

- Calcium chloride
- Dithiothreitol (DTT)
- DNA

- Formamide
- RNA
- Sodium chloride
- Sodium citrate

⇝ **AR quality**
⇝ **Use only for *in situ* hybridization.**

⇝ Molecular biology quality
⇝ Salmon or herring sperm in solution; sonicated
⇝ Preferably in ready-to-use solution
⇝ Deionized
⇝ Transfer, in solution, ready to use
⇝ AR quality, anhydrous
⇝ AR quality

5.5.2.3   Solutions

- 50X Denhardt's solution

- Deionized formamide
- Dithiothreitol (DTT): 1 *M*

- DNA: 10 mg/mL
- Probes
  – Antigenic probe

  – Radioactive probe

- 10 N Sodium hydroxide
- 20X SSC; pH 7.0

- Sterile water

- tRNA: 10 mg/mL

⇝ Molecular biology quality
⇝ *See* Appendix B2.19.
⇝ *See* Appendix B2.14.
⇝ *See* Appendix B2.11. Store at –20°C in 50 to 100 µL aliquots. Do not freeze a second time. The strong odor indicates that the chemical is functional. Do not use if the smell is absent or changed in any way.
⇝ *See* Appendix B2.3.

⇝ cDNA, oligonucleotides (*see* Section 5.5.4.2)

⇝ cDNA, oligonucleotides (*see* Section 5.5.4.1)

⇝ Store at 4°C or at room temperature (*see* Appendix B3.5).
⇝ *See* Appendix B1.1. Use only once or use DEPC water.
⇝ *See* Appendix B2.5.

5.5.3 Hybridization Buffer

1. Place in a sterile Eppendorf tube in the following order:

Solutions	Final concentration
• 20X SSC; pH 7.0	**4X**
• Deionized formamide	**50%**
• tRNA: 10 mg/mL	**100–250 μg/mL**
• DNA 10 mg /mL	**100–400 μg/mL**
• 50X Denhardt's solution	**1–2X**
• 1 *M* DTT	**10 m*M***

⇨ The hybridization buffer may be kept at –20°C with or without the probe.

⇨ Hybrids will form at a neutral pH in the presence of Na⁺ ions (ionic forces).
⇨ The concentration of Na⁺ ions stabilizes the hybrids.
⇨ **Indispensable**
⇨ Addition of formamide diminishes the hybridization temperature and facilitates the penetration of the different constituents.
⇨ **Important** for the efficiency and the specificity of the hybridization and diminishes the temperature necessary for hybridization and preserves the morphology of the tissue.
⇨ **Useful.** Formamide must be good quality (very pure) and deionized (to maintain the pH of the buffer).
⇨ Competitively saturates nonspecific binding to proteins at high concentrations
⇨ **Useful**
⇨ Saturates nonspecific binding.
⇨ **Useful** for cDNA probes
⇨ Competitively saturates nonspecific binding to macromolecules; the concentration may be increased.
⇨ **Useful**
⇨ Use only with ³⁵S labeled probes (antioxidant).
⇨ **Important,** do not add until after denaturation as it is destroyed at 100°C.

2. Mix by vortex.
3. Centrifuge.

⇨ Gently
⇨ Eliminates bubbles, collects all drops of solution at the bottom of tube

5.5.4 Reaction Mixture

⇨ Hybridization mixture containing the probe

5.5.4.1 Radioactive Probe

1. In an Eppendorf tube, resuspend the labeled probe in the hybridization buffer:
   • Labeled probes:
     – cDNA **0.1–1 ng/mL hybridization buffer**
     – Oligonucleotides **1–5 pmoles/mL hybridization buffer**

⇨ Dried probe
⇨ Denaturation is **indispensable.**

⇨ Do not denature.

2. Mix by vortex.   ⇀ Gently

3. Centrifuge.

4. Denature.   **3 min**   ⇀ *See* Chapter 1, Figure 1.17.
        at 100°C   ⇀ **Indispensable.** Use only for cDNA probes.

5. Chill rapidly on ice.   ⇀ To prevent rehybridization of the two denatured strands; at 0°C, the single strands are stable.

6. Add:
   - DTT   **10 mM**   ⇀ Use only for ^{35}S labeled probes.

7. Rapidly hybridize to floating sections.   ⇀ *See* Section 5.5.5.

5.5.4.2   Antigenic Probe

1. In an Eppendorf tube, resuspend the labeled probe in the hybridization buffer:
   - Labeled probes   ⇀ Dried probe
     - cDNA   **5–25 ng/mL hybridization buffer**   ⇀ Denaturation is **indispensable.**
     - Oligonucleotides   **5–200 pmoles/ mL hybridization buffer**   ⇀ Do not denature.

2. Mix by vortex.

3. Centrifuge.

4. Denature.   **3 min**   ⇀ *See* Chapter 1, Figure 1.17.
        at 100°C   ⇀ **Indispensable.** Use only for cDNA probes.

5. Chill rapidly on ice.   ⇀ To prevent rehybridization of the two denatured strands; at 0°C, the single strands are stable.

6. Rapidly hybridize to floating sections.   ⇀ *See* Section 5.5.5.

5.5.5   Protocol

⇀ **Wear gloves**
⇀ Radioactive probe
⇀ Antigenic probe

1. Transfer the floating sections to the reaction mixture.   ⇀ Remove as much of the prehybridization buffer as possible without drying the sections.

2. Incubate:
   - Radioactive probe   **3–16 h**   ⇀ A short hybridization time preserves the morphology without greatly reducing the signal.
          37–40°C   ⇀ The hybridization temperature may be raised to increase the specificity or lowered to decrease the signal.
   - Antigenic probe   **overnight 37°C**   ⇀ The penetration of the probes is slower due to the attached marker proteins.

❑ *Next stage:*
   - Post-hybridization treatment   ⇀ *See* Section 5.6.

5.6   POST-HYBRIDIZATION TREATMENT

5.6.1   Aims

This treatment specifically removes any non-hybridized probe (i.e., removes all the reaction mixture and excess probe and denatures all non-specific hybrids).

⇨ Present in the reaction mixture, adsorbed onto tissue structures or nonspecifically hybridized

Figure 5.6   Post-hybridization treatment.

5.6.2   Materials/Products/Solutions

1. Materials:
   • Oven
   • Tissue culture plates

⇨ 37°C
⇨ Sterile (*see* Figure 5.7).
⇨ 12 wells (diameter ≈ 1 cm)

2. Solutions:
   • Buffers
     – PBS
     – 20X SSC; pH 7.0

⇨ *See* Appendix B3.4.3.
⇨ Store at 4°C or at room temperature (*see* Appendix B3.5).

5.6.3   Protocols

❏ *Parameters:*

   • Duration of the washes
   • Quantity of buffer to prepare

⇨ ≈ 5 hours
⇨ Prepare a minimum of 2 mL for each wash
⇨ Numerous washes are necessary if the probe is radioactive.

- Conditions
  - Gloves do not need to be worn.

➾ RNase-free conditions are not necessary.

➾ Hybrids are double-stranded and cannot be degraded, unlike the nonhybridized probe which may be easily broken down.

- Washes are carried out in SSC, at room temperature or 37°C

➾ If the wash temperature has already been optimized for a light microscope *in situ* hybridization, the same conditions may be used for floating sections.

❏ *Protocol:*

1. Wash in a succession of changes of SSC:

➾ Use sterile tissue culture plates (*see* Figure 5.7)

- Radioactive probe:
  - 2X SSC   **2 × 15 min at RT**
  - 2X SSC   **3 × 15 min at 37°C**
  - 1X SSC   **2 × 60 min**
  - 0.5X SSC   **2 × 60 min**

➾ As the antibodies do not penetrate to the center of the section (*see* Figure 5.18), it is better to maintain the integrity of the tissue at the surface of the section than to eliminate nonspecific hybrids that will never be visible.

➾ The duration of the washes is reduced when an antigenic probe is used.

- Antigenic probe
  - 2X SSC   **5 × 15 min at RT**
  - 1X SSC   **2 × 60 min**

2. Immerse:
   - PBS

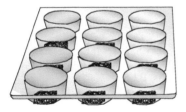

❏ *Next stages:*

➾ *See* Table 5.1.

- Fixation and embedding:
  - Vibratome sections in epoxy resin
  - Vibratome sections in acrylic resin

➾ *See* Section 5.7.4.

- Visualization of antigenic hybrids by enzymatic detection on vibratome sections

➾ *See* Section 5.8.3.1.

Figure 5.7 Washing floating sections.

Table 5.1 Summary of the Possible Stages after Post-Hybridization Treatment

Radioactive Hybrids		Antigenic Hybrids		
Post-hybridization Treatment				
Embedding thick sections in epoxy resin (§5.7.5.1)		Enzymatic development on thick sections (§5.8.3.1)		Embedding thick sections in acrylic resin* (§5.7.5.2)
Semithin sections	Ultrathin sections	Embedding thick sections in epoxy resin		Ultrathin sections
Developing autoradiography	Developing ultrastructural autoradiography	Semithin sections	Ultrathin sections	Immunocytochemistry (colloidal gold)
Light microscope	Transmission electron microscope	Light microscope	Transmission electron microscope	
* Embedding in acrylic resin is described in Chapter 4, Section 4.3.				

5.7   EMBEDDING MEDIA

5.7.1   Choice of Resin

In transmission electron microscopy, embedding resins can be divided into two major groups according to their polarity and hydrophobicity:

- Epoxy bases
- Acrylic bases

The choice of embedding media is determined by the way in which the hybrids are visualized (i.e., before or after embedding vibratome sections).

1. Visualization prior to embedding: After visualization of hybrids, it is necessary to maintain the reaction products in place. A fixative that preserves the ultrastructure without modifying the reaction product (e.g., glutaraldehyde and/or osmium tetroxide) may be used under standard conditions.

↪ Epoxy or acrylic

↪ Very hydrophobic
↪ Hydrophilic

↪ Hydrophobic resin

↪ Glutaraldehyde preserves nuclear structures well.
↪ Osmium tetroxide is a fixative which also gives contrast to the sections (the heavy atoms of osmium stop electrons).
↪ Used as a marker for the enzyme reaction (e.g., peroxidase)

2. Post-embedding visualization: It is possible to embed tissue after the post-hybridization washes and to carry out the visualization on ultrathin sections. In this case, embedding should maintain hybrids and not alter the marker.

⇨ Hydrophilic resin

Table 5.2 Summary of the Possible Stages According to Radioactive or Antigenic Probes

Label	Radioactive	Antigenic
Visualization		
Before embedding: thick sections		Weak fixation Immunocytochemistry Epoxy resin
After embedding: ultrathin sections	Strong fixation Epoxy resin Autoradiography	Weak fixation Acrylic resin Immunocytochemistry

5.7.2 Types of Epoxy Resin

Epoxy resins are used for ultrastructural observations. They are not recommended for immunocytochemical studies due to their hydrophobic nature and the possibility of copolymerization with biological material.

⇨ They form covalent bonds with biological material and this copolymerization makes it difficult for the probes to access the target nucleic acid sequence.

Different epoxy resins may be used after *in situ* hybridization. These are:
- Epon
- Spurr (ERL)
- Araldite
- Epon–Araldite

⇨ Difficult to obtain commercially

5.7.2.1 Characteristics
Epoxy resins are composed of a mixture of different monomers, hardeners, and accelerators:

1. Monomers:
   - Araldite
     - Chain with aromatic rings
   - Glycidyl ether 100
     - Aliphatic carbon chain

2. Epoxy hardeners:
   - Dodecenyl succinic anhydride
   - Methyl nadic anhydride:
     - Nonenyl succinic anhydride

⇨ Araldite M or CY 212
⇨ Polyarylethers of glycerol
⇨ Epikote 812
⇨ A mixture of glycidyl ethers of glycerol mono- to tri-substituted
⇨ Polymerization of the resin
⇨ DDSA or HY964
⇨ MNA or HY906
⇨ NSA

– Hexahydrophthalic anhydride

⇝ Epikote HPA

3. Epoxy accelerators:
   • 2,4,6-tridimethylamine methyl phenol
   • Dimethylaminoethanol

⇝ DMP 30 or DCY 264
⇝ DMAE or S-1

4. Additive:
   • Dibutyl phthalate

⇝ Plasticizer

The viscosity of epoxy resins is high due to the large size of the monomers. It increases with increasing size of the monomers.

⇝ ≈ 150–1650 centipoises; ERL is the least viscous.
⇝ Infiltration is very slow.

The advantage of epoxy resins is that the mixture can be adjusted, resulting in variable hardness.

⇝ Facilitates embedding and sectioning of very hard tissues such as bone or cartilage

Polymerization by heat is assured by the presence of a heat stable amine (DDSA).

⇝ Hardening of the epoxy resin is obtained at 60°C.

Shrinkage is low.

⇝ Only 4% loss of volume during polymerization; this is between 10 and 20% for acrylic resins.

These resins are not soluble in water but are soluble in certain organic solvents.

⇝ Embedding of biological samples requires total dehydration.
⇝ For example, acetone, ethanol, and propylene oxide

They are very stable under the electron beam.

⇝ Acrylic resins are less stable and require a support film.

5.7.2.1.1 EPON

⇝ Viscous before polymerization.

1. Epon is made up of a mixture of Epon A and Epon B, to which an accelerator is added:

⇝ Soluble in ethanol, acetone, and propylene oxide, (epoxy 1-2 propane).
⇝ Equal volumes of Epon A and B give a resin of medium hardness.

   • Epon A:
     – Glycidyl ether 100
     – Dodecenyl succinic anhydride

⇝ Replaces Epikote 812
⇝ DDSA (hardener)

   • Epon B:
     – Glycidyl ether 100
     – Methyl nadic anhydride

⇝ MNA (hardener)

   • Accelerator:
     – 2,4,6-tridimethylamine methyl phenol

⇝ DMP 30

2. The resin is more fluid than Araldite.

⇝ The monomers are shorter.

3. Heat polymerization

⇝ 60°C

5.7.2.1.2 SPURR (ERL)

⇝ Fluid before polymerization

1. ERL is composed of the following products:
   • Vinylcyclohexene dioxide
   • Nonenyl anhydride succinic
   • Dow epoxy resins
   • S-1

⇝ Soluble in ethanol and acetone
⇝ ERL 4206
⇝ NSA (hardener)
⇝ DER 736
⇝ Accelerator

2. The resin is very fluid and consequently penetrates biological samples very quickly. It is very hard and stable under the electron beam.
3. Highly toxic   ⇨ Carcinogenic
4. Heat polymerization   ⇨ 60°C

5.7.2.1.3 ARALDITE

1. Araldite is made up of the following products:
   - Araldite M or CY 212
   - Dodecenyl succinic anhydride
   - Dibutyl phthalate
   - 2,4,6-tridimethylamine methyl phenol

   ⇨ Viscous before polymerization.
   ⇨ Soluble in ethanol, acetone, and propylene oxide (epoxy 1-2 propane)

   ⇨ DDSA (hardener)
   ⇨ Plasticizer
   ⇨ DMP 30 (accelerator)

2. The resin is stable under the beam and the viscosity is high due to the large size of the monomers
3. Heat polymerization   ⇨ 60°C

5.7.2.1.4 EPON–ARALDITE

1. Epon–Araldite is made up of the following products:
   - Araldite CY 212
   - Glycidyl ether 100
   - Dodecenyl succinic anhydride
   - 2,4,6-tridimethylamine methyl phenol

   ⇨ Soluble in ethanol, acetone, and propylene oxide (epoxy 1-2 propane).

   ⇨ Replaces Epikote 812
   ⇨ DDSA (hardener)
   ⇨ DMP 30 (accelerator)
   ⇨ Viscous before polymerization

2. The resin is more fluid than Epon and consequently penetrates tissue more rapidly.
3. Medium hardness

   ⇨ Sections may be unfolded and spread using xylene vapors from a soaked filter paper.

4. Heat polymerization   ⇨ 60°C

Table 5.3 Summary of the Characteristics of Epoxy Resins

Type of Epoxy Resin	Characteristics				
	Penetration	Viscosity (before polymerization)	Hardness	Solvents	Stability
Epon	Rapid	Fluid	+++	Ethanol, acetone, propylene oxide	Very good
Spurr (ERL)	Very rapid	Fluid	+++	Ethanol, acetone	Very good
Araldite	Very slow	Viscous	+	Ethanol, acetone, propylene oxide	Very good
Epon–Araldite	Rapid	Viscous	++	Ethanol, acetone, propylene oxide	Good

5.7.2.2   Advantages/Disadvantages

❏ *Advantages*

* Easy to cut ultrathin sections

* Stability of sections under the electron beam

* Preservation of ultrastructure

* Storage of blocks

↪ These resins are very hard and permit ultrathin sections to be cut.

↪ Sections can be placed directly on the grid without a support film.

↪ Equivalent to that obtained with acrylic resins

↪ At room temperature for an unlimited time

❏ *Disadvantages*

* Nonspecific labeling

* Weak sensitivity

↪ Epoxy resins have a strong affinity for serum constituents.

↪ The absence of water often modifies antigenic sites.

5.7.3   Materials/Products/Solutions

1. Materials:
   * Disposable pipettes
   * Dry gelatin capsules

   ↪ Using small diameter (size 00) gelatin capsules makes embedding more rapid.
   ↪ Embed in very dry gelatin capsules (≥ 30 min at 37°C).

   * Embedding tubes and rubber stoppers
   * Flat embedding molds
   * Gloves
   * Oven
   * Pasteur pipettes
   * Siliconized coverslip
   * Vortex

   ↪ To orient the samples
   ↪ **Indispensable**
   ↪ 60°C to polymerize the resin

   ↪ *See* Appendix B3.3.
   ↪ Powerful; to homogenize the mixture

2. Products:
   * Epoxy embedding mixture:
     – Hexahydrophthalic anhydride
     – Dodecenyl succinic anhydride
     – Methyl nadic anhydride
     – 2,4,6-tridimethylamine methyl phenol

   ↪ *See* Appendix B5.2.
   ↪ Epikote 812.
   ↪ DDSA (hardener)
   ↪ MNA (hardener)
   ↪ DMP 30 (accelerator)

3. Solutions:
   * Buffers
     – 100 mM phosphate buffer; pH 7.4
     – 100 mM cacodylate buffer; pH 7.4
     – PBS; pH 7.4
   * Fixative
     – 2.5% glutaraldehyde in buffer; pH 7.4

     – 1% osmium tetroxide in buffer; pH 7.4

   * Infiltration solutions

   ↪ *See* Appendix B3.
   ↪ *See* Appendix B3.4.1.
   ↪ *See* Appendix B3.1.
   ↪ *See* Appendix B3.4.3.

   ↪ *See* Appendix B4.2.
   ↪ Salt buffer
   ↪ *See* Appendix B4.1.
   ↪ Salt buffer
   ↪ Prepare the different infiltration solutions.

- Resin
  - Epoxy resin

  - Acrylic resin

  ↝ Such as Epon (preparation, *see* Appendix B5.2.2).
  ↝ Such as LR White medium (preparation, *see* Appendix B5.3.2)

- Solvents
  - 30%, 50%, 70%, 80%, 90%, 95% and 100% ethanol
  - Propylene oxide

5.7.4   Fixation of Vibratome Sections

5.7.4.1   Principles

Vibratome sections may be embedded before or after visualization of hybrids.

Depending on the protocol used, sections may be post-fixed before embedding:

- Strong fixation before embedding in an epoxy resin

  ↝ A radioactive marker or the precipitate of the immunocytological reaction must be visualized.

- Weak fixation before embedding in an acrylic resin

  ↝ Hybrid must be accessible to the visualization system.

5.7.4.2   Protocols

After hybridization and washes, the thick sections are fixed.

↝ The time may be defined for 50-μm sections.

1. Before embedding in epoxy resin
   - Incubate:
     - 2.5% glutaraldehyde   **≥ 200 μL/**
       in buffer   **thick sections**
                                 60 min

     ↝ **Strong fixation**
     ↝ *See* Appendix B4.2.
     ↝ Use sterile culture dishes

   - Wash:
     - Buffer   **2 × 30 min**

     ↝ Same buffer as fixation

   - Post-fixation:
     - 1% osmium tetroxide   **≥ 200 μL/**
       in buffer   **thick section**
                                 60 min

     ↝ *See* Appendix B4.1.
     ↝ **Dangerous vapors: avoid contact**

   - Wash:
     - Buffer   **2 × 30 min**

❑ *Next stage:*
   - Embedding in epoxy resin

   ↝ Such as Epon (*see* Section 5.7.5.1).

2. Before embedding in a hydrophilic resin
   - Incubate:
     - 4% paraformaldehyde   **≥ 200 μL/**
       in buffer   **thick section**
                                 2–3 h
                                 at 4°C

     ↝ **Weak fixation**
     ↝ *See* Appendix B4.3.1.
     ↝ Use sterile culture dishes.
     ↝ Depending on the thickness of the section

- Wash:
  - Buffer   **2 × 30 min**   ↩ Same buffer as fixation

❏ *Next stage:*
- Embedding in acrylic resin   ↩ Such as LR White medium (*see* Section 5.7.5.2).

5.7.5   Embedding of Vibratome Sections

5.7.5.1   Embedding in an Epoxy Resin

↩ We will show the protocol for embedding in epoxy resin.

↩ All of the following steps are carried out at room temperature.

1. Dehydration

↩ The times may be changed depending on the thickness of the vibratome sections.

- 30% ethanol   **15–30 min**
- 50% ethanol   **15–30 min**

↩ At each change of solution, aspirate the liquid with a disposable pipette and replace with the next solution.

- 70% ethanol   **15–30 min**
- 80% ethanol   **15–30 min**
- 90% ethanol   **3 × 15–30 min**
- 95% ethanol   **3 × 15–30 min**
- 100% ethanol   **3 × 15–30 min**

↩ Total dehydration must be achieved. Spending too long in this solution is not good for the tissue and results in problems with sectioning.

2. Infiltration
   - Infiltrate the vibratome sections in the solvent with increasing concentrations of resin:

↩ Prepare the different infiltration solutions.

↩ In closed containers

↩ The time of infiltration may vary from 45 minutes to several hours depending on the thickness of the sections.

    – Propylene oxide + Epon   **2 vol + 1 vol**
                                 45 min
                                 at RT

↩ Toxic; use in a fume hood.

    – Propylene oxide + Epon   **1 vol + 1 vol**
                                 45 min
                                 at RT

    – Propylene oxide + Epon   **1 vol + 2 vol**
                                 45 min
                                 at RT

- Replace the substitution solution with Epon:

↩ Use freshly prepared resin.

↩ Do not stopper containers to allow complete evaporation of solvents.

    – 1st change: **1–2 h**
       Epon + DMP 30   **at RT**

↩ **Toxic; use in a fume hood.**

    – 2nd Change   **overnight**
       Epon + DMP 30   **at 4°C**

↩ Avoid air bubbles

3. Embedding

- Lift the thick sections and remove the maximum amount of resin.

- Place each thick section:
  - At the bottom of a flat embedding mold

  - On a siliconized coverslip

- Orient.

4. Polymerization
   - Duration   **48 h**
                    at 60°C

❏ *Next stage:*
- Ultramicrotomy

➼ Make up 300 μL of resin for each flat embedding mold.
➼ To help with the removal of samples, the resin is softened in an oven at 60°C.
➼ **Toxic; use in a fume hood.**

➼ For flat embedding, avoid any bumps which may distort the ultrastructure.
➼ Flat embedding permits the orientation of the sample and allows sectioning in the correct plane (*see* Figure 5.8).
➼ Preparation (*see* Appendix A3.3, Figure A2).
➼ If the vibratome section is very large

➼ To help in removing the mold, it should be chilled at 4°C for 15 to 30 min.

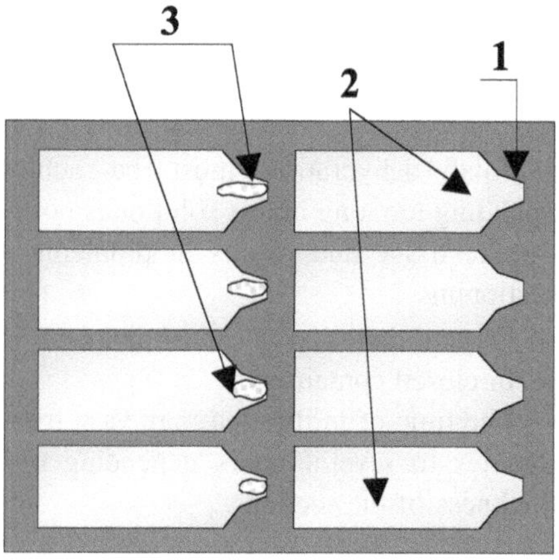

1 = **Extremity of the cavity**

2 = **Filling the cavity with the resin**

3 = **Orienting the thick sections**

Figure 5.8 Mold for flat embedding.

➼ Cutting semithin sections (*see* Chapter 7, Section 7.4)
➼ Cutting ultrathin sections (*see* Chapter 4, Section 4.4.3.3)

5.7.5.2   Embedding in Acrylic Resin

1. Dehydration
   - 30% ethanol   **45 min**

   - 55% ethanol   **45 min**
   - 70% ethanol   **3 × 30–40 min**
   - 90% ethanol   **3 × 30–40 min**
   - 100% ethanol   **3 × 30–40 min**

➼ We will show the protocol of embedding in LR White medium.
➼ **At room temperature (RT)**
➼ At each change of solution, aspirate the liquid with a disposable pipette and replace with the next solution.

2. Infiltration

- Infiltrate the vibratome sections in the solvent/resin with increasing concentrations of resin:
  - 100% ethanol **2 vol + 1 vol**
    + LR White **1–2 h**
  - 100% ethanol **1 vol + 1 vol**
    + LR White **2–4 h**
  - 100% ethanol **1 vol + 2 vol**
    + LR White **2–4 h**
- Replace the substitution mixture with pure resin:
  - 1st solution/pure LR White **1 h**
  - 2nd solution/pure **overnight**
    LR White
3. Embedding
- Lift out the thick sections carefully to remove most of the resin.
- Place each thick section in the base of a gelatin capsule or a plastic mold and fill with resin:
  - Pure LR White

- Orient.
- Completely fill the gelatin capsules with the resin mixture at 4°C.
- Close the gelatin capsules or molds.
4. Polymerization
- Temperature **60°C**
- Length of polymerization **≥ 48 h**

❏ *Next stage:*
- Ultramicrotomy

↝ **At room temperature**
↝ Prepare the different infiltration solutions.
↝ The time of infiltration may vary from 1 hour to several hours depending on the thickness of the sections.
↝ Do not increase the time of the first substitution solution to more than several hours.

↝ May be prolonged

↝ Resin without accelerator

↝ The duration of the impregnation may be prolonged for several hours at 4°C.
↝ Each embedding must be noted.
↝ **Toxic; use in a fume hood.**

↝ Without accelerator if polymerization at 60°C

↝ **Important:** do not trap air on top of the resin as oxygen inhibits polymerization.

↝ Cutting semithin sections (*see* Chapter 7, Section 7.4)
↝ Cutting ultrathin sections (*see* Chapter 4, Section 4.4.3.3)

5.8 VISUALIZING HYBRIDS

5.8.1 Choice of Method

The method of visualization depends on the labeling of the probe:

- Radioactivity

- Antigenic

↝ Detection of radioactivity incorporated into hybrids
↝ Detection of an antigen in antigenic hybrids

1. Radioactive hybrids:
   By autoradiography:

   Autoradiography is carried out after the embedding of vibratome sections and cutting of ultrathin sections.

⤳ Autoradiography involves placing a photographic emulsion on the section to register the radioactive emissions.

⤳ Autoradiography may also be carried out on semithin sections.

2. Antigenic hybrids:
   By immunocytochemistry:
   Visualization of antigenic hybrids is carried out after hybridization and washes by the following:

⤳ Immunocytochemistry involving an antigen–antibody reaction which is then visualized by:
   • An enzymatic marker
   • A specific marker

   • On vibratome sections before embedding

   • On ultrathin sections after embedding vibratome sections in acrylic resin

⤳ The antigen is incorporated into the probe.

⤳ Enzymatic detection (the most commonly used enzyme is peroxidase)

⤳ Vibratome sections are embedded in an acrylic resin and immunocytochemistry is carried out on ultrathin sections.

The visualization stage depends on the marker used and two variables:

   • Sensitivity
   • Resolution

⤳ The visualization stage reveals where the probe has hybridized. A lot of effort has been expended to make this technique as reliable as other forms of morphological analysis.

5.8.2   Radioactive Hybrids

⤳ On semithin or ultrathin sections

5.8.2.1   Principles

The principles of autoradiography are similar for both light and electron microscopy. However, the increased resolution of the electron microscope makes certain points more important:

⤳ *See* Chapter 7, Section 7.6.1 (autoradiography on semithin sections).

   • The size of the silver bromide crystals

   • The thickness of the emulsion

   • The size of the metallic silver grains

⤳ The size of the silver grain depends on the size of the silver bromide crystal.

⤳ Only a monolayer permits high resolution and an estimate of the signal.

⤳ The size depends on the chemical developer used.

⤳ Resolution does not depend on the size.

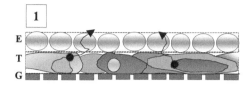

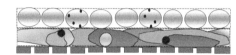

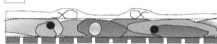

1 = Emissions

Radioactive hybrids (∗) form areas which have characteristic emissions depending on the radioactive element used. Autoradiography is the recording of these emissions on a photographic emulsion.

(E) Emulsion

(T) Tissue or cell section

(G) Support (grid)

2 = Exposure

The emulsion stays in contact with the hybrids, registering the emissions in the form of latent images (•) during the course of the exposure.

3 = Development

The latent images are transformed into visible silver grains during the development process. A number of parameters must be taken into account in the interpretation of the results.

Figure 5.9 Principles of microautoradiography.

❏ *Advantages*

• Cellular and subcellular resolution

⇝ Depending on the isotope and the emulsion used, it is possible to define subcellular localization.

• Association between signal and structure

⇝ In contrast to macroautoradiography

• Quantification

⇝ Necessary to control the thickness of the emulsion

⇝ By counting silver grains followed by statistical analysis

❏ *Disadvantages*

• Thickness of the emulsion

⇝ The thickness of the emulsion is the principal limitation of this technique.

• Background

⇝ This is proportional to the length of the exposure.

• Complex quantification

⇝ It is necessary to perform statistical analysis to determine the origin of the emission from within the tissue.

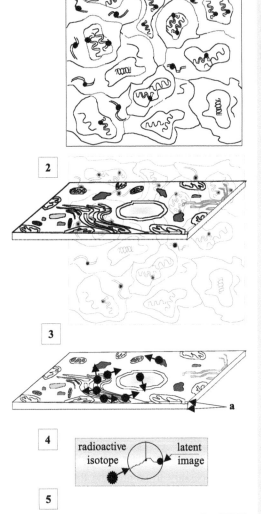

1 = **To visualize specific hybrids** (nonspecific hybrids are removed by washing). Only specific and stable hybrids are maintained.

2 = **Cell corresponding to the projection of section area**

3 = **Radiation emits from the radioactive isotope.**
(**a**) thickness of the section

4 = **Exposure**
The section is covered by a thin layer of photographic emulsion and is stored in darkness for a period of time.

5 = **In** *in situ* **hybridization,** the creation of an image on an autoradiographic emulsion by radiation reveals the position of a nucleic acid.

Figure 5.10 Principles of visualization of radioactive hybrids.

5.8.2.2 Choice of Autoradiography Emulsion

Different types of autoradiography emulsion are available; each one has different characteristics. An emulsion is chosen based on the degree of sensitivity required.

➯ This is dependent on the size of the silver bromide grains.
➯ For electron microscopy, the size of the silver bromide grains is less than that used for light microscopy.

Size of silver bromide grains:

- Ilford L4 emulsion   **0.18 μm**
- NTB5   **≈ 0.15 μm**
- EM   **0.13 μm**

5.8.2.3   Choice of Developer

The type and size of the metallic silver particle depends on the developer used:

	➥ *See* Table 5.4.
• Microdol X	➥ Tangled grains.
• D19	➥ Latent images are bigger than with Microdol X.
	➥ Tangled grains are more complex than those formed with Microdol X.
	➥ There is a risk of contamination of the sections due to the high pH of the developer.
• Ascorbic acid/Elon	➥ Small spherical grains with or without intensification
• Diamine-paraphenylene	➥ Fine spherical grains
	➥ Latent images are smaller than with Microdol X and D19

❏ *Parameters*

• Sensitivity	➥ D19 > Microdol > diamine-paraphenylene
• Length of development	➥ D19 < Microdol < diamine-paraphenylene
• Size of particles	➥ Increases with the duration of development
• Type of silver particles:	➥ *See* Table 5.4.
– Spherical grains	➥ Ascorbic acid/Elon and diamine-paraphenylene
– Tangled grains	➥ D19 and Microdol X

Table 5.4 Sensitivity and Type of Signal Depend on the Developer Used

Developers	Sensitivity	Type of Signal (Tangled Grains or Spherical Grains)
Microdol X	+++	Small, dissociated, tangled grains
D19	++++	Large tangled grains (complex)
Ascorbic acid/Elon	++	Small, fine, spherical grains
Diamine-paraphenylene	+	Fine spherical grains

5.8.2.4 Summary of Different Stages

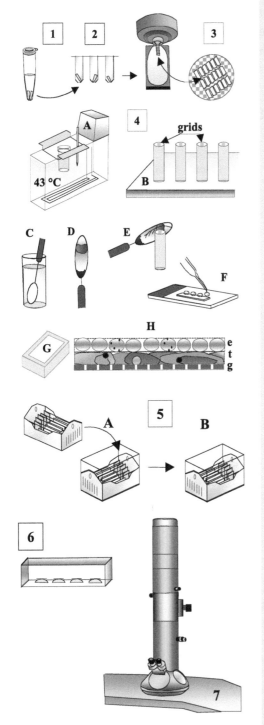

⇝ **Safe light**

1 = **Fixation of floating sections**

2 = **Embedding in an epoxy resin**

3 = **Ultramicrotomy**
 Ultrathin sections on grids

4 = **Coating with emulsion**
 (A) Dilution of emulsion (43°C)
 (B) Placing grids on a support
 (C) Preparation of a monolayer of emulsion
 (D) Checking the thickness of the emulsion: golden zone
 (E) Placing the emulsion monolayer on the grid
 (F) Sticking the grids to a glass slide
 (G) Storing the grids (4°C)
 (H) Exposure
  (e) Emulsion
  (t) Tissue or cell section
  (g) Grid

5 = **Developing autoradiography**
 (A) Development (17°C): washing
 (B) Fixation: washing

6 = **Staining**

7 = **Observation**

Figure 5.11 Protocol for ultrastructural autoradiography.

5.8.2.5   Materials/Products/Solutions

1. Location

   All stages must be carried out in a dark room, using a safe light placed 1.50 m from the workbench; humidity ≈ 60 to 80%.
   - ⇨ These stages do not have to be sterile.
   - ⇨ Wratten n°2 Kodak safe light screen

2. Materials/Small items
   - Aluminum foil
     - ⇨ To wrap the slide boxes and the bottle of emulsion.
   - Autoradiography boxes
     - ⇨ Clean and light-tight
   - Black tape
     - ⇨ To seal the slide boxes
   - Clean glass slides
     - ⇨ Test bubbles at the surface of the emulsion to verify the homogeneity.
   - Double-sided sticky tape
     - ⇨ To stick the grids to the slides
   - Glass slides for exposing the coated grids
     - ⇨ *See* Appendix A3.2.
   - Metal ring
     - ⇨ *See* Appendix A6.2, Figure A12.
     - ⇨ Formation of monolayer
     - ⇨ Platinum
   - Nickel or gold grids ≥ 200 mesh
     - ⇨ Diameter 3.05 mm.
   - Nickel or gold grids ≥ 200 mesh covered with a collodion
     - ⇨ Or formvar film (*see* Appendix A4.3/A4.4).
     - ⇨ To be determined by the monolayer of emulsion.
   - Oval and round glass tubes/emulsion tubes
     - ⇨ To dilute the emulsion
   - Porcelain spoon
     - ⇨ To mix the emulsion
   - Staining dishes
     - ⇨ Nonsterile
   - Support for the grids
     - ⇨ *See* Appendix A6.2, Figure A13. The grids are placed on upright metal rods of the same diameter as the grid (3.05 mm).
   - Thermostatic water bath at 43°C
     - ⇨ For melting the autoradiography emulsion

3. Products
   - Desiccant
     - ⇨ Silica gel wrapped in filter paper
   - Fixative
     - Sodium thiosulfate
       - ⇨ Or sodium hyposulfite
   - Nonaqueous mounting medium
     - ⇨ *See* Appendix B8.2.
   - Nuclear emulsion
     - L4 (Ilford)
     - EM1 (Amersham)
     - NTB5 (Kodak)
   - Potassium hydroxide
     - ⇨ Pellets
   - Sodium hydroxide
     - ⇨ Pellets
   - Stains
     - Uranyl acetate
       - ⇨ **Radioactive** (α emitter) (**wear a mask while weighing out**)
       - ⇨ Crystals
     - Sodium citrate
     - Lead nitrate

4. Solutions
 • Developer

   – Ascorbic acid
   – D19
   – Microdol X
 • Distilled water
 • 95% ethanol
 • Fixer
   – 30% sodium thiosulfate

 • Stains
   – 2.5% ethanolic uranyl acetate

   – Lead citrate

⇝ Depending on the emulsion used
⇝ According to the sensitivity required
⇝ Instructions for the use of the emulsion

⇝ Nonsterile
⇝ For diluting the 5% uranyl acetate
⇝ Fixer solution
⇝ Base solution of all fixatives
⇝ There are some commercially available solutions not suitable for electron microscopy.

⇝ *See* Appendix B7.2.1.1.
⇝ Dilute 5% aqueous uranyl acetate in 95% ethanol just before use ($^v/_v$) (*see* Appendix B7.2.1.1).
⇝ *See* Appendix B7.2.2.

5.8.2.6 Protocols

5.8.2.6.1 CUTTING ULTRATHIN SECTIONS

Cutting sections of tissue embedded in an epoxy resin is easier than cutting tissue embedded in an acrylic resin.

The ultrathin sections are mounted directly on copper or nickel grids without a support film.

⇝ *See* Chapter 4, Section 4.4.3.3.3 (cutting ultrathin sections).
⇝ The knife angle should be greater than 5°.
⇝ The speed of sectioning is around 1 to 2 mm/s.
⇝ The thickness of the sections is ≈ 80–100 nm.
⇝ Epoxy resin is very stable under the electron beam.

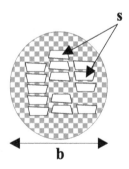

(s) ultrathin sections
(b) 3 mm

Figure 5.12 Mounting ultrathin sections on nickel grids.

5.8.2.6.2 COATING WITH AUTORADIOGRAPHY
        EMULSION
 1. Dilution of the emulsion:
    Use a specially designed container or a round glass tube. Mark with two lines: the first corresponding to the volume of emulsion required and the second to the total volume (water + emulsion) to complete the dilution:

⇝ **In a dark room with a safe light**

⇝ To be determined

⇝ The lines must be visible under the safe light.

- Ilford L4   **1:4**   ↝ Emulsion–sterile distilled water ($^v/_v$)
- EM1   **1:1**   ↝ Emulsion–sterile distilled water ($^v/_v$)
- NTB5   **1:1**   ↝ Emulsion–sterile distilled water ($^v/_v$)

2. Melting the emulsion:
   - Measure out the emulsion with a porcelain spoon.

   ↝ Move the lamp to the bench to see clearly; then return the lamp to the wall when the emulsion has melted.

   ↝ Dilute all the emulsion and store away from the light at 4°C.

   - Fill the container with emulsion up to the lower mark.
   - Place the emulsion in the water bath at 43°C for 5 to 10 min before dilution.
   - Add water up to the higher mark.
   - Mix gently for 20 minutes.

   ↝ Mix slowly (do not make bubbles); use a glass slide or rod; avoid using anything metallic.

   - Melt.   **1 h at 43°C**

   ↝ The emulsion must be perfectly homogeneous.

   ↝ Check the temperature of the emulsion, which should be 43°C. Too high a temperature denatures the emulsion and results in background.

   ↝ *See* Appendix A6.2, Figure A13 (preparation of materials for ultrastructural autoradiography).

3. Place the grids on the support.

   ↝ The grids must be perfectly positioned so that they do not fall when the film is placed on them.

   Place the grids on the metal rods with the sections at the top.

   ↝ The diameter of the rods (metal or Plexiglas) should be 3 mm.

   ↝ The height of the rods may be larger than 20 mm to break the film on either side of the grid.

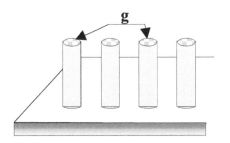

Figure 5.13 Positioning the grids (g).

↝ *See* Appendix A6.2, Figure A12 (preparation of materials for ultrastructural autoradiography).

4. Monolayer:

   ↝ Very delicately

   ↝ The room temperature should be 30–40°C and the humidity higher than 60%; this is very important in making gelatin film.

   - Dip the ring in the liquid emulsion.

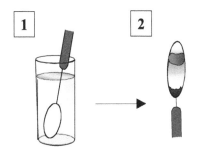

1 = **Dipping the loop**
2 = **Drying**

Figure 5.14 Formation of a monolayer.

- Hold the loop vertically to dry the emulsion.

- Check the monolayer of emulsion.

5. Placing the film on a grid
   - Place the monolayer of film on the grid by placing the zone over the rod, breaking the film by passing the loop to the base of the rod
   - Breathe on the grid to help the autoradiography film stick.

↪ The film forms from the top to the bottom. Only the zone that looks like a mustache is suitable. This zone is a monolayer of film.
↪ Place the membrane on a grid covered with a collodion or formvar film and observe with the electron microscope (*see* Appendix A4).

↪ If the film bursts too quickly, the room temperature is too high, the humidity is too low, and the dilution of the emulsion is too high.
↪ If the film does not dry, the room temperature is too low and the concentration of the emulsion too high.
↪ A monolayer film should form in 1 min.

Figure 5.15 Placing the monolayer on a grid.

6. Sticking the grids to a glass slide:
   Place the emulsion-covered grids on a piece of slide covered in double-sided sticky tape.

↪ *See* Appendix A3.2 (preparation of slides for exposure of grids for ultrastructural autoradiography).
↪ All the grids stuck to the same slide must be developed at the same time.

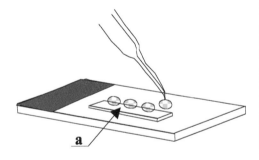

(**a**) double-sided sticky tape

Figure 5.16 Sticking grids to the slide.

7. Storage:
   Store the slides in watertight boxes containing desiccant.
   - Storage   **4°C**

8. Exposure time
   - Duration   **3–5 months at 4°C**

↪ The desiccant must be renewed every time the box is opened.
↪ To limit the growth of bacteria in the emulsion

↪ Prepare several sets of grids to be developed at different exposure times. It is best to keep all the same grids together in one box since some may come unstuck and fall to the bottom of the box.

❑ *Next stage:*
- Developing ultrastructural autoradiography ⮡ *See* Section 5.8.2.6.3.

5.8.2.6.3   DEVELOPING AUTORADIOGRAPHY

1. Allow the box of slides to come to room temperature for at least 1 hour.
2. Place the slides in a slide rack and place in the following solutions:

⮡ **In a dark room with a safe light**
⮡ *See* Figure 5.11.
⮡ Condensation may be a problem; the contents of the box must remain dry.
⮡ This stage is very delicate. Check that the grids are stuck firmly by pressing on the contact region.
⮡ Position the slides one by one. If a grid comes unstuck, it will fall between the two slides and it will be possible to save it.

- Development
  - D19   **4 min**
                           at 17°C

⮡ The temperature may be raised if the labeling is very weak.
⮡ The temperature may be lowered if the background is high.
⮡ The developer may be diluted to lower the signal.

  - Washes   **1 min**
- Fixation
  - 30% sodium thiosulfate   **5 min**

⮡ Without agitation

⮡ The section is embedded in resin, there is no risk of damaging the tissue.

  - Washes   **1 min**
  - Distilled water   **3 × 5 min**

⮡ Without agitation
⮡ The light may be switched on.

❑ *Next stage:*
- Staining

⮡ *See* Section 5.8.2.6.4

5.8.2.6.4   STAINING

1. Staining with uranyl acetate
   - Place the grids with sections against the stain:
     - 2.5% ethanolic   **50 min**
       uranyl acetate   **in the dark**
                              at RT

⮡ For sections in epoxy resin.

⮡ *See* Figure 5.21.

⮡ Uranyl is light sensitive.
⮡ Dilute 5% aqueous uranyl acetate in 95% ethanol just before use (*see* Appendix B7.2.1.1).

   - Wash the grids one after the other:
     - Distilled water   **≈ 40–50 μL/drop**
                              5 × 4 min
     - Distilled water   **Wash bottle**

⮡ Transfer between different drops.

⮡ Wash freely in a jet of distilled water from a wash bottle (*see* Figure 4.36).

   - Dry on filter paper.   **at RT**
2. Staining with lead citrate:
   - Place several pellets in a small container in a Petri dish with a hydrophobic film in the base.

⮡ Away from dust

⮡ Sodium or potassium hydroxide

- Keep the dish closed.   **5–10 min**
- Rapidly place the drops of lead citrate on the film

⇨ To eliminate carbon dioxide
⇨ The number of drops depends on the number of grids to be stained.
⇨ It is best to limit the number of grids to be stained (8 maximum).

- Place the dry grids on the drops of stain:
  – Lead citrate   **40–50 µL/grid**

  – Duration   **8 min
                         at RT**
- Wash the grids one after the other:
  – Distilled water   **Wash bottle**
- Dry on filter paper.   **≤ 37°C**

⇨ Place the sections face down against the stain.
⇨ Avoid opening the Petri dish because carbon dioxide will enter.
⇨ For sections in epoxy resin

⇨ *See* Figure 5.21.

❏ *Next stage:*
- Observation

⇨ *See* Examples of Results (Figures 2 and 14).

5.8.3 Antigenic Hybrids

⇨ On vibratome sections
⇨ On ultrathin sections embedded in acrylic resin

An antigenic probe has one or more sites (i.e., hapten) which form antigen–antibody complexes during the immunocytochemical reaction.
   Antigenic hybrids are visualized:

⇨ Antigenic markers (*See* Chapter 1, Section 1.2.2)

- On vibratome sections before embedding
- On ultrathin sections after embedding in acrylic resin.

⇨ *See* Section 5.8.3.1.
⇨ *See* Section 5.8.3.2.

The haptens that are generally used are:

⇨ *See* Figure 5.17. These molecules have a substitution of the radical $-CH_3$ of a thymidine. In this way they are transformed to deoxyuracil (dUTP) (*see* Chapter 1, Section 1.2.2.1).

- Digoxigenin
- Fluorescein
- Biotin

⇨ Better for animal tissue
⇨ Good for plant and animal tissues
⇨ Depending on the presence of endogenous biotin

1 = **Digoxigenin**
2 = **Fluorescein**
3 = **Biotin**
   (A) Probe
   (B) Target

Figure 5.17 Antigenic hybrid.

5.8.3.1   Visualization of Vibratome Sections

5.8.3.1.1   PRINCIPLES

The antigen is present in the hybrids formed throughout the vibratome section. The detection of the antigen by immunocytochemistry depends on the penetration of the tissue by different components of the reaction (antibody, marker, chromogen).

Two types of immunocytochemical reaction may be used after hybridization and washing:

• Direct

• Indirect

Enzymatic markers are the best for pre-embedding hybridization.

Peroxidase is the enzyme most commonly used. The peroxidase reaction results in a colored insoluble product (precipitate) (*see* Figure 5.19).

The chromogen used for peroxidase is DAB made electron opaque by osmic acid.

↝ The penetration of the tissue by different components of the reaction is the principal drawback of this technique (*see* Figure 5.18).

↝ The best system is the indirect system.

↝ Antigenic hybrids are detected by a primary antibody (IgG) or by antibody fragments conjugated to a marker.
↝ The first stage forms a complex between the hapten and a primary antibody. The second stage uses a conjugated secondary antibody (IgG or other).

↝ *See* Chapter 1, Section 1.2.2 (antigenic labels).
↝ This marker is absorbed by IgG.
↝ Oxidation of an appropriate substrate
↝ Endogenous peroxidase activity may be inhibited (*see* Appendix B6.1.3).
↝ 3′-diaminobenzidine tetrachloride

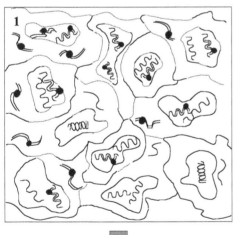

1 = **To visualize specific hybrids** (nonspecific hybrids are removed by washing).

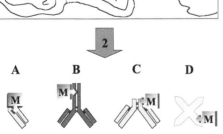

2 = **Conjugated antibodies are used to visualize the hybrids**:
(**A**) Fab
(**B**) IgG
(**C**) F(ab′)$_2$
(**D**) Streptavidin

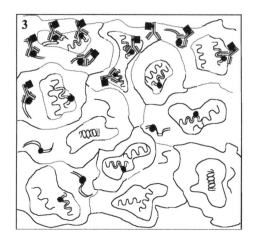

3 = **Immunocytochemical visualization.**

Figure 5.18 Principles of visualization of antigenic hybrids.

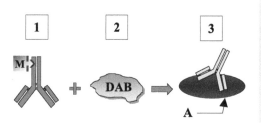

1 = **IgG conjugated to peroxidase**
  (**M**) peroxidase
2 = **Chromogen for peroxidase**
  e.g., DAB
3 = **Precipitation of the chromogen**
  (**A**) Colored product or electron-dense product

Figure 5.19 Principle of visualizing peroxidase activity.

5.8.3.1.2 SUMMARY OF DIFFERENT STAGES

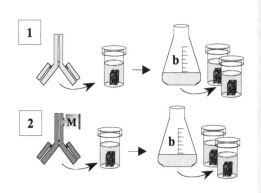

1 = **Incubation with the anti-hapten antibody** (species X)
  Washes in Tris–HCl/NaCl buffer
  (**b**) buffer

2 = **Incubation with anti-species X antibody conjugated with peroxidase**
  Washes in Tris–HCl/NaCl buffer
  (**M**) marker

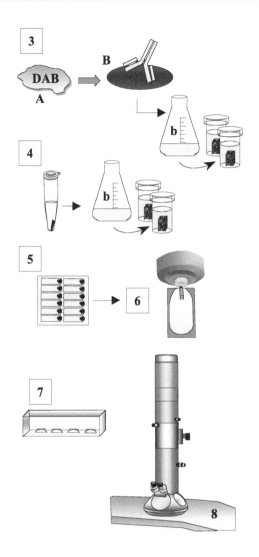

3 = **Visualization of enzymatic activity**
   (**A**) DAB
   (**B**) Chromogen precipitate
   Washes in Tris–HCl/NaCl buffer

4 = **Staining with osmium tetroxide**
   Washes in PBS

5 = **Embedding in epoxy resin**
   • Dehydration
   • Flat embedding
6 = **Ultramicrotomy**

7 = **Staining**

8 = **Observation**

Figure 5.20 Indirect immunocytochemical reaction.

5.8.3.1.3 SOLUTIONS

1. Antibodies:

• Biotinylated antibody anti-avidin	⇝ Indirect reaction
• Complex ABC-peroxidase	⇝ Indirect reaction
• IgG anti-hapten conjugated	⇝ IgG monoclonal or polyclonal; Fab fragments
• IgG anti-hapten nonconjugated (species X)	⇝ IgG monoclonal or polyclonal
	⇝ Indirect reaction
• IgG, F(ab′)$_2$, Fab	⇝ Indirect reaction
– Anti-species X conjugated	⇝ Peroxidase

2. Buffers:
- Blocking buffers
  - 20mM Tris–HCl buffer/300 mM NaCl/ 1% serum albumin; pH 7.6
  - 20mM Tris–HCl buffer/300 mM NaCl/ 2% goat serum/0.3–0.01% Triton X-100; pH 7.6
- Buffer PBS
- 20 mM Tris–HCl buffer/300 mM NaCl; pH 7.6

➠ *See* Appendix B6.1.
➠ Other agents may be used in the blocking buffer. These are:
- Fish gelatin
- Ovalbumin
- Nonfat milk powder

➠ *See* Appendix B3.4.3.
➠ *See* Appendix B3.7.5.

3. Blocking solution:
- Endogenous peroxidase

➠ *See* Appendix B6.1.3.

4. Fixative:
- 1% osmium tetroxide in PBS

➠ *See* Appendix B4.1.

5. Hydrogen peroxidase

➠ 110 V

6. Visualization solution:
- 0.025% DAB in 20 mM Tris–HCl buffer/ 0.006% hydrogen peroxidase; pH 7.6

➠ Solution for visualizing peroxidase activity (*see* Appendix B6.2.2.1)

5.8.3.1.4 INDIRECT REACTION PROTOCOL

1. Incubation of thick sections:
- 20 mM Tris–HCl buffer/300 mM NaCl; pH 7.6 — **≥ 200 µL/ thick section** — **10 min**

➠ After hybridization washes
➠ Equilibrate the osmolarity of the tissue after the post-hybridization washes in a buffer suitable for immunocytochemistry.
➠ *See* Appendix B3.7.5.

2. To block nonspecific sites:
- 20 mM Tris–HCl buffer/300 mM NaCl/1% serum albumin; pH 7.6 — **≥ 200 µL/ thick section** — **60 min** — **at RT**

➠ This stage is **indispensable** to remove any nonspecific reaction; the sections are pre-incubated with a nonspecific serum (BSA, fish gelatin, etc.) (*see* Appendix B6.1.1).
➠ The concentration of Triton X-100 may be varied between 0.01 and 0.3% according to the thickness of the section.
➠ Too high a concentration of Triton X-100 may cause background.

3. Inhibit endogenous enzyme activity.

➠ **Optional** (*see* Appendix B6.1.3)
➠ Check for enzyme activity prior to hybridization.

4. Incubate the anti-hapten antibody raised in species X:
- Dilute 1:100 in Tris–HCl buffer/300 mM NaCl; pH:7.6 — **≥ 200 µL/ thick section**

- Incubation — **overnight at 4°C**

➠ Formation of the hapten–antibody complex

➠ Primary antibody
➠ The dilution of the antibody is low (1:100 to 1:500 according to the density of the labeling).

5. Wash:
   - 20 m*M* Tris–HCl buffer/300 m*M* NaCl; pH:7.6   **3 × 10 min**

     ↬ The signal/background ratio is usually low.

6. Incubate with the conjugated anti-species X antibody:

   ↬ Conjugated secondary antibody or fragments diluted 1:200 (according to supplier's instructions)

   - Incubation   **≥ 200 µL 2 h at RT**

     ↬ Dilution is less than for the 1st antibody.
     ↬ Volume could be reduced.

7. Wash:
   - 20 m*M* Tris–HCl buffer/300 m*M* NaCl; pH 7.6   **3 × 10 min**

     ↬ **Important:** use a large volume of buffer.

8. Visualize peroxidase activity:
   - 0.025% DAB/20 m*M* Tris–HCl buffer/0.006% hydrogen peroxide; pH 7.6   **100 µL/ thick section**

     ↬ Preparation (*see* Appendix B6.2.2.1)
     ↬ **May be left out;** add an enzyme blocking agent (endogenous peroxidase, *see* Appendix B6.1.3).

9. Watch carefully while incubating.   **3–10 min at RT**

   ↬ Brown coloration
   ↬ Too long an incubation will cause a general brown coloration of the tissue.

10. Stop the reaction:
    - PBS

      ↬ *See* Appendix B3.4.3.

11. Post-fix the vibratome sections:
    - 1% osmium tetroxide in PBS   **30 min**

      ↬ *See* Appendix B4.1.
      ↬ The chromogen DAB is made electron dense by the reduction of the osmium atoms.
      ↬ Removes the fixative

12. Wash:
    - PBS   **3 × 10 min**

❑ *Next stage:*
   - Embedding in epoxy resin

     ↬ *See* Section 5.7.5.1.

5.8.3.2   Visualization of Ultrathin Sections

5.8.3.2.1   CUTTING ULTRATHIN SECTIONS

↬ *See* Chapter 4, Section 4.4.3.3 (ultramicrotomy).

Ultrathin sections of tissue embedded in an acrylic resin are mounted on collodion- and carbon-coated nickel grids.

5.8.3.2.2   VISUALIZATION

The protocol for immunocytochemistry on ultrathin sections after embedding in an acrylic resin is described in Chapter 4 (Post-Embedding Methods).

↬ *See* Chapter 4, Section 4.8.

↬ The marker incorporated into the probe (biotin, digoxigenin, or fluorescein) is detected by an indirect immunocytochemical reaction using an immunoglobulin (in general, an IgG) or an immunoglobulin fragment labeled with colloidal gold.

❑ *Next stage:*
  • Staining

↪ *See* Section 5.9.

5.9   STAINING

5.9.1   Principles

Heavy metal atoms (uranyl, lead) are dense to electrons and visualize subcellular structures when deposited on cellular structures.

↪ Positive staining

5.9.2   Summary of Different Stages

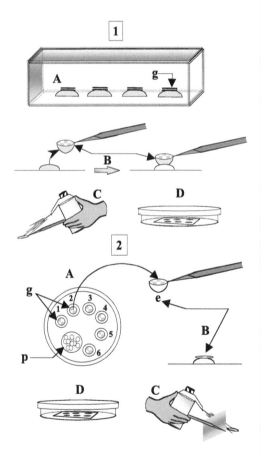

1 = Staining with uranyl acetate
  (A) Grid is placed on a drop of stain:
       • In the dark
       (g) grid
  (B) Transfer the stained grid to a drop of distilled water:
       • Transfer to different drops to wash
  (C) Wash under a jet of water.
  (D) Dry away from dust:
       • At room temperature

2 = Staining with lead citrate
  (A) Grid placed on a drop of stain
       (g) grid
       (p) pellets of sodium or potassium hydroxide
  (B) Transfer the grid from the stain to a drop of distilled water.
       (e) excess liquid
  (C) Wash under a jet of water.
  (D) Dry away from dust:
       • At room temperature
       • At or below 37°C

Figure 5.21   Staining ultrathin sections.

5.9.3   Materials/Products/Solutions

1. Materials

   * Container for staining   ⇝ Large enough for a container of sodium hydroxide pellets

   * Filter paper
   * Hydrophobic film like Parafilm   ⇝ For the drops of uranyl acetate

2. Products   ⇝ AR quality

   * Lead nitrate
   * Potassium hydroxide   ⇝ Pellets
   * Sodium citrate
   * Sodium hydroxide   ⇝ Pellets

3. Solutions

   * Distilled water
   * Stains   ⇝ Heavy metals
     – Aqueous uranyl acetate   ⇝ *See* Appendix B7.2.1.2.
       · 2% uranyl acetate   ⇝ Staining acrylic resin sections
       · 5% uranyl acetate   ⇝ Staining epoxy resin sections
     – Lead citrate   ⇝ *See* Appendix B7.2.2.

5.9.4   Protocol

1. Staining with uranyl acetate

   * Float each grid on a drop of stain, section   ⇝ Place the same number of drops as there
     side down: are grids on a hydrophobic film.
     – Aqueous uranyl   **In the dark**   ⇝ Filter before use (0.22 µm).
       acetate   **40–50 µL/grid**   ⇝ Uranyl is light sensitive.
     – Duration:
       · 5% uranyl acetate:
         Epoxy resin sections   **2 h**
       · 2% uranyl acetate:
         Acrylic resin sections   **30 min**
   * Wash the grids one after the other
     – Distilled water   ≈ **40–50 µL/drop**   ⇝ **Optional;** stop staining on drop.
                             5 × 4 min
     – Distilled water   **Wash bottle**   ⇝ Wash freely in a jet of distilled water.
   * Dry on filter paper.   ⇝ At room temperature away from dust

2. Staining with lead citrate

   * Place sodium hydroxide pellets in a con-   ⇝ Or potassium hydroxide
     tainer in the staining box.
   * Keep the container closed.   **5–10 min**   ⇝ Keep out carbon dioxide.

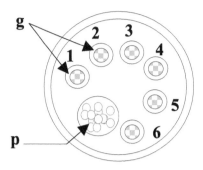

(g) grids
(p) pellets of sodium or potassium hydroxide

Figure 5.22 Staining ultrathin sections with lead citrate.

• Place the drops of lead citrate in the box as quickly as possible.

⮞ Use one drop per grid.
⮞ Do not stain too many grids at any one time (8 maximum).

• Rapidly place the dry grids on the lead citrate drops:
  – Lead citrate: **40–50 μL/grid**

⮞ Place the sections directly against the stain.

  · Duration:
    Epoxy resin sections   **10 min**
    Acrylic resin sections   **2 min**

⮞ Limit contact between carbon dioxide and the stain, as it precipitates the lead.

• Wash the grids one after the other:
  – Distilled water   ≈ **40–50 μL/drop**
                           5 × 2 min
  – Distilled water   **Wash bottle**
• Dry on filter paper.

⮞ Transfer the grids to different drops.

⮞ Wash freely in a jet of distilled water.
⮞ At room temperature away from dust or in an oven at or below 37°C

❑ *Next stage:*
  • Observation

⮞ *See* Example of Results (Figures 14 through 17).

5.10 PROTOCOL TYPES

⮞ Radioactive hybrids
⮞ Antigenic hybrids

• Fixation
• Cutting vibratome sections
• Pretreatments
  – Permeabilization
  – Deproteinization
  – Prehybridization
• Hybridization
• Post-hybridization treatment
• Visualization of hybrids
  – Developing ultrastructural autoradiography
  – Visualization of indirect immunocytochemistry

⮞ *See* Section 5.10.1 (radioactive hybrids).

⮞ *See* Section 5.10.2 (antigenic hybrids).

1. Fixation   ⮞ *See* Chapter 3.
   - 4% PF in 100 m*M*   **3–24 h**   ⮞ *See* Appendix B4.3.
     phosphate buffer   **at 4°C**
   - Washes in buffer   **4 × 15 min**
2. Cutting vibratome sections   ⮞ *See* Section 5.3.
   - Thick sections   **20–200 μm**
3. Pretreatments   ⮞ *See* Section 5.4.
   - Permeabilization   **30 min**   ⮞ *See* Section 5.4.3.
                               at RT
     – Triton X-100 in   **0.05%**
       phosphate buffer, or
     – Sarcosyl in   **0.2%**
       phosphate buffer, or
     – Washes in   **3 × 10 min**
       phosphate buffer
   - Deproteinization   ⮞ *See* Section 5.4.4.
     – Proteinase K, or   ⮞ *See* Section 5.4.4.2.1.
       Incubate floating sections
       in the following solutions:
       · 20 m*M* Tris–HCl   **15 min**   ⮞ *See* Appendix B3.7.2.
         buffer/2 m*M* CaCl$_2$;
         pH 7.6
       · Proteinase K in   **1 μg/mL**   ⮞ *See* Appendix B2.15.
         Tris–HCl/CaCl$_2$ buffer **1 h**
                                at RT
       · Phosphate buffer   **2 × 15 min**   ⮞ Removes the enzyme
       · 4% PF in phosphate   **5 min**   ⮞ *See* Appendix B4.3.
         buffer
       · Phosphate buffer   **3 × 5 min**   ⮞ Removes traces of fixative
     – Pronase   ⮞ *See* Section 5.4.4.2.2.
       Incubate floating sections
       in the following solutions
       · 100 m*M*   **10 min**
         phosphate buffer
       · Pronase B in   **25 μmoles/mL**   ⮞ Prepared just before use
         100 m*M*   **15 min**
         phosphate buffer;   **at RT**
         pH 7.4
       · Phosphate buffer   **3 × 10 min**   ⮞ Removes the enzyme
       · PF 4% in phosphate   **5 min**   ⮞ *See* Appendix B4.3.
         buffer
       · Phosphate buffer   **3 × 5 min**   ⮞ Removes traces of fixative
   - Prehybridization   ⮞ *See* Section 5.4.5.
     – Prehybridization buffer formula: ⮞ *See* Section 5.4.5.2.
                               Final concentration
       · 20X SSC   **4X**
       · 50X Denhardt's solution   **1X**
       · 10 mg/mL tRNA   **100 μg/mL**
       · 10 mg /mL DNA   **100 μg/mL**

· Sterile water	**to x mL**	
Vortex		
Centrifuge		➡ Removes bubbles
– Hybridization buffer	**≈ 100 µL**	
– Duration of incubation	**1 h**	
	at 37°C	

4. Hybridization — ➡ *See* Section 5.5.
   - Add the labeled probe to the hybridization buffer: — ➡ Dry probe
     - cDNA **5–25 ng/mL** — ➡ Must be denatured; **indispensable**
     - Oligonucleotide **5–200 pmoles/mL** — ➡ Do not denature.
   - Vortex.
   - Centrifuge.
   - Denature. **3 min at 100°C** — ➡ If necessary
   - Chill quickly on ice.
   - Incubate floating sections:
     - Hybridization **overnight at 37°C** — ➡ Transfer floating sections to the reaction mixture

5. Post-hybridization treatment — ➡ *See* Section 5.6.
   - 2X SSC **5 × 20 min at 37°C** — ➡ *See* Appendix B3.5.
   - 1X SSC **2 × 60 min**
   - 0.5X SSC **2 × 60 min**
   - PBS — ➡ *See* Appendix B3.4.3.

❏ *Next stages:*

- Fixation of floating sections — ➡ *See* Section 5.10.1 (radioactive hybrids). ➡ *See* Section 5.10.2.2 (antigenic hybrids).
- Indirect immunocytochemistry — ➡ *See* Section 5.10.2.1 (antigenic hybrids).

5.10.1 Radioactive Hybrids

- Fixation of floating sections — ➡ General example
- Embedding floating sections in epoxy resin
- Ultramicrotomy
- Coating with emulsion — ➡ Ultrathin sections
- Developing autoradiography
- Staining
- Observation

1. Fixation of floating sections — ➡ *See* Section 5.7.4.
   - Incubate: — ➡ **Strong fixation**
     - 2.5% glutaraldehyde **≥ 200 µL/** — ➡ *See* Appendix B4.2.
       in 100 m*M* **thick section** — ➡ Use sterile culture dishes.
       phosphate buffer **60 min**
   - Wash:
     - Buffer **2 × 30 min** — ➡ Same buffer as fixation

- Post-fix:
  - 1% osmium tetroxide **≥ 200 μL/**   ⇨ *See* Appendix B4.1.
    in 100 m*M*   **thick section**   ⇨ **Avoid skin contact, and vapors are**
    phosphate buffer   **60 min**   **dangerous.**
- Wash:
  - 100 m*M*   **2 × 30 min**
    phosphate buffer

2. Embedding floating sections in epoxy   ⇨ *See* Section 5.7.5.1.
   resin   ⇨ *See* Appendix B5.2.2.
   - Dehydrate: ⇨ **Room temperature**
     - 30% ethanol   **15–30 min**
     - 50% ethanol   **15–30 min**
     - 70% ethanol   **15–30 min**
     - 80% ethanol   **15–30 min**
     - 90% ethanol   **3 × 15–30 min**
     - 95% ethanol   **3 × 15–30 min**
     - 100% ethanol   **3 × 15–30 min**   ⇨ Total dehydration is **indispensable.**
   - Infiltrate: ⇨ **Room temperature**
     - Propylene   **2 vol + 1 vol**   ⇨ **Toxic; use in a fume hood** (*See* Appendix
       oxide + Epon   **45 min**   A1.2).
     - Propylene   **1 vol + 1 vol**
       oxide + Epon   **45 min**
     - Propylene   **1 vol + 2 vol**
       oxide + Epon   **45 min**
   - Substitute with Epon resin:
     - 1st mixture   **1–2 h**   ⇨ *See* Appendix B5.2.2.
       Epon + DMP 30   **at RT**   ⇨ **Toxic; use under a fume hood.**
     - 2nd mixture   **overnight**
       Epon + DMP 30   **at 4°C**
   - Embedding: ⇨ Each embedding must be recorded.
     - Epon + DMP 30   **300 μL/mold**   ⇨ Prepare the mixture just before use.
          ⇨ *See* Appendix B5.2.2.
   - Polymerize:
     - Duration   **48 h**
          at 60°C

3. Ultramicrotomy   ⇨ *See* Chapter 4, Section 4.4.3.3.
        ⇨ Semithin sections may be cut (*see* Chapter 7, Section 7.4).
   - Ultrathin sections/grid   **90–110 nm**
4. Coating with emulsion   ⇨ **Safe light/40°C/80% humidity**
        ⇨ *See* Section 5.8.2.6.2.
   - Dilute the emulsion:
     - Ilford L4   **1:4**   ⇨ Emulsion and sterile distilled water ($^v/_v$)
   - Melt.   **1 h**   ⇨ The emulsion must be perfectly
          at 43°C   homogenized.
   - Place the grids on the supports.
   - Prepare the emulsion monolayer.   ⇨ **Very delicate**
   - Check the thickness of the emulsion.   ⇨ Place the film on a collodion-coated grid
          and observe with the electron microscope.
   - Place the film on grids.

- Stick the grids to a microscope slide.   ⟿ *See* Appendix A3.2 (preparation of slides).
- Store grids.   **4°C**   ⟿ Water-tight boxes + desiccant.
- Expose.   **3–5 months at 4°C**

5. Developing autoradiography   ⟿ *See* Section 5.8.2.6.3.
   - D19   **4 min at 17°C**
   - Wash with distilled water   **1 min**
   - 30% sodium thiosulfate   **5 min**
   - Wash with distilled water   **3 × 5 min**

6. Staining   ⟿ *See* Section 5.8.2.6.4.
   - Staining with uranyl acetate   ⟿ **In the dark**
     - 5% aqueous   **40–50 μL/grid**   ⟿ *See* Appendix B7.2.1.2.
       uranyl acetate   **2 h**
     - Distilled water   **5 × 4 min**   ⟿ Wash **freely.**
                               Wash bottle
     - Drying   **at RT**
   - Staining with lead citrate   ⟿ **In the presence of sodium or potassium hydroxide**
     - Reynolds lead   **40–50 μL/grid**   ⟿ Keep the container closed.
       citrate solution   **2–5 min**
     - Distilled water   **5 × 4 min**   ⟿ Final wash in a jet of distilled water from a wash bottle

7. Observation   ⟿ *See* Examples of Results (Figures 2, 14, and 17).

5.10.2 Antigenic Hybrids

5.10.2.1 Visualization of Vibratome Sections

- Detection by indirect   ⟿ General example
  immunocytochemistry
- Post-fixation
- Embedding in epoxy resin
- Ultramicrotomy   ⟿ Ultrathin sections
- Staining
- Observation

1. Detection by indirect immunocytochemistry   ⟿ *See* Section 5.8.3.1.
   - Incubate thick sections: ⟿ After the hybridization washes
     - 20 mM Tris–HCl   **≥ 200 μL/**   ⟿ *See* Appendix B3.7.5.
       buffer/300 mM **thick section**
       NaCl; pH 7.6   **10 min**
   - Block nonspecific sites: ⟿ **Room temperature**
     - 20 mM Tris–HCl   **≥ 200 μL/**   ⟿ **Indispensable**
       buffer/300 mM **thick section** ⟿ *See* Appendix B6.1.1.
       NaCl/1% serum   **60 min**
       albumin; pH 7.6
   - Inhibit endogenous enzyme activity.   ⟿ **Optional** (*see* Appendix B6.1)

- Incubate floating sections with an anti-hapten antibody raised in species X:
  - Diluted 1:100 in **≥ 200 µL/** 20 m*M* Tris–HCl **thick section** buffer/300 m*M* **overnight** NaCl; pH 7.6 **at 4°C**

↝ Formation of a hapten–antibody complex

- Wash:
  - Tris–HCl buffer/NaCl **3 × 10 min**
- Incubate with a conjugated anti-species X-antibody

↝ Conjugated IgG, fragments diluted (1:200)

  - Diluted in 20 m*M* **≥ 200 µL** Tris–HCl buffer/300 m*M* **2 h** NaCl; pH 7.6

↝ **Room temperature**

- Wash:
  - Tris–HCl buffer/NaCl **3 × 10 min**
- Visualize peroxidase activity
  - 0.025% DAB in **100 µL/** 20 m*M* Tris–HCl **thick section** buffer/0.006% hydrogen peroxide; pH 7.6

↝ Preparation (*see* Appendix B6.2.2.1)

- Incubate while watching **3–10 min** carefully until the correct **at RT** level of coloration is obtained.

↝ Brown color

- Stop the reaction:
  - PBS **2 × 3 min**

2. Post-fixation

↝ *See* Section 5.7.4.

- 1% osmium tetroxide in PBS **60 min**

↝ The chromogen DAB is made electron dense.

- PBS **3 × 15 min**

3. Embedding in epoxy resin

↝ *See* Section 5.7.5.

- Dehydrate:

↝ **Room temperature**

  - 30% ethanol **15–30 min**
  - 50% ethanol **15–30 min**
  - 70% ethanol **15–30 min**
  - 80% ethanol **15–30 min**
  - 90% ethanol **3 × 15 min–30 min**
  - 95% ethanol **3 × 15 min–30 min**
  - 100% ethanol **3 × 15 min–30 min**
- Infiltrate:

↝ **Room temperature**
↝ *See* Appendix B5.2.2.
↝ **Toxic; use under a fume hood.**

  - Propylene **2 vol + 1 vol** oxide + Epon **45 min**
  - Propylene **1 vol + 1 vol** oxide + Epon **45 min**
  - Propylene **1 vol + 2 vol** oxide + Epon **45 min**
- Replace the substitution solution with Epon resin

↝ Use freshly prepared resin.

  - 1st mixture/Epon + **1–2 h** DMP 30

– 2nd mixture/Epon + DMP 30	**overnight at 4°C**	
• Embed the thick sections.		➯ Each embedding should be recorded.
• Polymerize:		➯ For flat embedding
– Duration	**48 h at 60°C**	
4. Ultramicrotomy		➯ *See* Chapter 4, Section 4.4.3.3.
• Ultrathin sections/grid	**90–110 nm**	
5. Staining		➯ *See* Section 5.9.
• Staining with uranyl acetate:		➯ **In the dark**
– 5% aqueous uranyl acetate	**40–50 µL/grid 2 h**	➯ *See* Appendix B7.2.1.2.
– Distilled water	**5 × 4 min Wash bottle**	➯ **Wash freely.**
– Drying	**at RT**	
• Staining with lead citrate:		➯ **In the presence of sodium or potassium hydroxide**
– Reynolds lead citrate	**40–50 µL/grid 10 min**	➯ *See* Appendix B7.2.2. ➯ Keep the container closed.
– Distilled water	**5 × 4 min**	➯ Final wash in jet of water from a wash bottle
6. Observation		➯ *See* Examples of Results (Figures 15 through 17).

5.10.2.2 Visualization of Ultrathin Sections

• Weak fixation
• Embedding in epoxy resin
• Ultramicrotomy
• Indirect immunocytochemistry
• Staining
• Observation

➯ Acrylic resin sections

➯ Ultrathin sections
➯ General example

1. Weak fixation		➯ *See* Chapter 3, Section 3.6.
• Fix:		➯ Weak fixation
– 4% PF in buffer	**≥ 200 µL/ thick section 2–3 h at 4°C**	➯ *See* Appendix B4.3. ➯ Use sterile culture dishes. ➯ According to the thickness of the section
• Wash:		
– Buffer	**2 × 30 min**	➯ Same buffer as fixation
2. Embedding in acrylic resin		➯ Such as LR White medium (*see* Chapter 4, Section 4.3.5) (*See* Appendix B5.3.2).
• Dehydrate:		➯ **At room temperature**
– 30% ethanol	**45 min**	
– 55% ethanol	**45 min**	
– 70% ethanol	**3 × 30–40 min**	
– 90% ethanol	**3 × 30–40 min**	
– 100% ethanol	**3 × 30–40 min**	

- Infiltrate of vibratome sections:
  - 100% ethanol   **2 vol/1 vol**
    + LR White   **1–2 h**
  - 100% ethanol   **vol/vol**
    + LR White   **2–4 h**
  - 100% ethanol   **1 vol/2 vol**
    + LR White   **2–4 h**
- Substitute with pure resin: ⇝ Resin without an accelerator
  - 1st mixture/pure   **1 h**
    LR White resin   **at 4°C**
  - 2nd mixture/pure   **overnight**
    LR White resin   **at 4°C**
- Embedding: ⇝ Close the gelatin capsules or plastic molds.
  - Temperature   **60°C**
  - Duration of polymerization   **≥ 48 h**

3. Ultramicrotomy   ⇝ *See* Chapter 4, Section 4.4.3.3.
- Ultrathin sections on grids   **90–110 nm**   ⇝ Grey–gold color.

4. Indirect immunocytochemistry   ⇝ *See* Section 5.8.3.2.2.
       ⇝ General example
       ⇝ Immunodetection is carried out on 40 to 50 µL drops on a hydrophobic film.

- Stabilize the structures:
  - 4% PF in 2X SSC   **5 min**   ⇝ *See* Appendix B4.3
- Wash:
  - 2X SSC   **3 × 5 min**   ⇝ *See* Appendix B3.5.

- Block nonspecific sites:
  - 100 mM phosphate   **15–30 min**   ⇝ *See* Appendix B6.1.1.
    buffer/300 mM NaCl/1%   ⇝ **Indispensable**
    serum albumin; pH 7.4
- Incubate with the 1st anti-hapten
  antibody, raised in species X:
  - Diluted 1:50 in   **≥ 40 µL/grid**
    100 mM phosphate
    buffer/300 mM NaCl;
    pH 7.4
  - Duration   **60 min**
         humid chamber
- Wash:
  - Phosphate buffer/NaCl   **2 × 5 min**   ⇝ *See* Appendix B3.4.2.
  - 20 mM Tris–HCl   **2 × 5 min**   ⇝ Change of buffer
    Buffer/300 mM NaCl;   ⇝ *See* Appendix B3.7.5.
    pH 7.6
- Anti-species X conjugated   ⇝ Detection of antibody complex.
  antibody:
  - Diluted 1:50 in   **≥ 20 µL/grid**   ⇝ Conjugated secondary antibody: IgG, Fab
    20 mM Tris–HCl   fragments.
    buffer/NaCl 300 mM; pH 7.6
  - Incubation   **60 min**   ⇝ Humid chamber

- Wash:
  - Tris–HCl/NaCl buffer **2 × 5 min**
  - 2X SSC **2 × 5 min** ⮞ Change of buffer.
- Fix:
  - 2.5% glutaraldehyde 2X SSC **5 min** ⮞ *See* Appendix B4.2.
- Wash:
  - 2X SSC **5 min**
  - Sterile water **5 min**

5. Staining ⮞ *See* Section 5.9.
  - Staining in uranyl acetate: ⮞ **In darkness**
    - 2% aqueous **40–50 μL/grid** ⮞ *See* Appendix B7.2.1.2.
      uranyl acetate **30 min**
    - Distilled water **5 × 4 min** ⮞ Wash freely.
      Wash bottle
    - Drying **Room**
      temperature
  - Staining with lead citrate: ⮞ **In presence of sodium or potassium hydroxide.**

    - Lead citrate **40–50 μL/grid** ⮞ *See* Appendix B7.2.2.
      2 min ⮞ Keep the container closed.
    - Distilled water **5 × 4 min** ⮞ Final wash in a jet of water from a wash bottle

6. Observation ⮞ *See* Examples of Results (Figures 14 through 17).

Chapter 6

Frozen Tissue Technique

CONTENTS

6.1   PRINCIPLES

To increase the accessibility of nucleic acids to a probe, embedding tissue can be avoided. In order to cut ultrathin sections, the tissue is hardened by freezing. Thus, the sections obtained by this method are called ultrathin sections of frozen tissue.

⮑ Semithin sections of frozen tissue can be cut (*see* Chapter 7, Section 7.4.1).
⮑ A cryo-ultramicrotome is required to cut frozen sections.

❏ *Advantages*

• Highly sensitive

⮑ Maximum preservation of nucleic acids due to the presence of water in the tissue
⮑ Good accessibility of nucleic acids and hybrids at the surface and in the interior of the section

• Good resolution
• Low quantity of probe necessary

⮑ Colloidal gold can be used.
⮑ The concentration of labeled probe is 5 to 20 times lower for the same detection compared with pre- and post-embedding methods.

• Preparation is easy

⮑ After fixation, the tissue is cryoprotected and frozen.

• Rapid technique

⮑ Freezing is a lot quicker than embedding.
⮑ Pretreatment is not necessary.

• Multiple labeling is possible

⮑ The preservation of antigens is also excellent.

❏ *Disadvantages*

• Equipment needed
• Preservation of ultrastructure
• Storage in liquid nitrogen

⮑ Freezing system and cryo-ultramicrotome
⮑ Inferior to that in embedded tissue
⮑ Storage is expensive and must be kept under constant control.

6.2 SUMMARY OF DIFFERENT STAGES

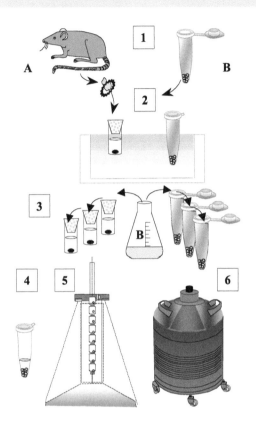

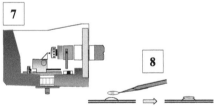

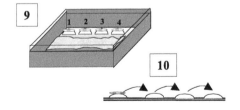

1 = **Specimen**
    (A) Dissection
    (B) Cell pellet

2 = **Fixation**
    • By super-cooling
    • By immersion

3 = **Washes**
    (B) Buffer

4 = **Cryoprotection**
    • Sucrose

5 = **Freezing**

6 = **Storage** (liquid nitrogen)

7 = **Cryo-ultramicrotomy**

8 = **Pretreatment** (optional)

9 = **Hybridization**
    • Humid chamber
    • Place grid on reaction mixture

10 = **Washes**
    • In SSC in decreasing concentrations

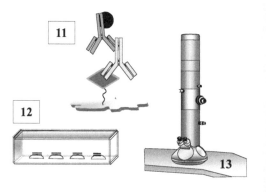

11 = **Detection by indirect immunocy-tochemistry**

12 = **Staining**

13 = **Observation**

Figure 6.1 Frozen tissue technique.

6.3 MATERIALS/PRODUCTS/SOLUTIONS

6.3.1 Materials

6.3.1.1 Small Items

• Nickel grids (200–400 mesh) collodion and carbon coated

⇝ Do not use copper; they will react chemically.
⇝ Ultrathin sections of frozen tissue must be mounted on coated sections.
⇝ Preparation of collodion-coated grid (*see* Appendix A4.3)

• Small laboratory items
  – Pipettes, forceps, etc.
  – Cryotubes, etc.

⇝ Material must be new or sterile and RNase free (*see* Appendices A1.1 and A2).
⇝ New

6.3.1.2 Apparatus

• Cryo-ultramicrotomy

⇝ Two brands of cryo-ultramicrotome are available, both with mechanical advance:
  • Leica
  • RMC
⇝ Monobloc cryogenic chamber (*see* Figure 6.17)

• Freezing system

⇝ Several freezing systems are available:
  1. Cold gradient (*see* Figure 6.2)
  2. Progressive cooling (*see* Figure 6.3)
  3. Immersion in cryogenic liquid (*see* Figure 6.4)
  4. Freezing on a block of chilled metal (*see* Figures 6.5, 6.6, and 6.7)
  5. Programmed freezing
  6. High pressure

- Knife maker

- Liquid nitrogen containers

- Refrigerator

1. Cold gradient

↬ For making glass knives. Diamond knives may be used for cryo-ultramicrotomy (*see* Appendix A5.1).
↬ For storing samples
↬ Liquid nitrogen reserve
↬ 4°C; storage of sections

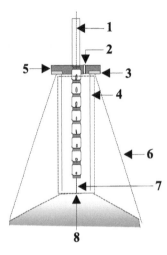

1 = **Mobile support**

2 = **Hole thermal probe**

3 = **Joint**

4 = **Collar**

5 = **Cover**

6 = **Cryobiological container**

7 = **Carrier**

8 = **Liquid nitrogen level**

Figure 6.2   Freezing system by cold gradient (Biogel).

2. Progressive cooling

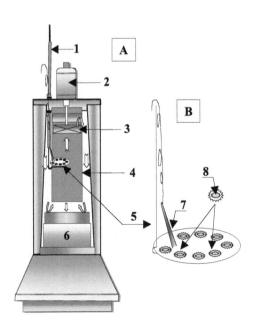

(A) **System**
(B) **Carrier for sample holders for cryo-ultramicrotomy**

1 = **Carrier for samples for cryo-ultra-microtomy**

2 = **Motor and speed control**

3 = **Motor**

4 = **Stainless steel cylinder**

5 = **Samples on the cryo-ultramicrotome holder**

6 = **Liquid nitrogen**

7 = **Thermal probe**

8 = **Sample**

Figure 6.3 Progressive cooling system (LM10).

3. Immersion in a cryogenic liquid

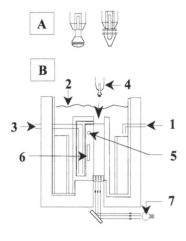

(A) **Injectors**
  • Tissue
  • Cell suspension
(B) **Freezing system**
1 = **Liquid nitrogen**
2 = **Nitrogen gas**
3 = **Cryogenic liquid**
4 = **Injector**
5 = **Temperature sensor**
6 = **Heat resistor**
7 = **Lamp**

Figure 6.4   Freezing system by immersion in a cryogenic liquid (KH 80).

4. Freezing on a chilled metal block

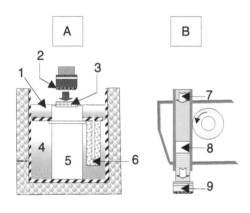

(A) **Metal mirror system**
(B) **Injector**
1 = Nitrogen gas
2 = Sample holder
3 = Copper block
4 = Liquid nitrogen
5 = Cryogenic agent
6 = Temperature sensor
7 = Piston
8 = Damping chamber
9 = Sample support

Figure 6.5   Principle of freezing metal mirror MM80.

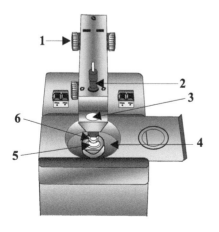

1 = **Pressure selector**
2 = **Piston**
3 = **Starting button**
4 = **Liquid nitrogen**
5 = **Metal mirror**
6 = **Sample holder**

Figure 6.6 Freezing system on a block of chilled metal.

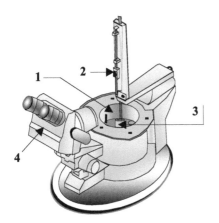

1 = **Well containing liquid nitrogen**

2 = **Injector**

3 = **Chilled metal mirror**

4 = **Stereo microscope**

Figure 6.7   Freezing system on a chilled block of metal.

6.3.2   Products

- Acetone
- Disodium phosphate
- DMSO

- Formvar powder
- Glycerol
- Isoamyl acetate
- Monosodium phosphate
- Paraformaldehyde

- Sucrose
- Solvents
  - Dichloroethane
  - Dioxan

⇀ To degrease grids
⇀ Powder
⇀ Dimethylsulfoxide
⇀ Cell culture quality

⇀ AR quality
⇀ AR quality
⇀ Powder
⇀ Powder
⇀ AR quality
⇀ AR quality
⇀ Formvar is soluble in dichloroethane or dioxan.

6.3.3   Solutions

- 2% collodion in isoamyl acetate

- DMSO 10%
- 0.15–0.25% formvar

- 5% glycerol
- 40% paraformaldehyde
- 10X phosphate buffer; pH 7.4

- Sucrose
  - 0.4 *M* sucrose
  - 2.3 *M* sucrose

⇀ *See* Appendix A4.3.
⇀ Ready-to-use solution (0.5%)

⇀ *See* Appendix A4.4.
⇀ Ready-to-use solution

⇀ *See* Appendix B4.3.
⇀ *See* Appendix B3.4.1. Can be stored at room temperature for 1 month after sterilization.
⇀ *See* Appendix B2.17.
⇀ In a phosphate buffer (100 m*M*)
⇀ Saturated solution

6.4   CRYOPROTECTION

☞ Indispensable

6.4.1   Principles

Cryoprotection limits the formation of crystals when tissue liquids freeze. Some molecules in solution become solid via an amorphous form (i.e., without crystallization).

☞ Water is the principal constituent of tissue.
☞ *See* Section 6.6.1.

6.4.2   Cryoprotective Agents

6.4.2.1   Sucrose or β-D-Fructofuranosyl α-D-Glucopyranoside

- Formula $C_{12}H_{22}O_{11}$
- Molecular weight   **342.3**
- Concentration   **0.4 *M* or 2.3 *M***

☞ Chemically neutral

6.4.2.2   Glycerol

- Formula $C_3H_8O_3$
- Molecular weight   **92.09**
- Melting point   **18°C**
- Concentration   **5–10%**

☞ **Toxic product;** highly penetrative
☞ Very good cryoprotective agent; establishes hydrogen bonds, but the tissue remains soft, and sectioning is more difficult

6.4.2.3   Dimethylsulfoxide (DMSO)

- Formula C_2H_6SO
- Molecular weight   **78.13**
- Melting point   **18°C**
- Concentration   **5–15%**

☞ **Toxic product**
☞ Good cryoprotective agent, but as an organic solvent, may alter membranes

6.4.3   Protocol

1. Incubate the samples in a buffered solution containing the cryoprotective agent:
   - 0.4 *M* sucrose   **≥ 90 min at 4°C**
   - 2.3 *M* sucrose   **16 h or more if necessary at 4°C**
   - 5% glycerol   **≥ 60 min at 4°C**
   - 10% DMSO   **≥ 90 min at 4°C**

2. Remove surplus cryoprotective agent.

☞ 100 m*M* phosphate buffer; pH 7.4

☞ Iso-osmolar solution

☞ Hyperosmolar solution
☞ The samples will float when first placed in this solution. When they become impregnated, the samples should fall to the bottom of the tube.
☞ This cryoprotective agent penetrates very rapidly. It may be used for large samples.
☞ This cryoprotective agent may be used for large samples.

☞ The samples must not dry out.

3. Incubate in 2.3 *M* sucrose.

- Duration ≤ **10 min at RT**

❏ *Next stage:*
- Freezing

⮞ After 10 min, the hyperosmolarity should become apparent.
⮞ This cryoprotective agent may be used for large samples.

⮞ This depends on the type of cryoprotective agent used.
⮞ *See* Section 6.6.6.

6.5 CRYOGENIC AGENTS

6.5.1 Principles

Freezing biological samples must be rapid and requires a large volume of cryogenic agent. The cryogenic agents used must allow ultra-rapid or the most rapid freezing possible. They must have a very low boiling point.

⮞ Ultra-rapid freezing is extremely difficult to obtain (*see* Section 6.6.6).

6.5.2 Liquid Nitrogen

⮞ N_2

Characteristics:

- Melting point −209.9°C
- Boiling point −195.7°C
- Cryogenic capacity **0.723 W/cm²/K**

❏ *Advantages*

- Easy to obtain

- Easy to store

- Used for storage of samples after freezing
- Skin contact dangerous

❏ *Disadvantages*

- Chelation

- Weak cryogenic capacity

⮞ It is the main component of the atmosphere.
⮞ Store in a well-ventilated area, otherwise it may reduce the oxygen level.
⮞ It is an inert gas.
⮞ May be frozen by contact with liquid nitrogen; also, metal that has been in contact with liquid nitrogen is potentially dangerous.

⮞ The boiling point of liquid nitrogen is very close to its melting point. Thus, a warm body placed in liquid nitrogen will initiate boiling. The nitrogen vapors form a layer of gas between the body and the liquid nitrogen, preventing further freezing. The layer of gas acts as an insulator.
⮞ It is necessary to raise the cryogenic capacity by using liquid nitrogen slush, metal chilled by nitrogen, nitrogen vapor, etc.).

6.5.3 Helium

➥ He

Characteristics:

- Melting point   **–271.4°C**
- Boiling point   **–268.6°C**
- Cryogenic capacity   **0.21 W/cm²/K**

❏ *Advantages*

- High cryogenic capacity
- Inert gas

➥ No chelation (*see* above)
➥ No chemical reaction between the samples and the gas

- Unlimited storage

➥ Its temperature is close to absolute zero and storage properties are unlimited.

❏ *Disadvantages*

- Rare gas
- Difficult to handle
- Difficult to store

➥ Expensive
➥ Special containers
➥ Special containers

6.6 FREEZING

6.6.1 Principles

This stage consists of preparing a sample that is hard enough for cutting ultrathin sections. The hardness is obtained by freezing the water in the tissue.

➥ The harder the tissue, the easier it is to cut ultrathin sections.
➥ Freezing involves a molecule changing from a liquid to a solid state.
➥ In animal tissue, around 70% of the weight is water. In plant cells, this value is around 90%.

Freezing occurs by chilling with a cryogenic agent. The chilling curve is shown in the next figure:

➥ Water is a pure liquid.

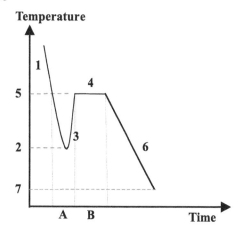

1 = **Liquid phase**
2 = **Temperature of super-cooling** (Ts)
3 = **Rising temperature phase**
4 = **Liquid–solid or freezing plateau phase**
5 = **Freezing temperature** (Tf)
6 = **Solid phase**
7 = **Storage temperature**

(A) **Period of super-cooling**
(B) **Crystallization period**

Figure 6.8 Cooling or freezing curve of a pure liquid (e.g., water).

Freezing is an exothermic reaction.

↬ During freezing, a liquid passes through an unstable state which becomes more stable by reducing its entropy (i.e., loss of heat).

❏ The characteristics of the freezing process are:

↬ At atmospheric pressure

• Super-cooling temperature (Ts)

↬ A lower temperature is obtained in the liquid phase.

↬ This temperature is not constant and corresponds to the appearance of the first crystal: **nucleation point**

↬ Water may reach temperatures much lower than its freezing temperature (0°C) without crystallization.

• Freezing temperature (Tf)

↬ Equilibrium temperature between the liquid phase and the solid phase

↬ This temperature is a constant and characterizes a pure substance.

↬ This corresponds to an equilibrium between the loss of heat during crystallization and the lowering of temperature by a cryogenic agent.

• The longer this phase, the higher the rise in temperature.

↬ The longer this phase is, the larger the number of ice crystals.

• The freezing phase

↬ The duration of the equilibrium between the liquid and solid phases depends on the size of the ice crystals.

• The storage temperature

↬ This temperature is lower than the recrystallization temperature.

• The recrystallization temperature

↬ Temperature at which the solid phase is modified by the appearance of hexagonal crystals (*see* Figures 6.9 and 6.10) or in place of amorphous ice (e.g., for pure water this temperature is around –100°C).

In nonhomogeneous liquids these parameters are modified:

• The super-cooling temperature

↬ The liquid phase is more stable and the first crystals appear much more quickly.

• The freezing temperature

↬ Usually this physical characteristic does not exist.

• The freezing plateau

↬ There is not a clearly defined stage as such, rather a succession of stages, each characterized by its constitutive elements. In the aqueous solutions found in animal tissues, the first crystals formed are made up of pure water. The last crystals formed are rich in ions, which lowers the freezing temperature.

• The recrystallization temperature

↬ The crystalline state is unstable, as the mixture is more complex. The temperature of the sample has to be maintained at a very low level to preserve the stability of the freezing process.

❑ The transformation of tissue water may be carried out in different ways:

• Crystallization in the form of hexagonal crystals

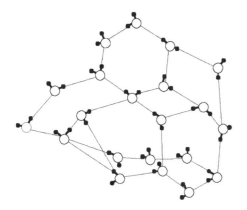

⇝ This is obtained by slow freezing or after recrystallization.
⇝ Crystal size is >1 μm

Figure 6.9 Hexagonal arrangements of water molecules.

• Crystallization in the form of cubic crystals

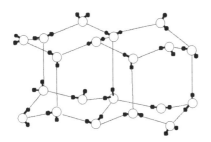

⇝ This is obtained by rapid freezing.
⇝ Crystal size is between 15 nm and 1 μm.

Figure 6.10 Cubic arrangements of water molecules.

• When crystallization takes place without the formation of crystals, the ice is said to be amorphous.

⇝ This is obtained by ultra-rapid freezing. This form of ice is evidenced by the rate of cooling being several hundred thousand degrees Celsius per minute over a thickness of a few micrometers.

❑ The parameters of freezing are:
• The structure of the sample

⇝ Heterogeneous structures are difficult to freeze uniformly.
⇝ The hardness of the sample is limited by the amount of ice included in it.

• The composition of the biological solution

⇝ This is never a pure liquid. Water is the principal component containing salts, proteins, carbohydrates, and lipids in varying concentrations, not only between tissues but within the same sample.

• The rate of cooling

⇝ The higher the speed, the smaller the ice crystals.
⇝ The speed of crystal formation may be 107 nm/s at −15°C and almost nothing at −100°C.

- The critical freezing points:
  - Appearance of the first ice crystals

    ➥ After this instant, the number increases rapidly.
  - Duration of the freezing plateau

    ➥ The duration is proportional to the size of the crystals.
  - Sample storage

    ➥ The more complex the solution the lower the recrystallization temperature.

The atmospheric pressure modifies certain parameters:

    ➥ This property is important when freezing at high pressures (*see* Section 6.6.2).

- The melting point

    ➥ For example, at 2000 bars, the Tf is –20°C for pure water.

- The super-cooling temperature

    ➥ For example, at 2000 bars, the Ts is –90°C for pure water.

6.6.2 Aim

Freezing is performed in order to obtain a sample hard enough for cutting ultrathin sections with a minimum number of crystals.

➥ The hardness of the sample is not a problem in and of itself.

➥ If the number and the size of the crystals are too high, it may prevent the cutting of sections because of heterogeneous and friable tissue.

Good freezing will reduce the number and size of crystals and preserve both the morphology and nucleic acids.

➥ The quality of the morphology is inversely proportional to the number and size of the crystals. However, sections containing a large number of small crystals can be examined at low magnifications under the electron microscope or be used for light microscopy.

The number and the size of the crystals are reduced by diminishing or bypassing the phases where the crystals appear and increase in size.

These phases are the nucleation point (i.e., the super-cooling temperature), the increasing temperature phase, and the freezing plateau.

If these three phases are reduced or bypassed, the number and size of the crystals will be reduced. For this to occur, it is necessary to know precisely the different parameters of the system used. Use of the appropriate rate of cooling can overcome the rising temperature phase and remove the freezing plateau.

Techniques to obtain this degree of cooling at the appropriate time are:

➥ The ideal situation is a biological liquid becoming an amorphous solid (i.e., no crystals).

➥ *See* Figure 6.6.

➥ The theoretical freezing curve for water is a function of the rate of cooling. This can be demonstrated by the absence of the three phases by obtaining a speed of 33.6 GJm^{-3}/s^{-1}. The freezing plateau disappears from this freezing speed half as fast.

- High-pressure freezing

➥ This technique is not appropriate for all samples. It is best to consult an expert. The protocol is not presented here.

- Freezing on a chilled metal block

➥ *See* Section 6.6.3.3.

- Programmed freezing

➥ *See* Section 6.6.3.4.

A cryoprotective agent replaces the complex cellular and extracellular mixture of the sample. The majority of the parameters thus become known and may be used to optimize the freezing procedure.

⇨ *See* Section 6.4.
⇨ Changes the complex aqueous biological medium to the simpler cryoprotective agent

The parameters of the cryoprotective solution used are known in the chosen freezing system and are applied to the sample.

⇨ This can also be applied directly without too many changes to fixed samples.

6.6.3   Freezing Methods

6.6.3.1   Immersion in Liquid Nitrogen

⇨ This is the simplest method, but can only be used if the samples are cryoprotected by prior incubation with 2.3 *M* sucrose (*see* Section 6.4.3).

6.6.3.1.1   SUMMARY OF DIFFERENT STAGES

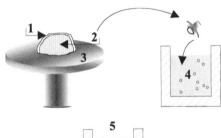

1 = **2.3 *M* sucrose**
2 = **Sample**
3 = **Cryo-ultramicrotome sample holder**
4 = **Immersion of sample holder in liquid nitrogen**
5 = **Freezing**

Figure 6.11   Mounting the sample on the cryo-ultramicrotome sample holder.

❏ *Advantages*

• Very simple method
• Identical results for all sample thicknesses

⇨ If the penetration of 2.3 *M* sucrose is identical in all samples

• No extra equipment is needed.

❏ *Disadvantages*

• Uses a single cryoprotective agent
• Hyperosmolarity of the cryoprotective agent
• Cannot be used on all tissues

⇨ 2.3 *M* sucrose
⇨ This phenomenon is less apparent in certain tissues (e.g., muscle).
⇨ The cryoprotective agent does not diffuse in certain tissues which cannot be frozen by this method (e.g., bone, adipose tissue).

• Difficult to cut sections
• Can result in artifacts

⇨ Heterogeneity of the hardness of the sample
⇨ Numerous small ice crystals are present.

6.6.3.1.2   MATERIALS

• Liquid nitrogen container
• Plastic forceps

⇨ For freezing the samples
⇨ For holding the sample and sample holder after freezing

6.6.3.1.3   PROTOCOL

1. Incubate the samples in   **16 h** 2.3 *M* sucrose.   **at 4°C**

⇒ *See* Section 6.4.3

2. Place each sample on a cryo-ultramicrotome support.

⇒ A thin layer of sucrose should cover the sample (*see* Figure 6.11).

3. Immerse the samples in the liquid nitrogen.

⇒ The samples must freeze before chelation is finished (*see* Section 6.5.2).
⇒ Rate of cooling is ≈ 2000°C/min.

4. Place the sample holder in the cryo-ultramicrotome.

⇒ The sample is ready for sectioning.
⇒ The sample may be stored.

❏ *Next stage:*
 • Cutting ultrathin sections

⇒ *See* Section 6.7.

6.6.3.2   Immersion in Nitrogen Slush

6.6.3.2.1   SUMMARY OF DIFFERENT STAGES

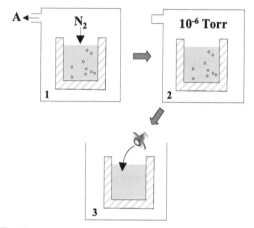

1 = **Liquid nitrogen at atmospheric pressure.**
    (A) Container under vacuum (≈ 10^{-6} Torr)

2 = **Solid nitrogen at 10^{-6} Torr pressure**

3 = **Immersion of sample and sample holder in nitrogen slush.**

Figure 6.12 Protocol for freezing in nitrogen slush.

❏ *Advantages*

 • Good reproducibility of results
 • No need for a particular cryoprotective agent
 • Rapid freezing speed

⇒ If the nitrogen remains as slush
⇒ High concentrations of cryoprotective agent give the best results.
⇒ Speed is slower than the vitrification of water.

❏ *Disadvantages*

 • A vacuum pump is necessary.
 • Slow method

 • Method is used solely for very small samples.

⇒ The nitrogen may evaporate before freezing.
⇒ Nitrogen slush remains in that state for a very short period of time.
⇒ Freezing the sample and the sample holder at the same time negates any advantage of this technique.

6.6.3.2.2   MATERIALS

 • Liquid nitrogen container

⇒ Small container with small opening to limit the evaporation of nitrogen, which will saturate the capacity of the pump

- Plastic forceps
⮑ For holding the sample and sample holder after freezing

- Secondary vacuum pump ($\approx 10^{-6}$ Torr)
- Vacuum container

6.6.3.2.3 PROTOCOL

1. Incubate the samples **≤ 10 min** in 2.3 *M* sucrose
⮑ *See* Section 6.4.3.
⮑ Or other cryoprotective agent

2. Position the liquid nitrogen container. Nitrogen vapor is removed under a vacuum.
⮑ The capacity of the pump must allow a secondary in 5–10 min.

3. Place each sample on a cryo-ultramicrotome sample holder.
⮑ A thin layer of sucrose should remain around the sample (*see* Figure. 6.11).

4. Break the vacuum and open rapidly.
⮑ Nitrogen should remain slush.

5. Place the samples in the nitrogen slush.
⮑ The sample should freeze without chelation (*see* Section 6.5.2).
⮑ Rate of cooling is > 2000°C/min.
⮑ Several samples can be frozen at one time.

6. The nitrogen reverts to liquid form.
⮑ More nitrogen can be added and the process recommenced.

7. Place the specimen holder in the cryo-ultramicrotome.
⮑ The specimen is ready to be sectioned.
⮑ The specimen may be stored.

❏ *Next stage:*
- Cutting ultrathin sections
⮑ *See* Section 6.7.

6.6.3.3 Projection on a Chilled Metal Block

6.6.3.3.1 SUMMARY OF DIFFERENT STAGES

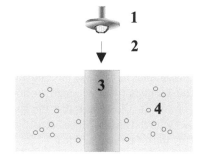

1 = **Sample/sample holder**

2 = **Drop the sample onto the metal block**

3 = **Metal block**

4 = **Liquid nitrogen**

Figure 6.13 Protocol for freezing on a chilled metal block.

❏ *Advantages*

- Good freezing
⮑ Few small crystals under optimal conditions

- It is possible to freeze small unfixed and uncryoprotected samples.
⮑ For example, cell suspension or monolayer

- Any cryoprotective agents may be used.
⮑ Nonspecific concentration

❏ *Disadvantages*

- Freezing occurs over a very thin layer.
⮑ A few μm
⮑ It is difficult to cut ultrathin sections of very thin specimens.

- Expensive

- Time consuming

⮑ The quality of the copper block is very important (composition, surface, oxidation).
⮑ The copper block must be cleaned between each freezing.

6.6.3.3.2 MATERIALS
- Freezing system

⮑ Different systems that use the same principles can be employed:
  1. Progressive system of chilling (*see* Figure 6.3)
  2. Freezing system using a chilled metal block (*see* Figures 6.5, 6.6, and 6.7)

- Liquid nitrogen container

⮑ For storing frozen samples and for keeping liquid nitrogen for freezing systems

- Plastic forceps

⮑ For holding the sample and sample holder after freezing

6.6.3.3.3 PROTOCOL
1. Clean the copper block.

⮑ The quality of freezing depends on the quality of thermal transfer. This is obtained between the chilled copper block (at the temperature of the cryogenic agent used) and the surface of the sample. All contaminants or oxidation on the surface of the block act as insulators and slow down the thermal exchange.

2. Chill the copper block.

⮑ Generally this is done by plunging the block into liquid nitrogen. More sophisticated systems exist, or helium can be used as the cryogenic agent.

3. Trim the sample to obtain a surface area that is large, homogeneous, and of minimum thickness.
4. Remove excess moisture from the samples.
5. Place the samples on a cryo-ultramicrotome sample holder with a thin layer of 2.3 *M* sucrose (*see* Figures 6.3 (B) and 6.11).

⮑ The surface of the section in contact with the block must be maximized.

⮑ The samples must never be completely dry.
⮑ The samples must be small to limit their weight and also to ensure that the surface in contact with the block will be correctly frozen.
⮑ This must be done rapidly.

6. Fix the sample holder and sample into the freezing apparatus.
7. Release the catch.

⮑ The speed of descent and the crushing force of the apparatus must be controlled depending on the samples to be frozen.

8. Immediately remove the sample holder from the freezing apparatus and immerse in liquid nitrogen.
9. Place the sample holder in the cryo-ultramicrotome.

⮑ This must be done rapidly.

⮑ The sample is ready to be sectioned.
⮑ The sample may be stored.

❏ *Next stage:*
 • Cutting ultrathin sections

↪ *See* Section 6.7.

6.6.3.4   Programmed Freezing

6.6.3.4.1   Summary of Different Stages

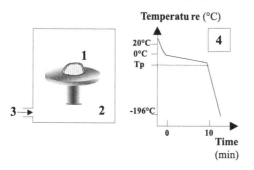

Temperature (°C)

20°C
0°C
Tp

-196°C

0   10   Time
            (min)

1 = **Sample/sample holder**

2 = **Chilled container**

3 = **Entrance for the cryogenic agent**

4 = **Freezing program**
     (**Tp**) Temperature plateau

Figure 6.14 Protocol for programmed freezing.

❏ *Advantages*

 • Good freezing throughout the sample

 • All cryoprotective agents may be used.

 • Optimal freezing with a low concentration of cryoprotective agent

 • Reproducible results

❏ *Disadvantages*

 • Method which requires small items of equipment
 • Method which requires careful optimization

↪ Numerous very small crystals under optimal conditions
↪ The characteristics of the cryoprotective agent used must be known (i.e., super-cooling temperature and freezing point (*see* Section 6.4.2).
↪ No risk of hyperosmolarity
↪ Limited risk of losing the molecules of interest
↪ After the parameters have been defined (see above)

↪ The cold gradient may be carried out in a polystyrene container.
↪ The temperature of the samples is compared with the room temperature:
 • If large crystals appear, the super-cooling temperature (Ts) is not compatible with the rate of chilling (*see* Figure 6.8).
 • If a number of small crystals are visible, the super-cooling temperature (Ts) has been passed before chilling. The rise in temperature to the freezing plateau (*see* Figure 6.8) causes an increase in the number of crystals.
↪ If numerous crystals are visible, the freezing of the tissue has taken place prior to chilling (*see* Figure 6.15).
↪ If the number of ice crystals increases, temperature increase owing to the freezing plateau has taken place (*see* Figure 6.15).

The figure below presents the freezing curve for samples cryoprotected in 0.4 *M* sucrose in 100 m*M* of phosphate buffer.

↪ For this iso-osmolar agent, the super-cooling temperature is –4.3°C. The freezing plateau is –4°C. With a slow rate of chilling prior to –4°C, the super-cooling temperature may be reached without risk before immersion in liquid nitrogen.

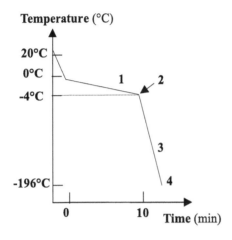

Temperature (°C)

1 = **Slow rate of chilling**

2 = **Temperature of freezing plateau** (Tp)

3 = **Immersion in liquid nitrogen**

4 = **Storage in liquid nitrogen**

Figure 6.15 Freezing curve for samples cryoprotected in 0.4 *M* sucrose.

6.6.3.4.2   MATERIALS

Digital thermometer

↪ Sensitivity ± 0.1°C

- Freezing system

↪ There are different systems at different levels of sophistication which control the rate of chilling.
↪ Freezing gradient system (*see* Figure 6.2)

- Liquid nitrogen container

↪ For the storage of frozen specimens

- Plastic forceps

↪ For handling the frozen tissue and tissue holder

- Thermal probe

↪ The type of thermocouple with cold reference or a ceramic probe has a small delay in response.

6.6.3.4.3   PROTOCOL

1. Trim the samples to a pyramidal surface ready to section.

↪ This surface is the correct shape for cutting ultrathin sections.

2. Incubate the specimens in 2.3 *M* sucrose.   **< 10 min at 4°C**

↪ This stage must be rapid (*see* Section 6.4.3).
↪ This stage aims to protect the specimens from desiccation during freezing in the gaseous phase.

3. Remove excess liquid from the specimens.

↪ The specimens should not be completely dry.

4. Place the specimen on a specimen holder for a cryo-ultramicrotome with a thin layer of 2.3 *M* sucrose (*see* Figure 6.11)

↪ The specimens should be small.

5. Place the specimen holder on a card with an identifying code on it (*see* Figure 6.3B).

↪ This must be done rapidly (*see* Section 6.4.3).

6. Place into the freezing apparatus.

↪ Different rates of cooling and freezing plateau should have been defined in advance.

7. Chill the samples down to the temperature of the freezing plateau (Tp).

↪ This must be done rapidly.

8. Chill the sample by injection of a large quantity of cold nitrogen gas, then immerse in liquid nitrogen.

9. Store the sample holder and sample in liquid nitrogen.

❏ *Next stage:*

• Cutting ultrathin sections

↪ *See* Section 6.7.

Table 6.1 Criteria for Choosing a Freezing Technique

Parameters	Freezing Methods			
	Immersion in Liquid Nitrogen	**Freezing on a Metal Block**	**Programmed Freezing**	**Freezing at High Pressure**
Simplicity	+	–	–	–
Materials	–	++	+	+++
Defining parameters	–	+	++	++
Choice of cryoprotector	2.3 *M* sucrose	+	+++	++
Duration	< 1 min	30 min/sample	20 min	60 min
Reproducibility	++	++	+++	+++
Quality	±	A few μm	All of the sample	All of the sample
Preservation • Morphology • Nucleic acid	 ± –	 ++ ±	 ++ ++	 +++ ++
Efficiency	–	±	+	±

6.7   CRYO-ULTRAMICROTOMY

6.7.1   Summary of Different Stages

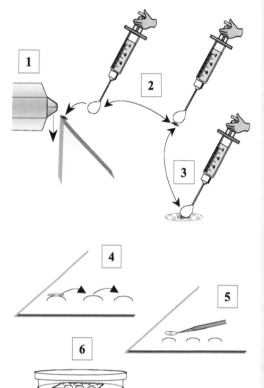

1 = **Cryosections** (80–100 nm).
    Ultramicrotomy of frozen tissue (cryo-ultramicrotome)

2 = **Picking up ultrathin sections.**
    Similar to the method for dry sections but using a drop of sucrose

3 = **Place the sections on a collodion and carbon-coated grid.**

4 = **Removal of sucrose**
5 = **Use**
6 = **Storage**
    At 4°C with coating (in a Petri dish) for several months without sucrose removal

Figure 6.16   Method for picking up cryosections.

6.7.2   Cutting Cryosections

6.7.2.1   Material

Cryo-ultramicrotomy involves cutting ultrathin sections of frozen tissue.

Temperature stability is very important in cutting sections and in sample preservation:

- The specimen holder within the cryo-ultramicrotome chamber is kept at a temperature lower than the temperature of recrystallization.
- The temperature of the glass knife is lower than that of the specimen.
- The temperature of the chamber for ultrathin sections must be kept at a lower temperature than that of recrystallization.

↪ Sections of 80 to100 nm in thickness

↪ To avoid deterioration of samples

↪ This difference in temperature favors the spreading of sections.
↪ To preserve the samples

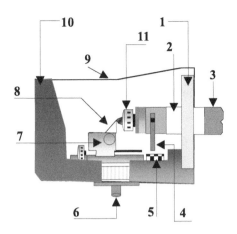

1 = **Support for the chamber insulation**
2 = **Insulation for the cutting arm**
3 = **Sectioning arm**
4 = **Heating plunger**
5 = **Liquid nitrogen container**
6 = **Fixation system for the ultramicro-tome**
7 = **Interchangeable knife holder**
8 = **Diamond or glass knife**
9 = **Incline to reduce the loss of cold nitrogen gas**
10 = **Insulation**
11 = **Specimen holder**
12 = **Specimen**

Figure 6.17 Monobloc cryogenic chamber.

The method is similar to that for dry sections with the sections cut directly onto the knife without floating on liquid.

↝ Dry sectioning method according to *Tokuyasu*

Sections may be cut with a diamond or a glass knife.

↝ Glass knives are better at low temperatures.

Sections are picked up with a drop of saturated sucrose solution at the end of a syringe. The drop is then placed on a collodion and carbon-coated nickel grid.

↝ 2.3 *M* sucrose solution

↝ *See* Appendix A4.3.

Cryosections may be treated immediately after removal of the sucrose (*see* Figure 6.16).

↝ Storage for several days after coating in sucrose is possible. The storage may be extended to several months at 4°C. The sucrose will not crystallize.

6.7.2.2   Section Parameters

Several parameters are important in cutting sections:

↝ The same parameters are applicable for cutting ultrathin sections of frozen tissue as for cutting ultrathin sections of tissues embedded in resin.

- Dimension of the block

↝ The block should be small to avoid breaking during sectioning.

- Preparation of the block

↝ The pyramidal sample shape (*see* Figure 6.11) puts a small zone of the section in contact with the cutting edge, which keeps section breakage to a minimum.

- Orientation of the knife

↝ The knife should be oriented so that the plane of the section is parallel to the cutting edge of the knife.

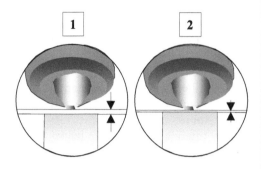

1 = **Plane of section parallel to the cutting edge of the knife**
Orientation of the knife by comparison with the block face

2 = **Final approach**

Figure 6.18 Positioning the sample.

• Hardness of the section	Direct consequence of the temperature of the specimen holder: the lower the temperature, the harder the block. ⇝ This can only be varied slightly.
• Angle of the glass knife ≈ 6°	

• Temperature of the biological sample
  – Soft tissue is –110°C
  – Hard tissue is –150°C
	⇝ Always higher than the recrystallization temperature
	⇝ For certain heterogeneous tissues with very hard components, low temperatures such as –150°C, must be used (e.g., bone).
	⇝ At very low temperatures, samples are fragile.

• Temperature of the knife – Generally lower than that of the specimen	⇝ Helps to spread the section on the knife.
• Temperature of the chamber • –90°C to –120°C	⇝ This is necessary to stabilize the specimen and the knife and limit the turbulence in the cryo-ultramicrotome.
• Speed of sectioning	⇝ This can be adjusted to improve the quality of the section. A high speed can heat the specimen, and too low a speed can cause crumbling of the surface.
• Static electricity	⇝ Sections roll at the knife surface due to repulsion of the sections. ⇝ May be overcome by the aid of a low current

6.7.2.3 Protocol

Cutting sections involves the following stages:

1. Place the specimen holder in the cryo-ultramicrotome.
2. Wait 15 min before sectioning.

⇝ All the temperatures used for cryo-ultramicrotomy must be stabilized prior to starting.
⇝ Preparation of the block is carried out prior to freezing (*see* Section 6.5).
⇝ Temperature of the specimen holder is that at which it is stored (≈ –196°C).
⇝ Temperature is allowed to equilibrate.

3. Position the frozen sample in front of the knife edge (*see* Figure 6.18).

➥ The right side of the knife edge is used to smooth the block and to approach the zone of interest.

➥ The left-hand side of the edge is used for cutting ultrathin sections (*see* Appendix A5.3).

4. Cut thick sections.

➥ 0.5 μm in thickness; thicker sections will damage the specimen.

5. Cut cryosections of around 100 nm in thickness with a glass knife at −100°C; the temperature of the block may vary between −110°C and −150°C, depending on the type of tissue.

➥ It is possible to use higher temperatures, but in that case, specimens cannot be used again.

6. Pick up the sections by contact with a drop of 2.3 *M* sucrose (*see* Figure 6.19).

7. Place the drop on a collodion and carbon-coated nickel grid.

➥ A good section looks like cellophane and has a yellow gold color (*see* Table 4.2)

➥ Outside the cryogenic chamber, the sucrose warms and spreads cryosections.

8. The grid may be inverted on a drop of buffer to eliminate the sucrose.

➥ The choice of buffer depends on the next stage (i.e., a buffer of Tris–HCl before a deproteinization and SSC 4X before hybridization).

➥ It is possible to store the grids with the drop of sucrose protecting them from desiccation and alteration (preferably at 4°C until use) (*see* Figure 6.16).

9. The sections must not be allowed to dry.

➥ Drying the sections results in a loss of morphology and soluble molecules.

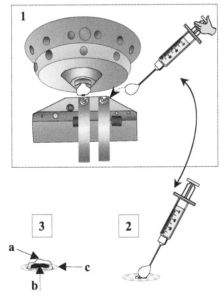

1 = Cutting cryosections

2 = Picking up sections using a drop of concentrated sucrose

3 = Placing the section on a collodion-carbon coated grid
   (**a**) sucrose
   (**b**) cryosection
   (**c**) grid

Figure 6.19 Picking up cryosections.

❑ *Next stages:*
 • Pretreatments

 • Hybridization

➥ **Optional**
➥ *See* Section 6.8.
➥ *See* Section 6.9.

6.8   PRETREATMENTS

⇨ **Optional**

Ultrathin sections of frozen tissue are permeable to all the molecules used for *in situ* hybridization. Thus, pretreatments are usually of little use. They may be used for the controls and for improving a weak signal:

⇨ It is always better to test a protocol for the first time without pretreatment.

- Deproteinization
- RNase/DNase treatment
- Prehybridization

⇨ It is used in exceptional cases.

6.8.1   Deproteinization

⇨ **Optional**

6.8.1.1   Aim

This treatment is performed to partially eliminate proteins, particularly those associated with nucleic acids, to allow better accessibility of the probe to the nucleic acid target.

This treatment may be done with a mixture or a single enzyme. The principal enzymatic treatments are:

⇨ This treatment must be carried out just prior to hybridization. Sections cannot be stored after deproteinization.

- Proteinase K

⇨ Removes proteins associated with nucleic acid

- Pronase
- Pepsin

⇨ Removes all proteins
⇨ Removes all proteins at an acid pH

Proteinase K is the pretreatment of choice.

⇨ If necessary

6.8.1.2   Pretreatment with Proteinase K

Proteinase K is generally considered selective for proteins associated with nucleic acids.

⇨ **Optional**
⇨ Its effect may be modulated by:
  - Concentration
  - Temperature
  - Buffer
  - Duration

❑ *Advantage*
- Raises the sensitivity

⇨ This stage is usually not indispensable.

❑ *Disadvantage*
- Destroys cellular structures

⇨ There is a loss of morphology.

6.8.1.2.1   PARAMETERS

1. Concentration:
   This varies depending on the sample type (cells, sections) or the tissue structure (homogeneous or not), and also according to the nature of the tissue and the nucleic acid target:

• Cells	**0.1–0.5 µg/mL**	⇒ A moderate digestion will preserve the morphology.
• Sections	**0.2–0.5 µg/mL**	
• Hard tissue	**0.4–1 µg/mL**	

2. Buffer:

Proteinase K (stock solution) is diluted in a buffer of 20 m*M* Tris–HCl/2 m*M* CaCl$_2$; pH 7.6.

⇒ This buffer contains Ca, which is the co-factor of proteinase K, but it is possible to use other buffers (e.g., Tris–EDTA) to limit the action of proteinase K.

3. Temperature:

The optimal temperature is 37°C, but activity occurs at ≈ 20°C.

⇒ It is best to use room temperature.

4. Duration:

Must be carefully regulated.   **2–15 min**

⇒ Washes must be very precise; treat the grids one at a time.

6.8.1.2.2   PRODUCTS/SOLUTIONS

⇒ **To be used solely for *in situ* hybridization**
⇒ Molecular biology quality

1. Products
   • Calcium chloride
   • Disodium phosphate
   • Monosodium phosphate
   • Paraformaldehyde

   • Proteinase K
   • Sodium chloride
   • Sodium citrate
   • Tris-hydroxymethyl-aminomethane

⇒ CaCl$_2$ · 2H$_2$O
⇒ Powder
⇒ Powder
⇒ Powder
⇒ AR quality
⇒ Best in solutions
⇒ NaCl
⇒ C$_6$ H$_5$Na$_3$O$_7$ · 2H$_2$O
⇒ Powder; keep dry

2. Solutions:
   • Buffers
     – 10X phosphate; pH 7.4

⇒ *See* Appendix B3.4.1. Store at room temperature for 1 month after sterilization.

     – 20X SSC; pH 7.0

⇒ *See* Appendix B3.5. Store at 4°C or room temperature.

     – Tris–HCl/CaCl$_2$ (20 m*M* Tris/2 m*M* CaCl$_2$); pH 7.6

⇒ *See* Appendix B3.7.2. Store at 4°C or room temperature.

   • 1 *M* calcium chloride

⇒ *See* Appendix B2.6.

   • 10 N hydrochloric acid
   • 4% paraformaldehyde (PF) in 100 m*M* phosphate buffer; pH 7.4

⇒ *See* Appendix B4.3. Store at –20°C.

   • Proteinase K (1 mg/mL) in sterile water

⇒ *See* Appendix B2.15. Store at –20°C in 50 µL aliquots.

   • 10 N sodium hydroxide
   • Sterile water

⇒ *See* Appendix B2.20.
⇒ *See* Appendix B1.1. Use only once or use DEPC water (*see* Appendix B1.2).

   • 1 *M* Tris

⇒ *See* Appendix B2.22.

6.8.1.2.3   PROTOCOL

1. Eliminate sucrose:
   - 100 mM phosphate buffer   **2 × 5 min at RT**
     - ➮ Place 50 µL drops on hydrophobic film at the bottom of a humid chamber (*see* Figure 6.1).
     - ➮ Sucrose is diluted by the buffer.

2. Wash the sections:
   - 20 mM Tris–HCl/2 mM CaCl$_2$; buffer pH 7.6   **5 min at RT**
     - ➮ The choice of buffer modifies the activity of proteinase K.

3. Enzymatic treatment:
   - Proteinase K 0.2 µg/mL   **5 min at RT**
     - ➮ A range of times of incubation should be tested.

4. Stopping the reaction:
   - 100 mM phosphate buffer   **5 min**
     - ➮ **Changing the buffer** stops the action of proteinase K by eliminating the NH$_2$ groups from the Tris buffer and the calcium chloride.

5. Post-fixation:
   - ➮ Restabilizes the tissue; **indispensable prior to hybridization.**
   - 4% PF in 100 mM phosphate buffer   **5 min**
     - ➮ This fixative offers the best compromise between the efficiency of hybridization and the preservation of morphology.

6. Washes:
   - Buffer phosphate   **2 × 5 min**
     - ➮ Removes traces of fixative
   - SSC 2X   **2 × 2 min**
     - ➮ Preparation for the next stage

❏ *Next stages:*

- Prehybridization
  - ➮ **Optional**
  - ➮ *See* Section 6.8.3.
- Hybridization
  - ➮ *See* Section 6.9.

6.8.2   Treatment with RNase/DNase

➮ **Optional**

6.8.2.1   Principles

Treatment destroys the nucleic acids present in the tissue sections: DNA by DNase I, RNA by RNase A, and also for:

➮ Permits selection of nucleic acid
➮ RNase A has the ability to destroy all single-stranded RNA.

- Preventing the formation of unwanted hybrids
- Controlling sections by destroying target nucleic acids

➮ For example, does not detect the expression of an exogenous gene, such as a virus
➮ Negative control of target nucleic acid

❏ *Advantage*
- Specifically eliminates one sort of nucleic acid

➮ This is a crude control but allows the expression of native DNA to be distinguished.

❏ *Disadvantages*

• Cellular structures are broken down.

⇨ Destruction of DNA modifies the structure of the nucleus, while destruction of RNA modifies the structure of the cytoplasm.

• Risk of increasing the nonspecific signal

⇨ If there are not enough washes, nonspecific hybridization may be increased.

6.8.2.2   Material/Products/Solutions

1. Material:
   • Oven

   ⇨ At 37°C

2. Products:

   ⇨ **To be used solely for *in situ* hybridization**
   ⇨ Molecular biology quality

   • DNase I

   ⇨ Destruction of all DNA

   • RNase A

   ⇨ Without DNase; also available in solution
   ⇨ Bovine pancreatic ribonuclease
   ⇨ Storage at –20°C
   ⇨ Destruction of all RNA

3. Solutions:

   ⇨ **The sterility of the solutions is of prime importance for hybridization (*see* Appendix A2)**

   • Buffers

     – Hybridization for a cDNA probe

     ⇨ *See* Section 6.9.2.1.

     – Hybridization for an oligonucleotide probe

     ⇨ *See* Section 6.9.3.1.

     – 10X phosphate; pH 7.4

     ⇨ *See* Appendix B3.4.1. Storage at room temperature for 1 month after sterilization

     – 20X SSC; pH 7.0

     ⇨ *See* Appendix B3.5. Storage at 4°C or room temperature

     – 10 mM Tris–HCl/5 mM MgCl$_2$; pH 7.3

     ⇨ *See* Appendix B3.7.4.
     ⇨ It is recommended to add 2% RNasin and 2 mM DTT to inhibit RNase activity.

     – 20 mM Tris–HCl/300 mM NaCl; pH 7.6

     ⇨ *See* Appendix B3.7.5. May be prepared just before use

   • DNase I (100 mg/mL) in sterile water

   ⇨ *See* Appendix B2.12. Storage at –20°C in 10 µL aliquots

   • 10 N hydrochloric acid

   • RNase A (100 mg/mL) sterile water

   ⇨ *See* Appendix B2.16. Storage at –20°C in 10 µL aliquots

   • Sterile water

   ⇨ *See* Appendix B1.1. Do not use once opened or use DEPC water (*see* Appendix B1.2).

   • 1 M Tris

   ⇨ *See* Appendix B2.22.

6.8.2.3 Protocol

The enzymes (DNase and RNase) must be of excellent quality and not be contaminated by any other enzyme.

↪ This is also important for the destruction of nucleic acid targets for controls.

Buffers:

↪ These should be specified by the companies supplying the enzymes.

* Tris–HCl/NaCl
* Phosphate

1. Incubate the tissue sections with:
   * DNase diluted in buffer   **1 mg/mL**
     30 min
     at 37°C

   ↪ The concentration must be tested.
   ↪ The duration may be increased or decreased.
   ↪ It is possible to work at room temperature.

   * RNase A diluted in buffer   **1 mg/mL**
     30 min
     at 37°C

   ↪ The concentration must be tested.
   ↪ The duration may be increased or decreased.
   ↪ It is possible to work at room temperature.

2. Wash freely in buffer.   **5 × 5 min**
   at RT

   ↪ **Indispensable**
   ↪ With buffer used for diluting the enzyme
   ↪ The enzymes cause breaks in nucleic acids. Washing is necessary to remove all nucleic acid fragments which may hybridize with the probe.

3. Do not let dry.

❏ *Next stages:*
* Prehybridization

↪ **Optional**
↪ *See* Section 6.8.3.

* Hybridization

↪ *See* Section 6.9.

6.8.3 Prehybridization

↪ **Optional**

6.8.3.1 Principle

Prehybridization is carried out prior to hybridization and consists of incubating sections in the hybridization medium in the absence of a probe.

↪ This stage is not indispensable but is useful for reducing the background.

❏ *Advantage*
* Reduces the nonspecific signal

❏ *Disadvantage*
* Reduces the sensitivity

↪ Risk of diluting the hybridization buffer and the probe during hybridization

6.8.3.2 Protocol

1. Incubate the sections with the hybridization medium in the absence of the probe:

↪ The composition of the hybridization medium differs depending on the probe used (cDNA or oligonucleotide).

   * Hybridization buffer   **≈ 50 µL/grid**

   ↪ *See* Section 6.9.2.1.
   ↪ It is possible to increase the concentrations of DNA, tRNA, and Denhardt's solution in relation to formamide.

• Duration of incubation	≈ **1 h** ➥ May be prolonged (several hours)
	at RT ➥ The temperature does not modify nonspecific binding (weak influence of formamide).

2. Remove most of the buffer before placing the grids on the reaction medium.
➥ The sections must not dry out. Leave the sections damp to help hybridization.

❏ *Next stage:*
  • Hybridization
➥ *See* Section 6.9.

6.9   HYBRIDIZATION

6.9.1   Materials/Products/Solutions

1. Materials
   • Centrifuge   ➥ More than 14,000 *g*
   • Eppendorf tubes   ➥ Sterile
   • Filter paper   ➥ To provide humidity in the Petri dish and for drying the grids

   • Gloves
   • Humid chamber   ➥ Petri dish with the base covered in hydrophobic film and containing a source of humidity (e.g., filter paper soaked in 5X SSC) for placing drops of different solutions (*see* Chapter 4, Figure 4.31)

   • Magnetic stirrer   ➥ To agitate the grids during the washes; the speed must be stable and slow.
   • Spotting dish   ➥ Sterile, for hybridization and for the washes; grids can be agitated lightly on a magnetic stirrer.

   • Vortex   ➥ Slow

2. Products
   • Dithiothreitol (DTT)   ➥ Molecular biology quality
   • DNA   ➥ Molecular biology quality; ready-to-use solutions are best.
   • Formamide   ➥ Deionized is available.
   • tRNA   ➥ Molecular biology quality; ready-to-use solutions are best.

3. Solutions
   • Buffers
     – Hybridization   ➥ Without the probe
     – Reaction buffer   ➥ Hybridization buffer containing the probe
     – 20X SSC; pH 7.0   ➥ *See* Appendix B3.5. Storage at 4°C or at room temperature

   • Deionized formamide   ➥ *See* Appendix B2.14. Storage at –20°C in 500 µL aliquots. If the product is not frozen at –20°C, do not use.

- 50X Denhardt's solution

  ➥ *See* Appendix B2.19. Storage at –20°C in aliquots. It is possible to thaw several times.

- 1 *M* dithiothreitol

  ➥ *See* Appendix B2.11. Storage at –20°C in 50 to 100 μL aliquots. May not be refrozen. It has an unpleasant odor. Do not use if the odor is absent or changes.

- 10 mg/mL DNA

  ➥ *See* Appendix B2.3. Sonicated; storage at –20°C. Repeated freezing/thawing results in breakages.

- Probes
  - Antigenic probes

    ➥ cDNA, oligonucleotide (*see* Chapter 1, Section 1.2.2).

  - Radioactive probes

    ➥ cDNA, oligonucleotide (*see* Chapter 1, Section 1.2.1).

- Sterile water

  ➥ *See* Appendix B1.1. Do not use after first opening or use DEPC water (*see* Appendix B1.2).

- 10 mg/mL tRNA

  ➥ *See* Appendix B2.5. Storage at –20°C. Repeated freezing/thawing results in breakages.

6.9.2   cDNA Probe

➥ **Wear gloves**

6.9.2.1   Hybridization Buffer

➥ This may be made just prior to use or stored at –20°C or –80°C.

1. In a sterile Eppendorf tube, place these products in the following order:

Solutions	Final concentration	
• 20X SSC	**4X**	➥ **Indispensable;** stabilizes the hybridization; source of Na^+
• Deionized formamide	**30%**	➥ **Useful;** denaturing agent that lowers the Tm
• Denhardt's solution	**2X**	➥ **Useful;** lowers the background by saturating the nonspecific binding sites.
• tRNA 10 mg/mL	**250 μg/mL**	➥ **Indispensable;** competes with proteins.
• DNA 10 mg/mL	**250 μg/mL**	➥ **Useful;** competes with proteins.
• 1 *M* DTT	**10 m*M***	➥ Solely for ^{35}S labeled probes
• Sterile water	**qsp × mL**	

2. Vortex.
   ➥ Gently
3. Centrifuge.
   ➥ To ensure all the liquid is at the bottom of the tube

6.9.2.2   Reaction Mixture

➥ **Wear gloves**

1. Add the probe to the hybridization buffer:
   - At a saturating concentration

     ➥ The addition of more probe does not increase the signal.
     ➥ Indispensable for quantifying the signal.

• At a non-saturating concentration

→ The background may be reduced by reducing the concentration of the probe.
→ Quantification is not possible.

  – Radioactive probe **10–100 ng/mL of the hybridization buffer**

  – Antigenic probe **0.1–2 µg/mL of the hybridization buffer**

→ After centrifugation and drying, the labeled probe is directly dissolved in the hybridization buffer.
→ Antigenic cDNA probes are used at 100 times higher concentrations.

2. Vortex.

→ Gently

3. Centrifuge.

→ To ensure all the liquid is at the bottom of the tube

4. Denature. **3 min at 100°C**

→ *See* Chapter 1, Figure 1.17.
→ **Indispensable;** double-stranded cDNA probes must be denatured.

5. Chill rapidly on ice.

→ To prevent the rehybridization of the two strands
→ At 0°C, the single strands are stable.

6. Add
  • DTT   **x µL**

→ Solely for ^{35}S-labeled probes
→ Final concentration is 10 mM

6.9.3   Oligonucleotide Probes

→ **Wear gloves**

6.9.3.1   Hybridization Buffer

→ This may be made just prior to use or stored at −20°C or −80°C

1. In a sterile Eppendorf tube, place these products in the following order:

Solutions	Final concentration	
• 20X SSC	**4X**	→ **Indispensable**
• Deionized formamide	**30%**	→ **Useful**
• Denhardt's solution	**2X**	→ **Useful**
• tRNA (10 mg/mL)	**400 µg/mL**	→ **Useful**
• DNA (10 mg/mL)	**250 µg/mL**	→ **Optional**
• DTT	**10 mM**	→ Solely for ^{35}S-labeled probes
• Sterile water	**qsp × mL**	

2. Vortex.

→ Gently

3. Centrifuge.

→ To ensure all the liquid is at the bottom of the tube

6.9.3.2   Reaction Mixture

→ **Wear gloves**

1. Add the probe to the hybridization buffer at a saturating concentration:

→ The addition of more probe does not increase the signal.
→ Indispensable for quantifying the signal
→ It is not always necessary to use this concentration. The background may be reduced by reducing the concentration of the probe.

- Radioactive probes **2–10 pmoles/mL of the hybridization buffer**

⇌ After centrifugation and drying, the labeled probe is directly dissolved in the hybridization buffer.

- Antigenic probes **20–100 pmoles/mL of the hybridization buffer**

⇌ Antigenic oligonucleotide probes are used at concentrations 10 times higher than radioactive oligonucleotide probes.

2. Vortex.

⇌ Gently

3. Centrifuge.

⇌ To ensure all the liquid is at the bottom of the tube

6.9.4 Hybridization Protocol

⇌ **Wear gloves**

1. Remove the sucrose by placing the grids on SSC

⇌ If the sections have not been pretreated

- 2X SSC: **3 × 5 min**

2. Incubate the grids on a drop of reaction mixture in a humid chamber:

⇌ A sterile spotting dish may be used (*see* Appendix A2).

⇌ The concentration of the liquid in the humid chamber must maintain a high humidity; several layers of filter paper soaked in 5X SSC are placed in the bottom of the chamber.

- Drop **≥ 30 µL/grid**

⇌ This quantity may be reduced by several µL per grid.

⇌ It is possible to place several grids on the same drop (100 µL of solution may support 5 grids).

- Incubation **3 h at RT humid chamber**

⇌ The temperature is chosen with the aim of forming specific hybrids and eliminating nonspecific hybrids during the washes.

⇌ The duration of the incubation is generally the shortest time (≤ 1h) that will give good results.

6.9.5 Post-Hybridization Treatments

6.9.5.1 Parameters

- Washes are carried out in SSC at room temperature.

⇌ SSC at pH 7.0

⇌ A light agitation of the grids can be obtained by using a porcelain spotting plate on a magnetic stirrer.

- Quantity of buffer
- Length of washes

⇌ A minimum volume of 50 µL for each grid

⇌ The duration of the washes can be adjusted, depending on the results obtained.

⇌ The washes are more numerous if the probes are radioactive.

- Conditions

⇌ The presence of RNase is not a problem; this enzyme destroys only single-stranded RNA. All hybrids, by definition, are double-stranded and are not destroyed by this enzyme.

All the washes are carried out by transferring the grids from one drop of solution to the next without removing the excess.

↝ The grids must not sink to the bottom of the drop.

6.9.5.2 Protocol

• 4X SSC	**5 min**	↝ **Necessary**
• 2X SSC	**3 × 5 min**	↝ Stop after this wash for the antigenic probes.
• 1X SSC	**3 × 5 min**	↝ Solely for radioactive probes

❏ *Next stages:*

• Visualization of radioactive hybrids

↝ *See* Section 6.10.1 and Chapter 5, Section 5.8.2.6.2.

• Visualization of antigenic hybrids

↝ *See* Section 6.10.2.

6.10 VISUALIZING HYBRIDS

6.10.1 Radioactive Hybrids

↝ *See* Chapter 5, Section 5.8.2 (visualization of radioactive hybrids).

After the final wash in 1X SSC, the grids are air dried.

↝ The morphological preservation will be diminished, but the signal will be more intense. A number of sections will be destroyed by coating in methylcellulose.

The development of the autoradiography is similar to that for sections of tissue embedded in acrylic or epoxy resin.

↝ *See* Chapter 5, Section 5.8.2.6.2.

❏ *Next stage:*
• Staining and coating in methylcellulose

↝ *See* Section 6.11.

6.10.2 Antigenic Hybrids

6.10.2.1 Solutions

1. Antibodies:
    • IgG anti-hapten nonconjugated (species X)

↝ IgG polyclonal

    • IgG, F(ab′)$_2$, or conjugated to biotin
        – Anti-species X conjugated

↝ Colloidal gold; 5, 10, or 15 nm in diameter

2. Buffers:
    – Blocking buffer
    – 100 mM phosphate buffer/ 300 mM NaCl/1% serum albumin; pH 7.4

↝ It is possible to use other agents in the blocking buffer (*see* Appendix B6.1.1).

    • 20X SSC; pH 7.0

↝ *See* Appendix B3.5. Storage at 4°C or at room temperature

    • 20 mM Tris–HCl/300 mM NaCl; pH 7.6

↝ *See* Appendix B3.7.5.

3. Fixatives:
- 1–2% glutaraldehyde in 2X SSC; pH 7.0
- 4% paraformaldehyde (PF) in 2X SSC; pH 7.0

⮡ *See* Appendix B4.2.
⮡ To maintain the salt concentration
⮡ *See* Appendix B4.3.

6.10.2.2 Indirect Reaction Protocol

⮡ For a direct reaction, the anti-hapten antibody must be labeled. The first stages are carried out under conditions identical to the protocol indirect reaction described below.

1. After the final wash in SSC, the ultrathin sections are post-fixed:

- 4% PF in 2X SSC   **5 min**

⮡ The buffer maintains the stringency of the washes and ensures that the hybrids are conserved.
⮡ To stabilize the hybrids; the immunocytochemical reaction is carried out in a low salt solution with the risk of losing the hybrids. Fixation stabilizes the cellular environment and preserves hybrids in a medium of high stringency.

2. Wash:
- 2X SSC   **3 × 5 min**
3. Block nonspecific sites:
- Blocking buffer   **10 min**

⮡ All paraformaldehyde must be removed.

⮡ **Indispensable;** to prevent all nonspecific binding, the sections are pre-incubated with nonspecific serum.

4. Anti-hapten-antibody (species X):

- Diluted 1:50 in Tris-HCl/NaCl   **≥ 40 µL/grid**
- Incubation   **1 h**

⮡ **Immunodetection:** formation of an antibody–hapten complex.
⮡ The dilution is between 1:50 and 1:200.

⮡ Humid chamber (water-soaked filter paper)
⮡ *See* Figure 6.1.

5. Wash:
- Tris-HCl/NaCl buffer   **3 × 5 min**
6. Conjugated antibody (anti-species X):
- Diluted 1:50 in Tris–HCl/NaCl buffer   **≥ 40 µL/grid**

- Incubation   **1 h**
7. Wash:
- Tris–HCl/NaCl buffer   **2 × 5 min**
- 2X SSC   **2 × 5 min**
8. Fixation:

- 1–2% glutaraldehyde in 2X SSC   **10 min**
9. Wash:
- 2X SSC   **3 × 5 min**
- Sterile water   **5 min**

⮡ Longer times can be used.
⮡ Detection of the complex
⮡ Secondary antibody: IgG, Fab fragments or biotin conjugated. Diluted (1:25 to 1:50) according to the manufacturer's instructions.
⮡ Humid chamber (water-soaked filter paper)

⮡ Longer times can be used.
⮡ Change of buffer
⮡ Stabilization of the immunocytochemical reaction and preservation of cellular structure
⮡ The time may be reduced.

⮡ To remove all traces of NaCl, which forms small crystals at the surface of the section and the membrane covering the grid

❏ *Next stage:*
- Staining and coating

↪ *See* Section 6.11.

6.11 STAINING/COATING

6.11.1 Aim

An ultrathin section of frozen tissue is very delicate. As the tissue dries, water, which constitutes a large part of the tissues, evaporates and structures that float in the liquid medium of the cell become distorted. To avoid this problem, sections are stained and then coated in a hydrophilic medium transparent to electrons.

↪ Problems that occur are often the result of the freezing process.
↪ Classical staining techniques for hydrophilic materials are used.

6.11.2 Principles

Several types of stain can be used to reveal the ultrastructure:

↪ The choice of stain depends on the aspect of the ultrastructure to be studied.

1. Positive staining:
   - 2% aqueous uranyl acetate
   - 4% aqueous uranyl acetate and 300 mM oxalic acid; pH 7.5

   - 0.5% aqueous ammonium molybdate
   - 1% buffered osmium tetroxide
   - Lead citrate

↪ 5–20 min (*see* Appendix B7.2.1.2)
↪ 10 min
↪ Neutralized with ammonia

↪ 1–5 min
↪ 5 min (*see* Appendix B4.1)
↪ 3 min
↪ May be used when the pH is very basic

2. Negative staining:
   - 2% aqueous phosphotungstic acid neutralized by potassium carbonate
   - 0.5% aqueous sodium silicotungstate

↪ 1 min

↪ 2–15 min

Coating ultrathin sections:

↪ This may contain a stain so that staining and coating may take place simultaneously.

- To limit artifacts due to drying of the section
- To visualize the ultrastructure

↪ To fill the gaps left by the liquid

Several methods can be used:

- Coating in methylcellulose
- Embedding in an epoxy or acrylic resin

↪ Staining may be positive or negative.
↪ Staining must be positive and carried out after embedding.
↪ Sections must be dehydrated. The grids must be covered in a film insoluble in ethanol (e.g., formvar).

6.11.3 Summary of Different Stages

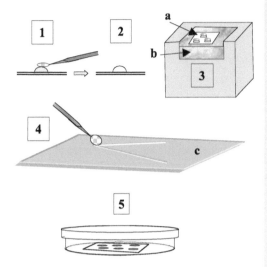

1 = **Uranyl acetate**
     In the dark
2 = **Washes**
3 = **Coating in methylcellulose**
     (a) Parafilm
     (b) ice

4 = **Removal of excess methyl cellulose**
     (c) filter paper

5 = **Drying**

Figure 6.20 Staining of ultrathin sections of frozen tissue.

6.11.4 Materials/Products/Solutions

1. Materials
   • Brown glass bottle

   • Filter paper

   • Glass Petri dish

   • Grid box
   • Hydrophobic film like Parafilm
   • Ice
   • Oven
   • Plastic wash bottle

   • Ultra-fine titanium anti-magnetic forceps
2. Products
   • Epoxy or acrylic resin
   • Lead nitrate
   • Methylcellulose
   • Phosphotungstic acid
   • Sodium citrate
   • Sodium hydroxide
   • Uranyl acetate

3. Solutions
   • 2% alcoholic uranyl acetate

⇨ For keeping the uranyl acetate solution (light sensitive)
⇨ For drying grids
⇨ For drying grids
⇨ For staining ultrathin sections
⇨ For storing grids
⇨ For staining ultrathin sections

⇨ 50°C (for polymerizing the epoxy resin)
⇨ Do not keep the water, or filter it before use.
⇨ To limit the problems of static

⇨ *See* Appendix B5.2 and 5.3.
⇨ Crystallized, $(PbNO_3)_2$
⇨ Tylose, MW = 300
⇨ $H_3O_{40}PW_{12} \cdot 2H_2O$
⇨ $C_6H_5Na_{36} \cdot 2H_2O$
⇨ NaOH (pellets)
⇨ $(CH_3COO)_2\ UO_2 \cdot 2H_2O$
⇨ Crystallized; **radioactive (wear a mask while weighing out)**

⇨ *See* Appendix B7.2.1.1.
⇨ Very sensitive to the light

- 2% aqueous phosphotungstic acid
- 0.5% aqueous sodium silicotungstate
- 50%, 80%, 95%, 100% ethanol
- Lead citrate

- 0.8% methylcellulose ± 2% neutral uranyl acetate
- 2% methylcellulose
- 2–4% neutral uranyl acetate

↪ *See* Appendix B7.2.4.

↪ According to Reynolds (*see* Appendix B7.2.2).

↪ *See* Appendix B7.2.3.2.

↪ *See* Appendix B7.2.3.1.
↪ *See* Appendix B7.2.1.3.

6.11.5   Protocols

1. Negative staining
   - After fixation and staining, cryosections are incubated in the following solutions:
     – 2% phosphotungstic acid   **1 min**
     – 0.5% sodium silicotungstate   ≈ **10 min**
   - Dry without washing.

   - Dry.

2. Coating in methylcellulose
   - After fixation and washing, the cryosections are stained
     – 4% neutral uranyl acetate   **10 min**
                                        In the dark
   - Incubate the cryosections on a drop of methylcellulose on a hydrophobic film on ice
     – 2% methylcellulose   **10 min**
       ± 20% of 4% neutral   **On ice**
       uranyl acetate

   - Dry.   **overnight**

3. Embedding in resin
   - After fixation and washes, the cryosections are dehydrated:
     – 50% ethanol   **2 min**

     – 80% ethanol   **2 min**
     – 95% ethanol   **2 min**

     – 100% ethanol   **2 × 2 min**

   - Incubate the sections in the following mixture:
     – 100% ethanol–resin $^v/_v$ **2 × 2 min**

↪ Staining is very important.
↪ Better filling of the cell and tissue gaps
↪ The stain fills the gaps occupied by the aqueous medium.

↪ Positive staining
↪ Uranyl is sensitive to the light.
↪ Move the grid with a dissecting needle to remove the liquid

↪ Remove excess with filter paper. Carefully absorb at the forceps.
↪ 10–20% of 4% neutral uranyl acetate may be added to the methylcellulose to obtain a positive/negative staining outlining the membranes.
↪ **Important;** the sections must be thoroughly dried prior to observation (≥ 1 hour).
↪ The support membrane must be formvar.
↪ The sections must not dry out between the different solutions.
↪ The grids are placed on the surface of a drop.

↪ May be enough for embedding in an acrylic resin (e.g., LR White).
↪ Indispensable for embedding in epoxy resin

↪ It is necessary to agitate the grids at the surface of the drop.

- Incubate the sections
  - 100% resin **15 min** ⇨ It is necessary to agitate the grids at the surface of the drop.

- Remove the maximum amount of resin from the grids
- Let polymerize **overnight at 50 °C** ⇨ The time may be reduced to several hours.

- Stain the sections
  - 2% alcoholic uranyl acetate **10 min**
  - Distilled water **5 × 4 min**
  - Lead citrate **5 min**

❏ *Next stage:*
- Observation ⇨ *See* Examples of Results (Figures 18 through 22).

6.12 PROTOCOL TYPE

⇨ Antigenic hybrids

- Fixation
- Cryoprotection
- Programmed freezing
- Storage in liquid nitrogen
- Cutting ultrathin sections
- Pretreatment ⇨ Optional
  - Deproteinization
  - Prehybridization
- Hybridization
- Post-hybridization treatments
- Visualization of hybrids
- Staining/Coating

1. Fixation: ⇨ *See* Chapter 3, Section 3.6.
   - 4% PF in 100 m*M* **3–24 h** ⇨ *See* Appendix B4.3.
     phosphate buffer **at 4°C**
   - Buffer washes **4 × 15 min** ⇨ *See* Appendix B3.4.
2. Cryoprotection: ⇨ *See* Section 6.4.
   - 0.4 *M* sucrose **> 90 min** ⇨ *See* Appendix B2.17.
     at 4°C ⇨ Depending on the size of the sample
   - 2.3 *M* sucrose **16 h** ⇨ *See* Appendix B2.17.
     at 4°C
3. Programmed freezing: ⇨ *See* Section 6.6.
   - Rate of cooling **Slow** ⇨ Less than 1°C/min below 0°C
     < 10 min ⇨ More than 10 min, and coating in sucrose may give rise to artifacts (*see* Section 6.6.3.4).
   - Freezing plateau **To be determined** ⇨ Temperature is slightly higher than Ts measured in the system used.
     ⇨ *See* Section 6.6.
   - Immersion in liquid nitrogen **−196°C** ⇨ As fast as possible
     ⇨ *See* Section 6.6.3.1.

4. Storage in liquid nitrogen ➭ No time limit
5. Cutting ultrathin sections: ➭ For soft tissue
   - Temperature parameters for the cryo-ultramicrotome: ➭ *See* Section 6.7.2.2.
     - Temperature of the sample **–110°C**
     - Temperature of the knife **–100°C**
     - Temperature of the chamber **–100°C** ➭ May be lower
   - According to the dry sectioning technique ➭ *See* Section 6.7.2.
     - Section thickness **80–100 nm**
   - Place the sections on a coated nickel grid. ➭ Preparation of a support film (*see* Appendix A4)
     ➭ Sections may be stored.
   - Remove sucrose by inverting the grids on the buffer surface: ➭ The sections must not dry out.
     ➭ The choice of buffer depends on the next stage.
     - Buffer **2 × 5 min at RT**

6. Pretreatments: ➭ **Optional** (*see* Section 6.8)
   - Deproteinization: ➭ *See* Section, 6.8.1.
     Incubate the sections in the following solutions:
     - 20 m*M* Tris buffer/ **2 × 5 min** ➭ *See* Appendix B3.7.2.
       2 m*M* CaCl$_2$; pH 7.6
     - Proteinase K in **0.2 µg/mL** ➭ *See* Appendix B2.15.
       Tris–HCl buffer/CaCl$_2$ **10 min** ➭ To be determined; the activity of the
       at RT enzyme may be reduced.
     - 20 m*M* Tris buffer/ **2 min**
       2 m*M* CaCl$_2$; pH 7.6
     - Phosphate buffer **2 × 5 min**
     - PF 4% in a phosphate buffer **5 min** ➭ **Indispensable** (post-fixation)
       ➭ *See* Appendix B4.3.
     - Phosphate buffer **3 × 5 min** ➭ Removes traces of fixative
   - Prehybridization ➭ *See* Seciton 6.8.3.
     - Preparation of the hybridization buffer ➭ *See* Section 6.9.3.1 (i.e., oligonucleotide probe).

	Final concentration	
· Buffer SSC 20X	**4X**	➭ **Indispensable**
· Deionized formamide	**30%**	➭ **Useful**
· 50X Denhardt's solution	**1X**	➭ This concentration may be increased.
· tRNA (10 mg/mL)	**400 µg/mL**	➭ This concentration may be increased.
· DNA (10 mg/mL)	**250 µg/mL**	➭ This concentration may be increased.
· Sterile water	**qsp x mL**	

   - Mix by vortex.
   - Centrifuge. ➭ Removes bubbles
   - Incubate the sections:
     · Drop **≤ 50 µL/grid** ➭ Or smaller volume

· Duration of incubation   **1 h**
      at RT
– Remove excess without drying and hybridize.

7. Hybridization: ↬ *See* Section 6.9.
   • Add the hapten-labeled probe to the hybridization buffer: ↬ Dry probe.
     – cDNA   **≈ 0.1–2 μg/mL**   ↬ **Denaturation is indispensable.**
     – Oligonucleotide   **≈ 50 pmoles/mL**   ↬ Without denaturation
   • Vortex.
   • Centrifuge.
   • Denature.   **3 min**   ↬ Solely for cDNA probes
         at 100°C
   • Rapidly chill on ice.
   • Incubate the grids.
     – Reaction mixture   **30 μL/grid**   ↬ Volume may be reduced to several μL.
           Humid chamber   ↬ 5X SSC (*see* Section 6.9.4)
     – Hybridization   **3 h**   ↬ Maximum duration
           at RT

8. Post-hybridization treatments: ↬ *See* Section 6.9.5.
   • Wash in the following solutions with increasing stringency: ↬ *See* Appendix B3.5.
     – 4X SSC   **10 min**
           at RT
     – 2X SSC   **2 × 10 min**
     – 1X SSC   **2 × 10 min**   ↬ **Optional**
   • Post-fix the ultrathin sections
     – 4% PF in 2X SSC   **5 min**
   • Wash
     – 2X SSC   **3 × 5 min**

9. Visualization of antigenic hybrids   ↬ Indirect immunocytochemical reaction
       ↬ *See* Section 6.10.2.

   • Block nonspecific sites:
     – Blocking buffer   **10 min**   ↬ **Indispensable** (*see* Appendix B6.1).
   • Incubate with the anti-hapten antibody: ↬ Raised in species X
     – Dilute 1:50 in   **≥ 40 μL/grid**   ↬ *See* Appendix B3.7.5.
     Tris–HCl/NaCl buffer
     – Incubation   **1 h**   ↬ Humid chamber (filter paper soaked in water)

   • Wash:
     – Tris–HCl/NaCl buffer   **3 × 5 min**   ↬ This time may be extended.
   • A conjugated anti-species X antibody is used   ↬ Generally conjugated to 10 nm colloidal gold
     – Diluted 1:50 in   **≥ 40 μL/grid**   ↬ According to the manufacturer's instructions
     Tris–HCl/NaCl buffer
     – Incubation   **1 h**   ↬ Humid chamber (filter paper soaked in water)

- Wash:
  - Tris–HCl/NaCl buffer   **2 × 5 min**   ⮌ This time may be extended.
  - 2X SSC buffer   **2 × 5 min**
- Fixation:
  - 1–2% glutaraldehyde   **10 min**   ⮌ *See* Appendix B4.2.
    in 2X SSC
- Wash:
  - 2X SSC   **3 × 5 min**
  - Sterile water   **5 min**   ⮌ To remove all traces of NaCl

10. Staining/Coating: ⮌ *See* Section 6.11.
    - Staining:
      - 4% neutral   **10 min**   ⮌ Positive staining (*see* Appendix B7.2.1.3)
        uranyl acetate   **In the dark**
    - Incubate the cryosections on a drop of   ⮌ Coating
      methylcellulose on a hydrophobic film   ⮌ Move the grid with a dissecting needle to
      on a bed of ice: remove all water.
      - 2% methylcellulose   **> 10 min**   ⮌ 10–20% of 4% neutral uranyl acetate may
        ± 20% of 4% neutral   **On ice**   be added to the methylcellulose to obtain a
        uranyl acetate   positive/negative staining, outlining mem-
        branes (*see* Appendix B7.2.3.2).
        ⮌ Remove the excess delicately with filter
        paper at the forceps (*see* Figure 6.20).
    - Dry.   **overnight**   ⮌ **Important.** The sections must be thor-
      oughly dried before observation (≥ 1 hour).

❏ *Next stage:*
  - Observation   ⮌ *See* Examples of Results (Figures 18,
    20–22, and 26).

Chapter 7

Semithin
Sections

CONTENTS

7.1 AIMS

The aim of this chapter is to demonstrate the potential uses of *in situ* hybridization at the level of the light microscope using the same techniques employed for electron microscopy.

☞ At the level of tissue and cellular structure

In situ hybridization on semithin sections can be considered a preliminary step to *in situ* hybridization on ultrathin sections (the intermediate step between light microscopy and electron microscopy).

☞ Less than or equal to 1 μm

Semithin sections may be cut before or after visualization of *in situ* hybridization, depending on the method chosen:

☞ At present, *in situ* hybridization on semithin sections embedded in acrylic or epoxy resins does not give satisfactory results.
☞ See Chapter 6.

• Unembedded tissue technique:
  In situ hybridization is carried out on semithin frozen sections.

• Pre-embedding technique:
  In situ hybridization is carried out prior to cutting semithin sections.

☞ See Chapter 5.
☞ Semithin sections are cut after visualization of the probe and embedding in resin.

❑ *Advantages*

• Simplicity
• Speed
• Results are better than those obtained using sections embedded in wax
• Localization of nucleic acids within tissues and cells
• Utilization of radioactive or antigenic probes
• Dual labeling may be carried out

☞ Similar to light microscopy
☞ Reduction in the number of steps

❑ *Disadvantages*

• Cutting semithin sections

☞ A cryostat is required for cutting frozen tissue.

• Sensitivity

☞ The absence of an *in situ* hybridization signal at this stage does not mean that there will be no signal when using electron microscopy.

• Tissue deterioration and loss of signal

☞ With the pre-embedding technique cutting one semithin section may remove any labeling at the surface of the tissue.

7.2   PRINCIPLES

This technique is comparable to *in situ* hybridization at the light microscope level. The only difference is in the thickness of the sections. The aim is to bind a specific **nucleic acid sequence** (target nucleic acid) to a complementary **probe**.

The target nucleic acid is present within the preserved tissue.

Hybridization is carried out on semithin sections *in situ* (i.e., there is penetration of the probe to the interior if the section is thin enough). Despite this, it is necessary to permeabilize the tissue to make the target nucleic acid accessible to the probe in the hybridization solution to allow specific hybridization to take place.

To allow detection of the hybrid, the probe is labeled with a marker that can be visualized.

⮞ Smaller than or equal to 1 μm
⮞ This phenomenon is identical for all molecular hybridization. Binding is specific when the hybrid formed is the result of a perfect match between the probe and the target nucleic acid.
⮞ This technique does not determine the characteristics of the nucleic acid visualized (e.g., molecular weight, sequence) within the tissue.
⮞ Difficulty in preserving morphology remains the greatest disadvantage associated with semithin sections.

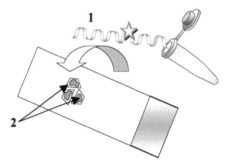

1 = Labeled probe ☆
- Antigenic
- Radioactive

2 = Semithin Sections
- Target nucleic acid

Figure 7.1 General principle.

This approach, which predates the electron microscope, is at the resolution limit of the light microscope.

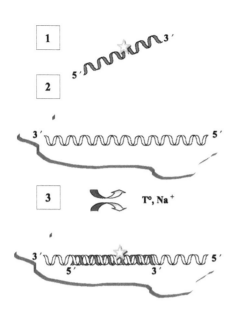

1 = Labeled probe

2 = Target nucleic acid

3 = Hybrid

Figure 7.2 Principles of hybridization.

7.2.1 Types of Hybrids

7.2.1.1 Stable Hybrids

These are characterized by the formation of the maximum number of hydrogen bonds possible in the new molecule.

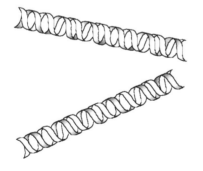

↪ Resin embedded and/or frozen sections mounted on a membrane adsorb partial or unstable hybrids that result in nonspecific background. Only stable hybrids should be conserved.
↪ The aim is to produce a perfect double strand, which is very stable and unlikely to be absorbed nonspecifically by the glass.

Figure 7.3 Stable hybrids.

RNA/riboprobes are the best probes for the formation of stable molecules.

Hybrids with large quantities of bases G and C are much more stable than those rich in bases A and T/U.

↪ In order of stability:
DNA–DNA < DNA–RNA < RNA–RNA.
↪ There are two hydrogen bonds between A and T (*see* Figure 7.4) and three between G and C (*see* Figure 7.5).

⮡ Too high a percentage of bases G and C leads to nonspecific bonds. The converse is true for hybrids containing a high percentage of bases A and T/U, which are easy to denature.

1. Diagram of bases adenine and thymine/uracil:

⮡ Bases **A and T** are linked by two hydrogen bonds:
- In DNA–DNA hybrids, the purine base adenine is complementary to the pyrimidine base thymine.
- In DNA–RNA hybrids, the purine base adenine is complementary to the pyrimidine base uracil.

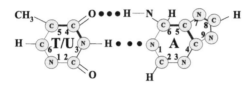

Figure 7.4   Diagram of A–T/U.

2. Diagram of bases cytosine and guanine:

⮡ Bases **G and C** are linked by three hydrogen bonds:
- When hybrids are formed, the purine base, guanine, is always bound to the pyrimidine base, cytosine.

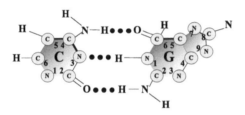

Figure 7.5   Diagram of C and G.

7.2.1.2   Unstable Hybrids

These are characterized by the lack of binding between noncomplementary bases placed opposite one another in the two nucleic acid molecules.

⮡ Partial homology

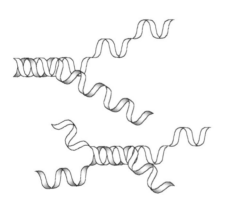

Figure 7.6   Unstable hybrids.

If the hybridization temperature is well below the Tm, a large number of the hybrids formed will be unstable (*see* Figure 7.6).

7.2.1.3   Partial Hybrids

When there is low homology between the sequence of two nucleic acid molecules, bonding will be incomplete.

↬ Only a proportion of the hydrogen bonds possible are formed in the molecule (slightly unstable hybrids).

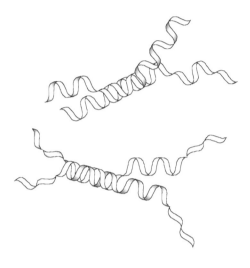

Figure 7.7   Partial hybrids.

7.2.2   Parameters

The hybridization conditions should ensure the formation of specific hybrids and limit the formation of nonspecific hybrids.

↬ However, it is often easier to form hybrids in the presence of nonspecific binding (hybrids or probe/protein interactions) and to remove the nonspecific binding later.

The specificity of the hybridization and its visualization *in situ* depends on a number of parameters:

1. Hybridization temperature
2. Salinity
3. Hybridization buffer
4. Probe type
5. Probe length

6. Label used

7. Sequence homology: probe–target nucleic acid

↬ Depends on the Tm (*see* Section 7.2.2.1).
↬ Concentration of Na⁺
↬ Ionic concentration
↬ *See* Chapter 1, Section 1.1
↬ The stability of the hybrid depends on the length of the probe.
↬ Radioactive or antigenic (*see* Chapter 1, Section 1.2).

7.2.2.1   Hybridization Temperature

The temperature at which the hybrids are formed depends on the specificity of the hybrids. The influence of temperature on the formation of hybrids is shown in Figure 7.8.

↬ The formation of hybrids is a spontaneous reaction which may be modulated by the hybridization temperature.

❑ *Determination of the Tm:*

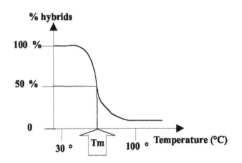

The temperature that corresponds to 50% hybrid formation is the fusion temperature or Tm. The Tm is constant for any given hybrid.

For each probe type there are several ways of calculating this value.

7.2.2.2   Na⁺ Concentration

The concentration of monovalent Na⁺ ions influences the stability of the hybrids. Raising the concentration by a factor of 10 increases the Tm by 16.6°C. This stabilization of the hybrids influences the speed of hybridization.

7.2.2.3   Hybridization Buffer

Large variations in pH destabilize the hybrids.

7.2.2.4   Type of Probe

DNA–DNA hybrids are less stable than DNA–RNA hybrids, which are less stable than RNA–RNA hybrids.

7.2.2.5   Length of Probe

The length of the probe influences the Tm and hence the stability of the hybrids.

7.2.2.6   Type of Marker Used

The marker used causes spatial and stereospecific modifications to the nucleotides with the exception of ^{33}P that are integrated into the molecule.

➯ Tm: melting temperature

➯ The Tm is the temperature at which 50% of strands are double and 50% are single and denatured.

Figure 7.8   Determination of the Tm.

➯ This temperature corresponds to the point of inflection of the sigmoid curve.
➯ Higher temperatures result in more denaturation of hybrids. The curve of denaturation vs. temperature is always sigmoidal.
➯ cDNA, oligonucleotides, and riboprobes

➯ Salinity
➯ The concentration of Na⁺ generally used is between 300 and 600 mM in SSC 2X to 4X (*see* Appendix B3.5).

➯ This parameter may inhibit the inter-base hydrogen bonds.

➯ Riboprobes form the most stable hybrids.

➯ The longer the probe, the more stable the hybrids; however, penetration into the tissue decreases with length.

➯ The substitution of oxygen with ^{35}S causes less perturbations than the addition of an antigen.

3′ extension of an oligonucleotide may give rise to a labeled oligonucleotide two to three times longer than the original.

The hybridization temperature is lowered as the number of antigenic-labeled nucleotides rises.

↪ The addition of a large number of radioactive nucleotides to the probe will inhibit the formation of hybrids.

↪ Antigenic markers consist of an antigenic molecule with an added carbon chain enabling the formation of hydrogen bonds. Labeling the probe at the 3′ end causes less structural modification.

7.2.2.7 Sequence Homology

The aim of hybridization is to detect a nucleic acid where the sequence is complementary to that of the probe. However, a percentage of heterologous bases is acceptable (each 1% reduces the hybridization temperature by 1°C).

↪ The degree of homology which exists between the two nucleic acid molecules influences the stability of the hybrid.

↪ The positions of any mismatches are important. Those situated at the extremities are less important to the stability of the hybrids.

7.2.3 Pretreatments

As in the case of ultrathin sections, the aim of pretreatment is to make the complementary nucleic acid sequences in the target nucleic acid accessible to the probe. The cellular constituents associated with nucleic acids are mainly proteins.

↪ **Optional**

↪ The treatments depend on the type of biological material and fixative used and the need to increase the signal.

7.2.3.1 Deproteinization

Proteinase K is generally considered a selective protease for proteins associated with nucleic acids.

Its activity is modulated by the following parameters:

• Concentration

• Buffer

• Temperature

• Duration

↪ This enzymatic treatment is a compromise between the intensity of the signal and the preservation of the tissue.

↪ The concentration of the enzyme can be varied according to the type of sample (frozen sections or embedded tissue), type of tissue and the target nucleic acid.

↪ 20 mM Tris–HCl/2 mM CaCl$_2$ buffer; pH 7.6, containing calcium, which is a cofactor of proteinase K. Other buffers can be used (e.g., Tris–EDTA) to limit the action of proteinase K.

↪ The optimal temperature is 37°C but the enzyme remains active at lower temperatures.

↪ This parameter may be varied, but avoid prolonged incubations because the tissue structure may be destroyed.

7.2.3.2 Prehybridization

Prehybridization prior to hybridization involves the incubation of sections in the hybridization mixture in the absence of the probe.

❏ *Advantage*
  • Reduces the nonspecific signal

❏ *Disadvantage*
  • Reduces sensitivity

⇒ Necessary to reduce background

⇒ Saturates nonspecific sites; however, the labeled probe may be absorbed onto the macromolecules contained in the buffer (dextran sulfate, Denhardt's solution, RNA, DNA).

⇒ Risk of diluting the hybridization buffer and hence the probe

7.2.4 Post-Hybridization Treatments

Washing removes nonspecific hybrids which mask specific hybrids and reduce the signal/background ratio. A number of washes under stringent conditions are carried out, removing all but the most stable hybrids.
  The stringency depends on:

  • The concentration of Na^+
  • The temperature

The stringency is higher when the salt concentration is lower and the temperature is higher.

⇒ The probe may bind nonspecifically to proteins.

⇒ The stringency may be weaker for semithin and ultrathin sections.

⇒ Start with a low stringency and increase if necessary.

7.2.5 Visualization of Hybrids

The choice of visualization technique depends on the probe label:

  • Radioactive isotope
  • Antigen

Visualization depends on two parameters:

  • Sensitivity
  • Resolution

Two approaches are used, depending on the label:

  • Autoradiography

  • Immunocytochemistry

⇒ *See* Chapter 1, Section 1.2.

⇒ Detection of radioactive decay
⇒ Immunocytochemical detection

⇒ Visualization of hybrids gives results that can be analyzed in the same way as other morphological techniques.

⇒ Autoradiography uses a liquid photographic emulsion overlaying the tissue and registering radioactive emissions in the form of silver grains (*see* Section 7.2.5.1).
⇒ Immunocytochemistry involves an antigen–antibody reaction that is visualized by a color reaction product (*see* Section 7.2.5.2).

7.2.5.1 Radioactive Hybrids

The isotopes most used are ^{35}S and ^{33}P; each has advantages and disadvantages according to its characteristics.

Radioactive hybrid decays are recorded by autoradiography (i.e., visualized on a photographic emulsion).

↬ Isotope characteristics (*see* Chapter 1, Section 1.2.1.1)

7.2.5.1.1 *IN SITU LOCALIZATION*

For *in situ* hybridization silver grains are produced on autoradiographic emulsion as a result of emitted radiation, revealing the location of the nucleic acids. Parameters to take into consideration with this technique are:

- Emission energy
- Size of emulsion grains (the precision of the labeling depends on the size of the crystals in the nuclear emulsion)
- Thickness of the emulsion

↬ The resolution is determined by statistical analysis, which calculates the distance between the source of the emission (i.e., the isotope) and the silver grain. This depends on the energy of the isotope and the size of the silver grain. The higher the energy of decay, the longer the particle will travel and the further from the hybrid the latent image will be formed. Also, the larger the silver grain the more difficult it is to draw a circle of probability around it containing the hybrid and the source of the radioactivity.

↬ The best results are those obtained with low energy emitters, which do not penetrate very far. Thus, the silver grains are localized close to the source of radiation. ^{35}S provides a good compromise between energy and efficiency.

7.2.5.1.2 CHOICE OF AUTORADIOGRAPHY EMULSION

Different types of autoradiography emulsion are available, each with different characteristics. The choice of emulsion depends on the sensitivity required.

↬ This depends on the size of the silver bromide crystals.

❏ *Size of silver bromide crystals*

- K5 Ilford   **0.2 μm**
- LM1 Amersham   **0.25 μm**
- NTB2 Kodak   **0.26 μm**

7.2.5.1.3 CHOICE OF DEVELOPER

The type of silver grains depends on the developer used:

- Microdol X
- D19

↬ *See* Chapter 5, Table 5.2.

↬ Fine twisted strands
↬ Latent images are larger than with Microdol X
↬ Twisted strands are more complex than those with Microdol X.
↬ Risk of contaminating sections with the high pH of this developer

❏ *Parameters*

- Sensitivity
- Duration of development
- Size of silver grains
- Type of silver grains

➯ D19 is more sensitive than Microdol.
➯ D19 is shorter than Microdol.
➯ Increases with the length of development
➯ *See* Chapter 5, Table 5.4.

7.2.5.2   Antigenic Hybrids

The aim of the immunocytochemical reaction is to form a high affinity complex between the antigen and a molecule that is specific for the antigen and may be visualized and, hence, reveal the location of the hybrid.

Molecules which have a high affinity for the hapten are used:

➯ These are molecules produced by an immunological reaction (antibodies) or that spontaneously form a complex with the hapten.

- Polyclonal or monoclonal immunoglobulins
- IgG, Fab and F(ab')₂ fragments

➯ Immunoglobulin G (IgG) is used most frequently.

Depending on the type of immunocytochemical reaction, the enzymatic markers used are:

➯ These markers are adsorbed onto the immunoglobulin.

- Horseradish peroxidase

➯ Small molecule of plant origin MW = 40 kDa.

- Alkaline phosphatase

➯ Enzyme extracted from intestine

Colloidal gold may be used.
Fluorescent markers are not often used, as they are less sensitive than other methods.

➯ Silver intensification is necessary.

7.2.5.2.1   PEROXIDASE

Peroxidase activity (oxidation of an appropriate substrate) produces an insoluble color precipitate that reveals the reaction site.

The chromogens (substrates) used for peroxidase are:

➯ The electron donor is hydrogen peroxide.

- 3'-diaminobenzidine tetrachloride (DAB)
- 3-amino-9-ethylcarbazole (AEC)
1. DAB (3'-diaminobenzidine tetrachloride)

➯ There are other chromogens.

➯ For use, *see* Section 7.6.2.4.1.

➯ The substrate is oxidized in the presence of peroxidase and produces a yellow-brown precipitate.

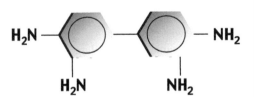

Figure 7.9   Formula of DAB.

❏ *Advantages*

- Yellow-brown precipitate
- Precipitate is insoluble in ethanol.

↝ May be intensified by nickel salts.
↝ The precipitate is stable on mounted sections.

- Allows counterstaining

↝ All types of counterstaining.
↝ Reduction of background

- Reaction is very intense.
- Double labeling is possible.

↝ Used for second reaction.

❏ *Disadvantages*

- Dangerous when in powder form
- Background

↝ Use tablet form if possible.
↝ May be reduced by pre-incubation with DAB in the absence of hydrogen peroxidase.

- Very pure hydrogen peroxide must be used

↝ The shelf life of hydrogen peroxide is short (a few weeks).

- Toxic waste products are generated.

2. AEC (3-amino-9-ethylcarbazole)

↝ This substrate is soluble in dimethylformamide.
↝ The reaction product is a red precipitate.

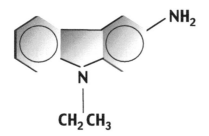

Figure 7.10 Formula of AEC.

❏ *Advantages*

- Red precipitate is clearly visible.
- Double labeling
- Counterstaining is possible.

❏ *Disadvantage*

- Toxic product

↝ The solvent is toxic by inhalation.

7.2.5.2.2 ALKALINE PHOSPHATASE

The chromogens (substrates) most used for alkaline phosphatase are:

↝ This enzyme is widespread in biological tissue. Its endogenous activity must be inhibited by levamisole or by heat treatment.

- NBT-BCIP
- Fast Red

1. NBT-BCIP
   - NBT or nitro blue tetrazolium
   - BCIP or 5-bromo-4-chloro-3-indolyl phosphate

↝ For use, *see* Section 7.6.2.4.2.
↝ $C_{40}H_{30}Cl_2N_{10}O_6$; MW = 817.70
↝ $C_8H_6NO_4BrCIP \times C_7H_9N$; MW = 433.60

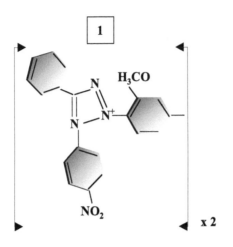

1 = NBT

⇨ This is a redox reaction (oxidation of BCIP and reduction of NBT) that results in two water-insoluble precipitates (soluble in ethanol). It gives a good signal/background ratio.

2 = BCIP

Figure 7.11 Structure of NBT-BCIP.

❑ *Advantages*

• Sensitivity
• Stability of the precipitate
• Compatible with multiple labeling

⇨ Due to the formation of two precipitates
⇨ **The precipitate is soluble in ethanol.**

❑ *Disadvantages*

• Reaction is time consuming
• Precipitate is soluble in ethanol

⇨ Visualization takes several hours.
⇨ If the labeling is very intense, dehydration should be rapid (reduces the background).

2. Fast Red

⇨ This method is particularly good for smears. The colored reaction product is soluble in ethanol. An aqueous counter-stain and mountant must be used. Ready to use solutions of Fast Red are available.

Figure 7.12 Formula of Fast Red.

❑ *Advantages*

• Sensitivity
• Stability of the precipitate
• Compatible with multiple labeling

⇨ **Soluble in ethanol**

❑ *Disadvantages*

- Reaction is time consuming.
- Precipitate is soluble in ethanol.

➥ Visualization takes several hours.
➥ If the labeling is very intense, dehydration should be rapid (reduces the background).

7.2.5.2.3   COLLOIDAL GOLD

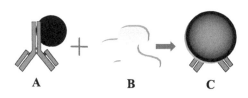

A   B   C

➥ The intensity of immunolabeling with colloidal gold depends on the size of the gold particles. Gold particles are below the resolution of the light microscope. To obtain a signal for light microscopy, it is necessary to increase the size of the particle by silver latensification.

(A)   IgG conjugated
(B)   Silver salts
(C)   Visible product

Figure 7.13   Intensification of colloidal gold signal.

❑ *Advantages*

- Highly specific
- Reaction product is stable in ethanol
- Multiple labeling
- Counterstaining possible

➥ Observation with polarized light
➥ May be kept for some time
➥ Must be carried out last

❑ *Disadvantages*

- Nonspecific binding
- Necessary to control conditions

➥ Background
➥ Temperature and duration must be adjusted for each reaction.

7.2.6   Methods of Observation

7.2.6.1   Bright Field

➥ Normal light microscopy is referred to as bright field.

PRINCIPLES
The light passes through the object before it enters the eye. The background of the preparation is clear and the object absorbs part of the transmitted light. The object is stained and appears on a bright background.

➥ The color of the section depends on the wavelengths of light absorbed, which depend on the stains used.

USES
Bright field microscopy is the basis for all microscopes. It is used for observing tissue morphology.

➥ In autoradiography, silver grains appear black.
➥ Observation of chromogen precipitates

❏ *Advantages*

- Easy to use
- Immunocytochemistry and autoradiography can be observed

❏ *Disadvantages*

- Low contrast

⇨ Only the contours of an object appear (e.g., colloidal gold gives a red color).
⇨ Small particles are difficult to see.

- Staining is required to observe structures
- Low definition

7.2.6.2   Epi-Polarization

⇨ Used because of diffraction effects in a transmitted light bright field.

PRINCIPLES
The use of polarized light aids observation by removing glare from off-axis light scattered by silver grains.

⇨ Natural light is composed of waves oscillating in all directions. Polarized light reduces glare by allowing the user to blank out all but the waves oscillating in one direction.

USES
Observation by epi-polarized light may be employed when:

- Silver grains appear bright on a colored background
- Colloidal gold is latensified by silver salts after immunocytological staining

⇨ Radioactives probes

⇨ Antigenic probes

❏ *Advantages*

- Sensitivity and contrast are increased
- Visualization of labeling and tissue at the same time

⇨ Good for weak labeling
⇨ Combined bright field and dark field images allow simultaneous visualization of tissue morphometry and grain concentration.

- Background is diminished

⇨ A weak signal cannot be intensified using only a transmitted light bright field.

- Images are of high quality

⇨ Black and white and, particularly, color

❏ *Disadvantages*

- Additional equipment is necessary (epi-illuminator, polarizer, analyzer).
- Precise microscope setup is required.
- Difficult to get good results at low magnifications

⇨ Important in the case of weak staining
⇨ Dark field gives better results.

7.3   SUMMARY OF DIFFERENT STAGES

7.3.1   Radioactive Hybrids

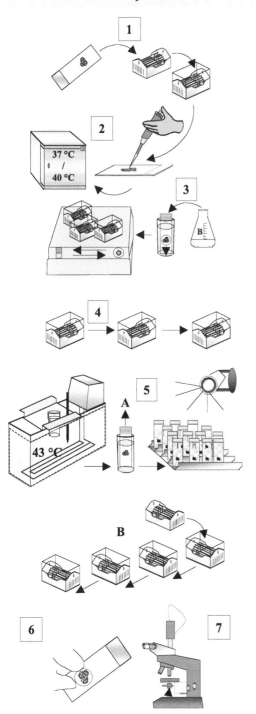

1 = **Pretreatments** (optional)
  • Deproteinization
  • Prehybridization
2 = **Hybridization**

3 = **Washes**
  (**B**) Buffer

4 = **Dehydration**

⤳ Safe light
5 = **Visualization of radioactive hybrids**
  (**A**) Coating in emulsion
    • Dilution of emulsion
    • Melting
    • Coating
    • Drying
    • Exposure time
    • Storage
  (**B**) Developing microautoradiography
    • Development
    • Washing
    • Fixing
    • Washing

6 = **Counterstaining and mounting**
7 = **Observation with the light microscope**
  (**A**) Bright field
  (**B**) Polarized light

Figure 7.14 Protocol for hybridization and visualization of radioactive hybrids.

7.3.2 Antigenic Hybrids

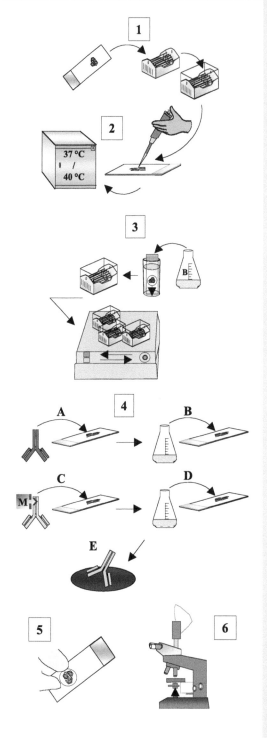

1 = **Pretreatments** (optional)
  • Deproteinization
  • Prehybridization

2 = **Hybridization**

3 = **Washes**
  (B) Buffer

4 = **Indirect immunocytochemistry**
  (A) Incubation:
    with the anti-hapten antibody (species X)
  (B) Washes:
    Tris–HCl/NaCl buffer
  (C) Incubation:
    with the enzyme conjugated antibody (i.e., alkaline phosphatase), anti-species X
  (D) Washes:
    Tris–HCl/NaCl buffer
  (E) Visualization of enzyme activity i.e., NBT-BCIP

5 = **Counterstaining** (optional) **and mounting**

6 = **Observation with light microscope**
  Bright field

Figure 7.15 **Protocol for hybridization and visualization of antigenic hybrids.**

7.4 CUTTING SEMITHIN SECTIONS

7.4.1 Frozen Sections

7.4.1.1 Aims

Semithin sections are obtained using the same protocol as ultrathin sections.

The thickness of the sections is no more than 1 μm.

The sections are picked up in a drop of sucrose, like ultrathin sections, and placed on a pretreated glass slide (*see* Figure 7.16).

⮎ Cryo-ultramicrotomy
⮎ Method without embedding

⮎ *See* Chapter 6, Section 6.7.2.

⮎ To increase the thickness of the sections, the temperature must be increased.
⮎ It is best to cut semithin sections after ultrathin sections (there is a risk of recrystallization after warming).
⮎ Pretreated slides (*see* Appendix A3.1)

7.4.1.2 Summary of Different Stages

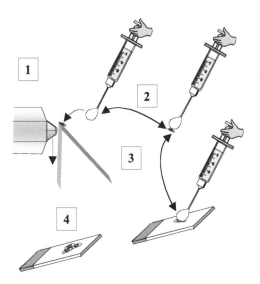

1 = **Semithin section**

2 = **Picking up the section in a drop of sucrose**

3 = **Placing the drop containing the section on the slide**

4 = **Sticking semithin sections to the slide**

Figure 7.16 Technique for picking up semithin frozen sections.

7.4.1.3 Materials/Products/Solutions

1. Materials:
   - Cryo-ultramicrotome
   - Knife-maker
2. Small items:
   - Coverslips
   - Diamond knife

   - Pipettes, tubes, forceps, etc.
   - Syringes with plastic tips

⮎ For cutting thin sections of frozen tissues
⮎ For making glass knives

⮎ For mounting semithin sections
⮎ Better than glass knives; allows larger sections to be cut
⮎ New and/or sterilized (*see* Appendix A2)
⮎ Picking up semithin sections

3. Pretreated material:
   • Pretreated glass slides

➭ Treatment is necessary for the adhesion of semithin sections to the slide (*see* Appendix A3.1).

4. Solutions:
   • 10X phosphate buffer; pH 7.4

➭ To remove sucrose
➭ *See* Appendix B3.4. May be stored at room temperature for 1 month after sterilization.

   • Sterile water
   • 2.3 *M* sucrose

➭ *See* Appendix B1.1.
➭ *See* Appendix B2.17.

7.4.1.4 Protocol

The preparation of the block for semithin sectioning involves the following stages:

1. The specimen is placed in the specimen holder.
2. Temperatures:
   • Object   **−70 to −100°C**

➭ Temperatures allow recrystallization to occur. The sample cannot be used for cutting ultrathin sections.

   • Knife   **−60 to −80°C**
   • Chamber   **−80°C**

3. The sample may be retrimmed if necessary.

➭ **Optional**

4. Smooth the block until the zone to be sectioned is found.
5. Cut sections.   **0.5–1 μm**

➭ Automatic or manual system

6. Pick up the sections with a drop of sucrose.
7. Place the drop and the section on a pretreated glass slide.

➭ Place the section on a glass slide covered with a drop of sucrose (*see* Figure 7.17).

(**A**) Sucrose
(**B**) Frozen sections

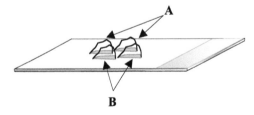

Figure 7.17 Placing semithin sections on a glass slide.

8. Let adhere.   **5 min**

➭ May be stored at 4°C

9. Remove the drops of sucrose by turning the slide surface down in a beaker of buffer (*see* Figure 7.18).

• Buffer   **1–5 min**

↪ Do not let the sections dry. Dehydration with ethanol is necessary.

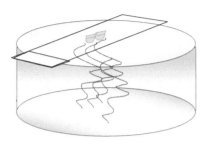

Figure 7.18   Removal of sucrose.

10. Dehydrate:

• 70% ethanol	**1 min**
• 95% ethanol	**1 min**
• 100% ethanol	**1 min**

↪ **Optional** stage; unnecessary if the next stage is part of the pretreatment
↪ Dehydration is fast since the section is thin

11. Dry.

❑ *Next stages:*

• Pretreatments

• Hybridization

↪ Storage for 1–3 h maximum

↪ *See* Section 7.6.1.1.2 (radioactive probe).
↪ *See* Section 7.6.2.1 (antigenic probe).
↪ *See* Section 7.6.1.2.2 (radioactive probe).
↪ *See* Section 7.6.2.2 (antigenic probe).

7.4.2   Embedded Tissue

↪ **Post-embedding method**
↪ Ultramicrotomy

7.4.2.1   Aims
Semithin sections are prepared according to the same protocol as ultrathin sections.

The sections are picked up with a wire loop and placed on pretreated slides (*see* Figure 7.19).

Depending on the resin used, it is recommended to let the embedded samples cool down after removing them from the oven or from under UV light for several hours at room temperature before trimming.

↪ *See* Chapter 4, Section 4.4.3
↪ The thickness of these sections is 1 µm or 0.5 µm.

↪ Pre-embedding method:
  • Epoxy resin (e.g., Epon)
↪ Post-embedding method:
  • Acrylic resin (e.g., Lowicryl K4M, Unicryl, LR White, etc.)

7.4.2.2 Summary of Different Stages

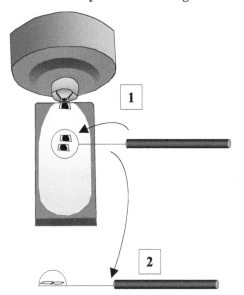

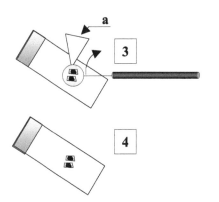

1 = **Float semithin sections on water and maneuver into loop.**

2 = **Pick up semithin sections with loop.**

3 = **Place the drop containing the sections on a glass slide and remove the water.** **(a)** filter paper

4 = **Adhesion of semithin sections to the glass slide** (i.e., hot plate)

Figure 7.19 Picking up semithin sections with a wire loop.

7.4.2.3 Materials/Products/Solutions

1. Materials
   - Hot plate
   - Knife-maker

   - Ultramicrotome

2. Small items
   - Coverslips
   - Dental wax or nail polish

   - Diamond or glass knife

⇨ For drying semithin sections
⇨ For making glass knives (*see* Appendix A.5.1)
⇨ For cutting resin-embedded tissue (*see* Chapter 4, Figure 4.23)

⇨ For mounting semithin sections
⇨ For mounting the boat onto the glass knife
⇨ A specially designed hot plate is available that keeps the wax liquid and the knife warm.
⇨ *See* Appendix A5.

- Dissecting needle

- Filter paper

- Plastic boats

- Single-edged razor blades

- Pipettes, tubes, fine paint brush, forceps, etc.
- Wire loop

⮌ *See* Figure 7.20. For picking up semithin sections: a fine point allows sections to be manipulated without breaking, both in the boat and on a water drop on the slide.

1 = Wire loop
2 = Dissecting needle

Figure 7.20 Tools for picking up semithin sections.

⮌ Filter paper is used to quickly remove the water drop.
⮌ Mounting the boat onto the knife (*see* Appendix A, Figure A10)
⮌ Sections do not stick to the plastic boat.
⮌ Or black sticky tape
⮌ Cleaned in ethanol; the blades have to be fine enough to trim the block prior to sectioning.
⮌ New and sterile (*see* Appendix A2).

⮌ *See* Figure 7.20 or comet for picking up semithin sections (*see* Appendix A6.1, making a comet).

7.4.2.4 Protocol

The preparation of the block prior to cutting semithin sections involves the following steps:

1. Fix the block in the specimen holder of the ultramicrotome.
2. Trim the block by removing the maximum amount of resin possible from around the specimen to give a pyramid-shaped surface with top and bottom edges parallel (*see* Figure 7.21).

⮌ *See* Chapter 4, Section 4.4.3 (post-embedding method).
⮌ Unlike blocks for ultrathin sectioning, there should be a margin of resin left around the tissue (*see* Chapter 4, Section 4.4.3.3).
⮌ Lowicryl K4M is difficult to trim and breaks more easily than epoxy resin.

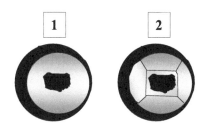

1 = Sample embedded in resin.
2 = Upper and lower edges of the trapezium are parallel.

Figure 7.21 Trimming a block for semithin sections.

3. Place the specimen holder with the block in the moving arm of the ultramicrotome.

4. Place the glass knife in the knife holder, and adjust the knife angle according to the hardness of the resin.

↪ A small plastic boat is fixed to the knife with either dental wax or nail polish.

↪ The knife angle varies according to the hardness of the block:
  - Greater than 5°, harder block = Lowicryl K4M, Unicryl
  - Less than 5°, softer block (LR White)
  - 6°–8°, Epon–araldite block
  - ≈ 5°, Epon block

5. Adjust the level of the water in the boat according to the resin to be cut.

↪ A very low level is used for acrylic resin (*see* Figure 7.22).

1 = **Epoxy resin**
  (a) The surface of the water is uniformly shiny.
2 = **Acrylic resin**
  (b) The level of the water is low with a dark area against the cutting edge of the knife.

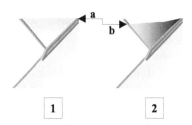

Figure 7.22 The level of water required for different types of resin.

6. Trim the block using the medium advance control (this allows sections of ≈ 1 μm) using the right-hand side of the cutting edge (*see* Appendix A, Figure A11).

↪ This is the least sharp part of the knife. Move the knife to the left as knife marks begin to appear.

7. Wait for the zone of interest to be sectioned.

↪ As the cutting arm descends, the block is sectioned by the knife and the sections float into the bath.

8. Move the knife holder at the appearance of the first knife marks.

↪ **Important:** reverse the block before moving the knife.

9. Cut semithin sections of up to ≤ 1 μm thickness

↪ Each graduation of the micrometric advance corresponds to 0.5 μm, and a section of maximum thickness 1.5 μm may be cut.

↪ If the section is too thick it will not stick to the slide.

10. Spread the sections.

↪ Epon–araldite sections may be spread and made thinner by xylene vapors from a paper filter or fine paint brush.

11. Pick up sections that are ≈ 1 μm with a wire loop (*see* Figure 7.19) or with a dissecting needle (*see* Figure 7.23).

↪ **Useful for acrylic resin sections:** a wire loop can pick up several sections at the same time. Place the loop under the sections and lift them in a drop of water, placing them immediately on a glass slide.

↪ Place the point of the dissecting needle under a section and lift. Place the section immediately in water.

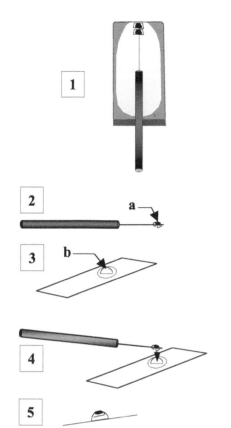

1 = **Dissecting needle under section**

2 = **Lifting the section (a)**

3 = **Placing drop of water on slide (b)**

4 = **Placing section on a drop of water**

5 = **Semithin section floating on water**

Figure 7.23 Picking up semithin sections with a fine point.

12. Immediately place the semithin sections on a pretreated glass slide.

❏ *Next stages:*

- Pretreatments

- Hybridization

- Observation

↝ Mark around the sections with a diamond pen (*see* Appendix A3.1, pretreated slides).

↝ *See* Section 7.6.1.1 (radioactive hybrids).
↝ *See* Section 7.6.2.1 (antigenic hybrids).
↝ *See* Section 7.6.1.2 (radioactive hybrids).
↝ *See* Section 7.6.2.2 (antigenic hybrids).
↝ If vibratome sections are embedded after the probe is visualized (*see* Chapter 5)

7.5 MATERIALS/PRODUCTS/SOLUTIONS

7.5.1 Pretreatments/Hybridization

1. Materials:
   - Centrifuge
   - Humid chamber

↝ Radioactive probes
↝ Antigenic probes

↝ More than 14,000 *g*
↝ A Petri dish (≈ 24 × 24 cm) may be used.

- Micropipettes
- Oven
- Rubber cement
- Slide forceps

- Slide holder
- Staining baths and slide holders

- Sterile coverslips
- Thermostatic hot plate

- Vortex
- Water bath with stirrer

2. Products:

- Bipotassium phosphate
- Calcium chloride
- Dextran sulfate
- Disodium phosphate
- Dithiothreitol (DTT)
- DNA

- Formamide
- Hydrochloric acid
- Monosodium phosphate
- Paraformaldehyde (PF)
- Proteinase K
- RNA
- Sodium chloride
- Tris-hydroxymethyl-aminomethane

3. Solutions:
- Buffers
  - 1X or 10X PBS

  - Hybridization buffer
  - Reaction buffer
  - 10X phosphate buffer; pH 7.4

  - 20 mM Tris–HCl buffer/2 mM CaCl$_2$; pH 7.6
  - 20X SSC; pH 7.0

- Deionized formamide

⇌ 1000 μL
⇌ 37–40°C
⇌ For sealing the coverslips (optional)
⇌ Sterile forceps wrapped in aluminum foil and kept at room temperature

⇌ Pyrex (may be baked at 180°C); keep these solely for *in situ* hybridization or use disposable baths.
⇌ Sterilization (*see* Appendix A2)
≈ 92–100°C. The temperature is important for denaturing the probe but preserving the tissue structure.

⇌ 37°C
⇌ **To be used only for *in situ* hybridization**
⇌ AR quality
⇌ Powder

⇌ Powder or solution
⇌ Powder
⇌ Molecular biology quality
⇌ Salmon or herring sperm, in solution/sonicated
⇌ Ready-to-use solution is best.
⇌ Deionized

⇌ Powder
⇌ Powder
⇌ Molecular biology quality
⇌ Transfer, in ready-to-use solution.
⇌ Anhydrous
⇌ Powder; keep dry.

⇌ Store at room temperature after sterilization (*see* Appendix B3.4.3).
⇌ Without the probe
⇌ Hybridization buffer containing the probe
⇌ Store at room temperature for 1 month after sterilization (*see* Appendix B3.4.1).
⇌ Store at room temperature for 1 month after sterilization (*see* Appendix B3.7.2).
⇌ Store at 4°C or at room temperature (*see* Appendix B3.5).
⇌ Store at –20°C in 500 μL aliquots. If the solution does not freeze at –20°C, do not use (*see* Appendix B2.14).

- 50X Denhardt's solution

↝ Store at –20°C in aliquots. It can be thawed and refrozen several times without losing its properties (*see* Appendix B2.19).

- 50% dextran sulfate

↝ *See* Appendix B2.21.

- Dithiothreitol (1 *M*)

↝ *See* Appendix B2.11. Store at –20°C in 50 to 100 µl aliquots. It must not be thawed and refrozen. Its strong smell indicates that it is still usable. Do not use if the smell has changed or is absent.

- 10 mg/mL DNA

↝ Sonicated and stored at –20°C. Repeated freezing and thawing causes breaks in the structure (*see* Appendix B2.3).

- 95%, 100% ethanol
- 4% paraformaldehyde (PF) in phosphate buffer

↝ Store at –20°C (*see* Appendix B4.3).

- Probes
  - Antigenic probes

  ↝ cDNA, oligonucleotides (*see* Chapter 1, Section 1.2.2)

  - Radioactive probes

  ↝ cDNA, oligonucleotides (*see* Chapter 1, Section 1.2.1)

- 1 mg/mL proteinase K in sterile water

↝ Store at –20°C in 250 µL aliquots (*see* Appendix B2.15).

- 150 m*M* sodium chloride

↝ Store at room temperature after sterilization (*see* Appendix B2.9).

- Sterile water

↝ *See* Appendix B1.1. Do not use more than once or use DEPC-treated water (*see* Appendix B1.2).

- 10 mg/mL yeast tRNA

↝ Store at –20°C. Repeated freezing and thawing causes breaks in the structure (*see* Appendix B2.5).

7.5.2   Post-Hybridization Treatments

- 95%, 100% ethanol

↝ Dehydrate semithin sections before applying autoradiography emulsion.

- 150 m*M* sodium chloride

↝ *See* Appendix B2.9.

- 20X SSC; pH 7.0

↝ *See* Appendix B3.5. Store at 4°C or room temperature.

↝ Washes are carried out in SSC; pH 7.0 at room temperature (antigenic hybrids) or with gentle agitation at different temperatures in a water bath (radioactive hybrids).

7.5.3   Visualization of Hybrids

7.5.3.1   Radioactive Hybrids

1. Location:

   All of the following stages must be carried out in a dark room with a safe light placed 1.50 m from the bench; the humidity should be between ≈ 20 and 40%.

   ↪ This stage does not need to be carried out under sterile conditions.

2. Materials/small items:
   - Aluminum foil
     ↪ To wrap the autoradiography boxes
   - Autoradiography boxes
     ↪ Clean and light-tight.
   - Black sticky tape
     ↪ To seal the autoradiography boxes light-tight.
   - Clean glass slides
     ↪ To test for bubbles and to hold the desiccant in the autoradiography boxes
   - Filter paper
     ↪ To dry the backs of the slides
   - Hot plate
     ↪ **Useful** if counterstaining resin-embedded semithin sections
   - Porcelain spoon
     ↪ To mix the emulsion
   - Round and oval glass tubes/emulsion tubes
     ↪ To dilute the emulsion
   - Staining baths
     ↪ Nonsterile
   - Support for drying slides vertically
     ↪ Always use the same support to keep the angle constant at which the slides are dried.
   - Thermostatic water bath
     ↪ At 43°C; to melt the autoradiographic emulsion

3. Products:
   - Desiccant
     ↪ Silica gel wrapped in filter paper

4. Solutions:
   - 100% ethanol
     ↪ Standard
   - 1% aqueous toluidine blue
     ↪ If counterstaining (*see* Appendix B7.1.1)
   - Distilled water
     ↪ Nonsterile
   - Emulsions
     – LM1 Amersham
     – NTB2 Kodak
     – K5 Ilford
   - Fixer
     – 30% aqueous sodium thiosulfate
     ↪ Or sodium hyposulfite
     ↪ Fixer solution
   - Nonaqueous mounting medium
     ↪ *See* Appendix B8.2
   - Developer
     – D19
     ↪ Developer solution
     ↪ According to the manufacturer's instructions

7.5.3.2 **Antigenic Hybrids**

1. Products:
   - Dimethylformamide ⮡ $(CH_3)_2NOCH$
   - Substrates
     - Diaminobenzidine tetrahydrochloride ⮡ $C_{12}H_{14}N_{14} \cdot HCl$ (DAB)
       ⮡ Peroxidase substrate
     - NBT–BCIP ⮡ Alkaline phosphatase substrate; these two components are soluble in dimethylformamide.
       · NBT or nitro blue tetrazolium ⮡ $C_{40}H_{30}C_{12}N_{10}O_6$
       · BCIP or 5-bromo-4-chloro-3-indolyl phosphate ⮡ $C_8H_6NO_4Br \times C_7H_9N$
2. Solutions:
   - Antibodies
     - Complex ABC - peroxidase ⮡ Indirect reaction
     - IgG anti-hapten nonconjugated (raised in species X) ⮡ IgG monoclonal or polyclonal
       ⮡ Indirect reaction
     - IgG, F(ab′)$_2$, Fab
       · Anti-hapten conjugated ⮡ Direct immunocytochemical reaction
       · Anti-species X conjugated ⮡ Indirect immunocytochemical reaction
   - Buffers
     - Blocking buffer:
       · 20 mM Tris–HCl buffer/300 mM NaCl/1% serum albumin ⮡ *See* Appendix B6.1.
         ⮡ Other blocking agents may also be used:
         • Nonspecific IgG
         • 2% goat serum
         • 1% fish gelatin
         • 1% ovalbumin
         • 2% nonfat milk powder
     - 20 mM Tris–HCl buffer; pH 7.6 ⮡ *See* Appendix B3.7.1.
     - 20 mM Tris–HCl buffer/300 mM NaCl; pH 7.6 ⮡ *See* Appendix B3.7.5. 50 mM Tris–HCl/600 mM NaCl to better stabilize the hybrids
   - Color development buffers
     - 20 mM Tris–HCl buffer/300 mM NaCl/50 mM MgCl$_2$; pH 7.6 ⮡ Color development buffer for peroxidase (*see* Appendix B3.7.6)
     - 20 mM Tris–HCl buffer/300 mM NaCl/50 mM MgCl$_2$; pH 9.5 ⮡ Color development buffer for alkaline phosphatase (*see* Appendix B3.7.6)
   - Color development solution
     - 0.025% DAB in 20 mM Tris–HCl buffer/300 mM NaCl/0.006% hydrogen peroxidase; pH 7.6 ⮡ Color development solution for peroxidase (*see* Appendix B6.2.2.1)
       ⮡ The precipitate is insoluble in ethanol.
   - Counterstain
     - 2% aqueous methyl green ⮡ *See* Appendix B7.1.2.
   - Distilled water
   - 70%, 90%, 100% ethanol
   - 30% hydrogen peroxide ⮡ 110 V
     ⮡ Peroxidase substrate and inhibitor of endogenous peroxidase.

- Inhibition of endogenous enzymatic activity
  - Peroxidase:
    Hydrogen peroxide
  - Phosphatase:
    1 *M* levamisole

  ⇨ *See* Appendix B6.1.3.

  ⇨ *See* Appendix B6.1.2.
  ⇨ $C_{11}H_{12}N_2S$
  ⇨ Enzyme blocking agent

- Mounting medium
  - Aqueous
  - Permanent
- Xylene

  ⇨ *See* Appendix B8.1.
  ⇨ *See* Appendix B8.2.
  ⇨ Permits a better infiltration of mounting medium

7.6   PROTOCOLS

7.6.1   Radioactive Hybrids

⇨ **Wear gloves**
⇨ Oligonucleotide or cDNA probe labeled with ^{35}S or ^{33}P

- Pretreatments
  - Deproteinization

  ⇨ *See* Section 7.6.1.1.
  ⇨ Generally of no use for semithin sections and particularly not for frozen sections

  - Prehybridization
- Hybridization
  - Hybridization buffer
  - Reaction mixture
  - Hybridization
- Post-hybridization treatment
- Dehydration
- Visualization of radioactive hybrids
- Coating with emulsion
  - Dilution of emulsion
  - Melting
  - Coating
  - Drying
  - Exposure time
  - Storage
  - Microautoradiography development
- Counterstaining
- Observation
  - Bright field
  - Epi-polarization

  ⇨ *See* Section 7.6.1.2.

  ⇨ *See* Section 7.6.1.3.

  ⇨ *See* Section 7.6.1.4.

  ⇨ *See* Section 7.6.1.5.
  ⇨ *See* Section 7.6.1.6.

7.6.1.1   Pretreatments

7.6.1.1.1   SEMITHIN FROZEN SECTIONS

1. Deproteinization:
   - Warm the Tris–HCl/CaCl$_2$ buffer in the water bath (37°C) and place the slides into the warmed buffer
     - Proteinase K   **1 µg/mL**
           5–10 min
           at RT

     Add the proteinase K at the last moment and agitate the slide holder.
   - Wash:
     - 100 m*M* phosphate   **3 min**
     - 150 m*M* NaCl   **3 min**
   - Post-fix:

     - 4% PF in phosphate buffer   **5 min**

   - Wash:
     - 100 m*M* phosphate buffer   **3 × 5 min**
     - 150 m*M* NaCl   **3 min**
   - Dehydrate:
     - 95%, 100% ethanol   **2 min/bath**

   - Dry:
     - Under a fume hood   **> 15 min**

❑ *Next stages:*

   - Prehybridization
   - Hybridization

2. Prehybridization:

   - Sections are incubated with the hybridization mixture without the probe:
     - Hybridization buffer   **≈ 100 µL**
     - Incubation duration   **1–2 h**
           at RT
   - Remove as much of the buffer as possible before placing the probe mixture onto the sections.

↝ **Wear gloves.**
↝ **Optional stages**

↝ **Optional;** to be determined

↝ Proteinase K is the enzyme used.
↝ Different buffers modify the activity of proteinase K.

↝ To be determined since the sections are extremely fragile.

↝ To avoid precipitation in ethanol
↝ Restabilization of the tissue; **indispensable prior to hybridization** for semithin sections and frozen tissue
↝ This fixative gives the best compromise between hybridization signal and preservation of morphology.
↝ To fix sections to the slides during hybridization

↝ Removes traces of fixative

↝ To avoid precipitation in ethanol

↝ Dry carefully around the section and the back of the slide.

↝ The enzymes (RNases, DNases) are inactive in the absence of water.

↝ *See* Section 7.6.1.1.1.
↝ *See* Section 7.6.1.2.1.

↝ **Wear gloves.**
↝ **Optional stage**

↝ *See* Section 7.6.1.2.1.

↝ The sections must not be dried. The sections should remain wet prior to hybridization.

❏ *Next stage:*
- Hybridization

↪ *See* Section 7.6.1.2.1.

7.6.1.1.2 SEMITHIN RESIN SECTIONS

1. Deproteinization

↪ Proteinase K is the enzyme used most often, but pepsin or pronase may also be used.
↪ Different buffers modify the activity of proteinase K.

- Warm the Tris–HCl/CaCl$_2$ buffer in the water bath (37°C) and place the slides into the warmed buffer.
- Enzymatic treatment:
  - Proteinase K   **5 μg/mL**
    10 min
    at RT

↪ A water bath at 37°C with agitation should be used.

  Add the proteinase K at the last moment and agitate the slide holder.

↪ The parameters described above modulate the action of the enzyme.

- Stop the reaction:
  - Tris–HCl/CaCl$_2$ **2 min**
  - Phosphate buffer   **5 min**

↪ Rapid washing
↪ The action of proteinase K is stopped by eliminating NH$_2$ groups in Tris buffer with calcium chloride by changing the buffer.

  - 150 mM NaCl   **3 min**
- Dehydration
  - 95%, 100% ethanol   **2 min/bath**

↪ To avoid precipitation in ethanol
↪ Dry carefully around the section and the back of the slide.

- Drying
  - Under the fume hood   **> 15 min**

↪ The enzymes (RNases and DNases) are inactive in the absence of water

❏ *Next stages:*
- Prehybridization
- Hybridization

↪ *See* Section 7.6.1.1.2.
↪ *See* Section 7.6.1.2.2.

2. Prehybridization

↪ **Wear gloves**
↪ **Optional stage**

- After dehydration and drying the sections are incubated with the hybridization buffer without the probe:
  - Hybridization buffer   **≈ 100 μL**

↪ *See* Section 7.6.1.2.2. The concentrations of DNA, RNA, and Denhardt's solution may be increased and the concentration of formamide reduced or left out.

  - Incubation duration   **1–2 h**
    at RT

↪ This stage may be prolonged for several hours.
↪ The temperature does not modify the levels of nonspecific binding (weak influence of formamide).

- Remove as much of the buffer as possible before placing the probe mixture on the sections.

↪ The sections must not be dried. The sections should remain wet prior to hybridization.

❏ *Next stage:*
- Hybridization

↪ *See* Section 7.6.1.2.2

7.6.1.2   Hybridization

↪ **Wear gloves.**

7.6.1.2.1   SEMITHIN FROZEN SECTIONS
1. Hybridization buffer:

↪ Prepare just before use or can be kept at −20°C/−80°C.

- Add:

Solutions	Final concentration
− 20X SSC	**4X**

↪ **Indispensable;** stabilizes the hybridization and is a source of Na$^+$

− Deionized formamide	**50%**

↪ **Useful;** denaturing agent that lowers the Tm, and permits hybridization to take place at 40°C.

− 10 mg/mL DNA	**400 µg/mL**
− 10 mg/mL yeast tRNA	**400 µg/mL**
− 50X Denhardt's solution	**2X**

↪ **Useful;** competes with proteins
↪ **Indispensable;** competes with proteins
↪ **Useful;** reduces the background and binds to nonspecific sites

− 1 *M* DTT	**10 m*M***

↪ Solely for ^{35}S probes
↪ Antioxidant acts against radiolysis due to ^{35}S.
↪ Add after denaturation.

− Sterile water	**to x mL**

- Vortex.
- Centrifuge.

↪ Gently, or agitate by hand

2. Reaction mixture:
   - Suspend the labeled probe in the hybridization buffer in an Eppendorf tube:

↪ **Wear gloves.**
↪ After centrifugation, the labeled probe is dissolved directly in hybridization buffer.

− Oligonucleotide probe	**2–5 pmoles/mL hybridization buffer**
− cDNA probe	**10–20 ng/mL hybridization buffer**

↪ Labeled by 3′ extension (*see* Chapter 1, Section 1.3.4)
↪ Labeled by random priming or PCR (*see* Chapter 1, Sections 1.3.1 and 1.3.2)

- Vortex.
- Centrifuge.
- Denature.   **3 min at 100°C**

↪ Gently, or mix with a pipette

↪ **Indispensable** for double-stranded cDNA probes (*see* Chapter 1, Figure 1.17).
↪ Denaturation is not necessary for oligonucleotide probes.

- Chill rapidly on ice.

↪ To prevent rehybridization of the two denatured strands; at 0°C, the two single strands are stable.

- Add
  − 1 *M* DTT   **x µL**

↪ Final concentration is 10 m*M*.
↪ Solely for ^{35}S probes

3. Hybridization protocol:
   - Place the reaction mixture on the frozen semithin sections:

↪ **Wear gloves.**

– Drop size **30 μL**	➭ This volume can be reduced if the semithin sections are close together.
• Cover with a sterile coverslip.	➭ To avoid air bubbles, place the coverslip carefully.
• Seal the coverslips with rubber cement or DPX	➭ **Optional** (*see* Figure 7.25)
• Keeping the slides flat, place them in a humid chamber	➭ *See* Figure 7.19.
• Incubate. **overnight at 37– 40°C (humid chamber)**	➭ The hybridization temperature is defined depending on the characteristics of the probe. ➭ 35–45°C is the temperature range usually used to ensure formation of specific hybrids and to enable removal of nonspecific hybrids during washing.

❑ *Next stage:*
  • Post-hybridization treatments

➭ *See* Section 7.6.1.3.

7.6.1.2.2 SEMITHIN RESIN SECTIONS

1. Hybridization buffer

➭ It can be made just before use or stored at –20°C/-80°C.

  • Add

Solutions	Final concentration	
– 20X SSC	**4X**	➭ **Indispensable;** stabilizes the hybridization and is a source of Na$^+$
– Deionized formamide	**50%**	➭ **Useful,** denaturing agent that lowers the Tm and permits hybridization to take place at 40°C
– 50% Dextran sulfate	**10%**	➭ **Useful;** high-molecular-weight polymer that artificially concentrates the probe
– 10 mg/mL DNA	**250 μg/mL**	➭ **Useful;** competes for binding with proteins
– 10 mg/mL yeast tRNA	**250 μg/mL**	➭ **Indispensable,** competes for binding with proteins
– 50X Denhardt's solution	**2X**	➭ **Useful;** reduces background and saturates nonspecific binding sites
– 1 *M* DTT	**10 m*M***	➭ Solely for ^{35}S probes ➭ Antioxidant counters ^{35}S radiolysis. ➭ Add after denaturation.
– Sterile water	**to x mL**	

  • Vortex. ➭ Gently, or agitate by hand
  • Centrifuge.
2. Reaction mixture: ➭ **Wear gloves.**
  • Resuspend the labeled probe in hybridization buffer in an Eppendorf tube: ➭ After centrifugation, the labeled probe is dissolved directly in the hybridization buffer.
    – Oligonucleotide probe **5–10 pmoles/mL of hybridization buffer** ➭ Labeled by 3′ extension (*see* Chapter 1, Section 1.3.4)
    – cDNA probe **10–20 ng/mL of hybridization buffer** ➭ Labeled by random priming or PCR (*see* Chapter 1, Sections 1.3.1 and 1.3.2)
  • Vortex. ➭ Gently, or agitate using a pipette.

- Centrifuge.
- Denature. **3 min**
  at 100°C

➥ **Indispensable** for cDNA probes that are double-stranded (*see* Chapter 1, Figure 1.17).
➥ Denaturation is not necessary for oligonucleotide probes.
➥ The tube containing the mixture is floated on the surface of a water bath at 100°C for 3 min to denature the DNA (remember to pierce the lid if the volume is greater than 50 µL) (or use clips).

- Chill quickly on ice.

➥ To prevent rehybridization of the two denatured strands; at 0°C, single strands are stable.

- Add:
  - 1 *M* DTT **x µL**

➥ Final concentration is 10 m*M*.
➥ Solely for ^{35}S probe
➥ **Wear gloves.**

3. Hybridization protocol:
   - Place the reaction mixture on semithin sections:
     - Drop size **30 µL**

➥ The volume may be reduced if the sections are closely arranged.

   - Cover with a sterile coverslip.
   - Seal the coverslips with rubber cement or DPX.

➥ Carefully, to avoid air bubbles
➥ **Optional** (*see* Figure 7.25)

   - Place the slides flat in a humid chamber.

➥ *See* Figure 7.19.
➥ The concentration of liquid in the humid chamber should keep the humidity the same as that of the hybridization buffer: use several layers of filter paper soaked in 5X SSC in the bottom of the box.

   - Incubate. **overnight**
     at 37–40 °C
     (humid chamber)

➥ The hybridization temperature is defined by the characteristics of the probe.
➥ 35–45°C is the range of temperatures generally used to maximize specific and minimize nonspecific binding.

❏ *Next stage:*
   - Post-hybridization treatments

➥ *See* Section 7.6.1.3.

7.6.1.3 Post-Hybridization Treatments

➥ **Gloves do not have to be worn.**
➥ For semithin sections of frozen or embedded tissue
➥ Reduce the washes for antigenic probes.

   - Duration of washes
   - Quantity to prepare in advance

➥ Usually ≈ 2 h
➥ Prepare a minimum volume of 200 mL for each wash (bath containing 20 slides).

1. Wash the slides in the following solutions:
   - 4X SSC **5 min**
   - 2X SSC **30 min**
     at RT

➥ Soak off the coverslips in 4X SSC.
➥ **Necessary**

• 2X SSC	**30 min at 40°C**	⇨ **Optional** ⇨ In certain exceptional cases where background is excessive, a wash at 40°C is carried out.
• 1X SSC	**20 min at RT**	⇨ **Possible**
• 0.5X SSC	**20 min at RT**	⇨ **Solely in the case of high background**

2. Dehydrate
   • 95%, 100% ethanol   **2 min/bath**   ⇨ Gently dry around the sections and the back of the slide.

3. Dry
   • Under a fume hood   **30 min**   ⇨ Any liquid left on the slide will result in a nonspecific signal (chemoreactivity).

❏ *Next stage:*
   • Visualization of radioactive hybrids   ⇨ *See* Section 7.6.1.4.

7.6.1.4   Visualization of Radioactive Hybrids

1. Coating with emulsion   ⇨ **In a dark room with a safe light**
   • Dilution of emulsion: ⇨ *See* Section 7.5.3.1.
   Take a glass tube or specifically designed container. Draw two lines on the container: the first corresponds to the volume of water needed to make the dilution and the second corresponds to the total volume (water + emulsion).

   ⇨ For 20 to 30 slides, a volume of ≈ 30 mL is needed.
   ⇨ The lines must be clearly visible under the safe light.

– LM1 Amersham	**2:1**	⇨ Emulsion: sterile distilled water ($^v/_v$)
– NTB2 Kodak	**1:1**	⇨ Emulsion: sterile distilled water ($^v/_v$)
– K5 Ilford	**1:1**	⇨ Emulsion: sterile distilled water ($^v/_v$)

   • Melting:
   – Fill with water up to the lower line.   ⇨ Thermostatic water bath with a holder for the glass container

   – Place the tube with the emulsion at 43°C for 5 to 10 min before dilution   ⇨ To melt the emulsion

   – Use a porcelain spoon to measure out the emulsion, or pour the melted emulsion up to the higher mark.   ⇨ Use a safe light on the workbench. The lamp is turned toward the wall when the emulsion is melted.
   ⇨ The emulsion may be stored in the diluted form away from dust and light at 4°C.

   – Mix gently every 20 minutes.   ⇨ With a glass rod or glass slide, **mix slowly** (avoid making bubbles); do not use metal instruments.
   ⇨ The emulsion must be perfectly homogeneous.

   – Melt.   **1 h at 43°C**   ⇨ Check the temperature of the emulsion, which should remain constant. Too high a temperature will denature the emulsion, causing background.

- Coating:
  - Dip and throw away around 10 clean glass slides to clean the surface of the emulsion and to get rid of air bubbles.
  - Slowly dip the slides vertically into the emulsion. Remove with a slow and even motion letting the excess emulsion drain at the edge of the container for a few seconds before using filter paper.
- Dry vertically:
  - Duration   **2 h–overnight at RT**

⇝ Removes all bubbles

⇝ The slides are dipped up to 1 cm below sections.
⇝ Avoid changes in thickness; the film must be uniform.
⇝ Use one slide with no sections (control for assessing the background).

⇝ The vertical position assures the homogeneity of the film.

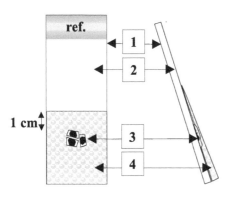

⇝ The slides are air dried on a slide support at an angle of 80°, section side up.
⇝ Wipe the under-side of the slide.
⇝ Turn the slides upside down after 15 to 30 min (optional) to dry the zone of contact between the slide and support.
1 = Glass slide
2 = Support
3 = Semithin section
4 = Autoradiography emulsion

Fig. 7.24 Drying emulsion-coated slides.

Important: the emulsion must be dry before the slides are stored.

- Storage:
  Store the slides in a watertight container with desiccant. The container should be wrapped in aluminum foil or in black plastic at 4°C
- Exposure time:
  - Duration   **1–3 weeks at 4°C**

2. Developing microautoradiography
   - Take the slide box out at least 1 hour before opening to equilibrate to the temperature of the laboratory.
   - Place the slides in a slide holder and transfer them to the following baths

⇝ If the tissue is wet, the water acts on the silver bromide crystals (BrAg), resulting in background.
⇝ Desiccant (e.g., silica gel) to absorb moisture.
⇝ Store at 4°C to limit the development of bacteria in the emulsion.

⇝ Set up several sets of slides to develop after different exposure times.
⇝ **In dark room with a safe light**
⇝ Condensation is a problem. The box must be warm before it is opened.

- Development:
  - Developer (D19)   **4 min at 17°C**

    ⇝ The temperature can be increased if the labeling is weak.
    ⇝ The temperature can be decreased if the background is high. This will also decrease the specific labeling.
    ⇝ The developer can be diluted if the labeling is too high.
  - Wash with distilled water   **30 s**   ⇝ Quickly
- Fix:
  - 30% sodium thiosulfate   **2 min**

    ⇝ The time can be reduced if the tissue is very fragile.
  - Distilled water   **3 × 10 min**

❏ *Next stage:*
- Counterstaining/mounting   ⇝ *See* Section 7.6.1.5.

7.6.1.5   Counterstaining/Mounting

1. Place a drop of stain on the semithin sections:
   - 1% aqueous toluidine blue   **20–30 s at RT or Hot plate**

     ⇝ Other stains may be used (*see* Appendix B7.1.1).
     ⇝ If over-staining occurs, the slide may be soaked in 95% ethanol for several seconds.
     ⇝ The hot plate is used for resin-embedded tissue.
2. Rinse well
   - Distilled water   **Wash bottle**
3. After dehydration, mount permanently.   ⇝ *See* Appendix B8.2.

7.6.1.6   Modes of Observation

- Light field

  ⇝ *See* Section 7.2.6.1.
  ⇝ Silver grains appear black (*see* Examples of Results, Figures 1 and 2).
- Epi-polarization

  ⇝ *See* Section 7.2.6.2
  ⇝ Silver grains appear shiny.

7.6.2   Antigenic Hybrids

⇝ Probes are labeled with biotin, digoxigenin, or fluorescein.
⇝ **Useful** mainly for semithin sections of frozen tissue
⇝ Disappointing results with acrylic resins
⇝ **Optional** (*see* Section 7.6.2.1).

- Pretreatment
  - Deproteinization
  - Prehybridization

- Hybridization   ⇨ *See* Section 7.6.2.2.
  - Hybridization buffer
  - Reaction mixture
  - Hybridization
- Post-hybridization treatments   ⇨ *See* Section 7.6.2.3.
- Immunocytochemical detection   ⇨ *See* Section 7.6.2.4.
  - Direct immunocytochemical reaction   ⇨ Only used if the target nucleic acid is present in large amounts (*see* Section 7.6.2.4.1)

  - Indirect immunocytochemical reaction   ⇨ *See* Section 7.6.2.4.2.
- Counterstaining   ⇨ **Optional** (*see* Section 7.6.2.5)
- Mounting   ⇨ *See* Section 7.6.2.6.
  - Aqueous
  - Permanent

7.6.2.1   Pretreatments

⇨ **Wear gloves.**
⇨ **Optional stage** (test without)
⇨ The enzyme used is proteinase K.

1. Deproteinization
   - Enzyme treatment:
     - Proteinase K   **1 µg/mL**
                                    10 min
                                    at RT

   - Stop the reaction:
     - Tris–HCl/CaCl$_2$ **2 min**   ⇨ Rapid rinse
     - Phosphate buffer   **5 min**   ⇨ Used to stop the action of proteinase K by removing NH$_2$ groups and calcium chloride present in the Tris buffer by changing the buffer

   - Post-fix: ⇨ Restabilization of the tissue; **indispensable prior to hybridization** for semithin sections of frozen tissue

     - 4% PF in phosphate buffer   **5 min**   ⇨ This fixative gives the best compromise between efficiency of hybridization and preservation of morphology.
                                                  ⇨ To prevent loss of sections from slides during hybridization

   - Washes:
     - Phosphate buffer   **3 × 5 min** ⇨ Removes traces of fixative
     - 150 m*M* NaCl   **3 min**   ⇨ To avoid precipitation in ethanol
   - Dehydrate:
     - 95%, 100% ethanol   **2 min/change**   ⇨ Carefully dry around the sections and the backs of the slides.

   - Dry:
     - Under the fume hood   **> 30 min**   ⇨ Enzymes (RNases and DNases) are inactive in the absence of water.

❑ *Next stages:*

- Prehybridization   ⇨ *See* Section 7.6.2.1.
- Hybridization   ⇨ *See* Section 7.6.2.2.

2. Prehybridization

↪ **Wear gloves.**
↪ **Optional stage**

- After dehydration and drying, sections are incubated with hybridization mixture without the probe:
  - Hybridization buffer   ≈ **100 μL**

↪ *See* Section 7.6.2.2. The concentrations of DNA and RNA in Denhardt's solution depend on formamide.

  - Duration of incubation   **1–2 h at RT**

↪ May be prolonged (several hours)
↪ The temperature does not modify nonspecific binding (weak influence of formamide).
↪ The sections must not dry out. Leave the sections wet.

- Remove as much buffer as possible before using the labeled probe.

❏ *Next stage:*
- Hybridization

↪ *See* Section 7.6.2.2.

7.6.2.2   Hybridization

↪ **Wear gloves.**

1. Hybridization buffer

↪ It can be made just before use or stored at –20°C/–80°C.

- Add

Solutions	Final concentration
– 20X SSC	**4X**

↪ **Indispensable,** stabilizes hybridization, is a source of Na⁺

| – Deionized formamide | **50%** |

↪ **Useful;** denaturing agent that lowers the Tm and permits hybridization to take place at 40°C

| – 50% dextran sulfate | **10%** |

↪ **Useful;** high-molecular-weight polymer that artificially concentrates the probe

| – DNA (10 mg/mL) | **250 μg/mL** |

↪ **Useful;** competes with proteins

| – Yeast tRNA (10 mg/mL) | **250–400 μg/mL** |

↪ Indispensable; competes with proteins

| – 50X Denhardt's solution | **2X** |

↪ Useful; diminishes the background and saturates nonspecific binding sites

| – Sterile water | **to x mL** |

- Vortex.
- Centrifuge.

2. Reaction mixture

↪ **Wear gloves.**
↪ After centrifugation, the labeled probe is dissolved directly in the hybridization buffer.
↪ The concentration depends on the type of labeling.

- Suspend the labeled probe in hybridization buffer in an Eppendorf tube:

| – Oligonucleotide probe | **20–50 pmoles/mL of hybridization buffer** |

↪ Labeling by 3′ extension (*see* Chapter 1, Section 1.3.4).

| – cDNA probe | **50–100 ng/mL of hybridization buffer** |

↪ Labeling by random priming or PCR (*see* Chapter 1, Sections 1.3.1 and 1.3.2).

- Vortex.

↪ Gently, or mix with a pipette

- Centrifuge.
- Denature.   **3 min**
                                at 100°C

- Chill quickly on ice.

3. Hybridization protocol
   - Place the reaction mixture on the semi-thin sections:
     – Drop size   **30 μL**

   - Cover with a sterile coverslip.
   - Seal the coverslip with rubber cement or DPX.

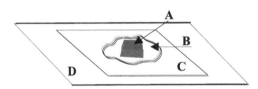

- Place the slides flat in a humid chamber.

❑ *Next stage:*
   - Post-hybridization treatments

7.6.2.3   Post-Hybridization Treatments

- Duration of washes
- Quantity to prepare

Wash the slides in the following solutions:

1. 4X SSC   **5 min**
                                  at RT
2. 2X SSC   **10 min**
                                  at RT
3. 1X SSC   **2 × 10 min**
                                  at RT

❑ *Next stages:*

- Direct immunocytochemistry reaction

➷ **Indispensable** for double-stranded cDNA probes (*see* Chapter 1, Figure 1.17).
➷ Denaturation is not necessary for oligonucleotide probes.
➷ To prevent rehybridization of the two denatured strands; at 0°C, the single strands are stable.
➷ **Wear gloves.**

➷ Depends on the area occupied by the sections
➷ **Carefully,** to avoid air bubbles
➷ **Optional**

(**A**) Section
(**B**) Probe
(**C**) Coverslip
(**D**) Slide

Fig. 7.25   Sealing the coverslip.

➷ *See* Figure 7.19.
➷ To keep the humidity in the humid chamber high, place several layers of filter paper soaked in 5X SSC in the bottom of the chamber.

➷ *See* Section 7.6.2.3.

➷ **Gloves do not have to be worn.**

➷ ≈ 1 h
➷ Prepare a minimum volume of 200 mL for each wash (bath of 20 slides).

➷ The coverslips will slide off the slides.

➷ **Necessary**

➷ The semithin sections must not be allowed to dry out prior to immunocytochemistry.
➷ Used when the target nucleic acid is present in large amounts (*see* Section 7.6.2.4.1)

• Indirect immunocytochemistry reaction	➥ Increases the sensitivity and the specificity (*see* Section 7.6.2.4.2)

7.6.2.4   Immunocytochemical Detection

7.6.2.4.1   DIRECT IMMUNOCYTOCHEMICAL REACTION

➥ Short protocol: ≈ 4 h
➥ All of the following stages are carried out at room temperature.

1. Wash:
   • Tris–HCl/NaCl buffer   **10 min**

➥ This step equilibrates the osmolarity of the tissue after the post-hybridization washes.
➥ *See* Appendix B3.7.5.

2. Block nonspecific sites:
   • Blocking buffer   **15–30 min**

➥ *See* Appendix B6.1.1.
➥ **Indispensable;** to remove all of the non-specific binding sites, the sections are incubated with nonimmune serum.

3. Remove excess buffer.

➥ Either by aspiration or by delicately wiping around the tissue with filter paper; a line of hydrophobic solution can be drawn around the sections to limit the dilution of the solution placed on the sections.

4. Inhibition of endogenous enzyme activity:
   • Peroxidase with hydrogen peroxide
   • Alkaline phosphatase with levamisole
5. Place the conjugated anti-hapten antibody on the slide (e.g., peroxidase):
   • Diluted in Tris–HCl/NaCl buffer   **≥ 20 μL**

➥ *See* Appendix B6.1.3.
➥ *See* Appendix B6.1.2.1.
➥ Forms a hapten–antibody complex
➥ Alkaline phosphatase may be used.
➥ IgG and Fab fragments conjugated to peroxidase
➥ The dilution of the antibody is low to compensate for the low sensitivity (between 1:10 to 1:100 depending on the intensity of labeling).

6. Incubate with the antibody. **2 h–overnight**
7. Wash
   • Tris–HCl/NaCl buffer   **3 × 10 min**

➥ Humid chamber (water-soaked filter paper)

➥ The ratio of signal/background is usually low.

8. Visualization of peroxidase:
   • Prepare the developing solution:

➥ **Prepare just before use** (*see* Appendix B6.2.2).
➥ NBT-BCIP for alkaline phosphatase (*see* Appendix B6.2.1.1)

   – DAB   **10 mg**
   – Buffer for developing the color product   **10 mL**
   • Filter then add:
   – 30% hydrogen peroxide   **100 μL**
   • Place the substrate on the sections.
   – Substrate   **100 μL**

➥ Or 1 tablet
➥ 20 mM Tris–HCl/300 mM NaCl/50 mM MgCl$_2$ buffer; pH 7.6 (*see* Appendix B3.7.6)

➥ Final concentration is 0.01%.

• Incubate while watching carefully.	**3–10 min** **at RT**	↪ Forms a brown precipitate
		↪ Too long a development leads to a general brown staining of the tissue (background).
• Stop the reaction by plunging the slide into distilled water:		↪ Counterstaining may be carried out.
– Distilled water	**1 min**	

❑ *Next stages:*
  • Counterstaining
  • Permanent mounting

↪ *See* Section 7.6.2.5.
↪ The precipitate is insoluble in ethanol.
↪ *See* Section 7.6.2.6.1.

7.6.2.4.2   INDIRECT IMMUNOCYTOCHEMICAL REACTION

↪ Duration is ≈ 6 h

↪ All of the following stages are carried out at room temperature. A line of hydrophobic solution can be drawn around the sections to limit the quantity of solution needed.

1. Wash:
   • Tris–HCl/NaCl buffer   **10 min**

↪ This step equilibrates the osmolarity of the tissue after the post-hybridization washes (*see* Appendix B3.7.5).

2. Block nonspecific sites:
   • Blocking buffer   **15–30 min**

↪ **Indispensable;** to remove all of the nonspecific binding sites, the sections are incubated with nonimmune serum (*see* Appendix B6.1).

3. Inhibition of endogenous enzyme activity:
   • Peroxidase by hydrogen peroxide
   • Alkaline phosphatase by levamisole

↪ *See* Appendix B6.1.3
↪ *See* Appendix B6.1.2.1

4. Wash:
   • Tris–HCl/NaCl buffer   **3 × 10 min**
5. Remove excess buffer.
6. Place the 1st nonconjugated anti-hapten antibody (raised in species X) on the sections:

↪ Risk of diluting the next solution
↪ Forms a hapten–antibody complex

   • Diluted in   **≥ 20 μL**
   Tris–HCl/NaCl buffer

↪ The dilution is between 1:50 and 1:500.

7. Incubate with antibody.   **60–90 min**

↪ Humid chamber (filter paper soaked in water)

8. Wash:
   • Tris–HCl/NaCl buffer   **3 × 10 min**
9. Remove excess buffer.

↪ Or Tris–HCl buffer

10. Place the 2nd antibody (anti-species X) conjugated with alkaline phosphatase:

↪ IgG, F(ab′)$_2$, Fab
↪ Peroxidase can be used (*see* Section 7.6.2.4.1).

- Diluted in Tris–HCl/NaCl ≥ **20 µL**
  buffer

↪ Secondary antibody: IgG or Fab fragments diluted 1:25–1:50 (according to the supplier's instructions)

↪ *Dilution is less than that of the 1st antibody.*

11. Incubate with antibody.   **60–90 min**

↪ Humid chamber (filter paper soaked in water)

12. Wash:
    - Tris–HCl/NaCl   **3 × 10 min**
      buffer

↪ On the slide (≥ 100 µL) or in a bath

13. Remove excess buffer.
14. Visualize the alkaline phosphatase:

↪ Substrate used: NBT-BCIP
↪ For peroxidase, use DAB (*see* Section 7.6.2.4.1)

- Prepare the solution for developing the color reaction:

↪ Prepare just before use (*see* Appendix B6.2.1.1).
↪ Ready-to-use solutions are best.

  - NBT   **0.34 mg or 4.5 µL**

↪ 75 mg/mL dimethylformamide

  - BCIP   **0.18 mg or 3.5 µL**

↪ 50 mg/mL dimethylformamide.

  - Buffer for visualizing   **to 1 mL**
    the color product

↪ Tris–HCl/NaCl/MgCl$_2$ buffer; pH 9.5 (*see* Appendix B3.7.6)

- Place the substrate on   **1:250**
  the section:
  - NBT-BCIP   **100 µL**

↪ **Optional;** add an enzyme-blocking agent (1 m*M* levamisole) (*see* Appendix B6.1.2.1).

- Incubate while watching carefully:
  - Duration   **10 min–24 h at RT in the dark**

↪ The reaction may be accelerated by incubating at 37°C.

- Stop the reaction by plunging the slide into distilled water:
  - Distilled water   **5 min**

↪ If the color is not dark enough, use another 100 µL of the NBT-BCIP solution.

❑ *Next stages:*
- Counterstaining

↪ *See* Section 7.6.2.5.
↪ It is **necessary** to counterstain if the substrate used is Fast Red.

- Mount in aqueous mountant

↪ The precipitate produced by the substrate NBT-BCIP is soluble in ethanol (*see* Section 7.6.2.6.2).

- Observation

↪ Use light field microscopy.
↪ *See* Examples of Results (Figures 1 and 2).

7.6.2.5   Counterstaining

↪ **Optional**

1. Drop onto section
   - 2% methyl green   **1–3 min**

↪ *See* Appendix B7.1.2

2. Wash:
   • Running water   **1 min**
3. Mount in aqueous mountant.   ➟ *See* Section 7.6.2.6.2

7.6.2.6   Mounting Media

7.6.2.6.1   PERMANENT

1. Dehydrate:
   • 70%, 90%, 100% ethanol   **3 × 5 min**
   • Xylene   **2 × 5 min**   ➟ These last two changes may be prolonged.
2. Mount:
   Place the mountant onto the slide directly   ➟ Mountants (e.g., Permount, Eukitt, or
   out of the xylene for best results; place   DPX); Use a hot plate for the best results.
   coverslip on immediately
3. Observe with the light microscope.

7.6.2.6.2   AQUEOUS   ➟ Use aqueous mountants if the precipitate
                                                              is soluble in ethanol.

   • Commercially available mountants
   • Buffered glycerine   ➟ *See* Appendix B8.1.1.
   • Moviol   ➟ *See* Appendix B8.1.2.

Chapter 8

Controls and Problems

CONTENTS

8.1   SIGNAL/BACKGROUND RATIO

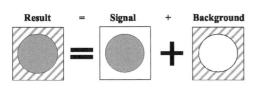

⮡ The observed result is the additive effect of the signal and the background. The ratio between these two parameters is critical.

8.1.1   Probes

Criteria

1. Type:
   - cDNA → → ⮡ Reference
   - Single-stranded DNA ↗ → ⮡ Limits the formation of interchain hybrids.
   - Oligonucleotides x oligo ⮡ A number of oligonucleotides may be specific for the same target.
   - cRNA ↗ ↘ ⮡ Due to a high specific activity and the stability of RNA/RNA hybrids

2. Homology:
   - 100% → → ⮡ Reference
   - Less than 100% ↘ ↗ ⮡ Loss of hybrid stability and nonspecific binding

3. Label:
   - Radioactive ↗ ↗ ⮡ Higher sensitivity than antigenic labels and can be quantified
   - Antigenic ↘ ↘ ⮡ Threshold effect limits detection.

4. Specific activity:
   - \+ ↘ ↘ ⮡ Threshold effect
   - \++ → → ⮡ Reference
   - \+++ ↗ ↗ ⮡ Possible to use washes of increasing stringency

5. Purification: ↗ ↘ ⮡ Diminution of nonspecific binding
   ⮡ Also necessary for sense probes

8.1.2   Samples

Criteria

1. Condition of sample: ↗ ⮡ Reduction of exogenous RNase
2. Preservation of nucleic acids: ⮡ Facilitates specific hybridization
   - Post-embedding ↘
   - Pre-embedding ↗
   - Frozen tissue ↗↗
   - Semithin sections ↗

8.1.3 Pretreatments

Criteria

Criteria	⬤	◯	
1. Fixation:			
• +	↗	↘	⇒ Preservation of the accessibility of the nucleic acids
• ++	→	→	⇒ Compromise between cell preservation and preservation of nucleic acids
• +++	↘	↗	⇒ Loss of endogenous nucleic acids (particularly cellular RNA) due to the formation of complexes between nucleic acids and proteins
• +++	↗ or ↘	↗ or ↘	⇒ Preservation of exogenous nucleic acids (e.g., virus)
2. Preparation of tissue:			
• Post-embedding			
– Acrylic resin	↘	↗	⇒ Extraction of RNA is greater than DNA.
– Epoxy resin	0	↗	⇒ Limits the hybridization possible
			⇒ Possibility of visualization
• Pre-embedding	↗	→	
• Frozen tissue	↗	↘ or →	⇒ Avoids dehydration and embedding
			⇒ Increases the sensitivity
• Semithin sections			
– Frozen	→	→	
– Epoxy resin	↘	↘	
3. Deproteinization:			⇒ Increases the accessibility of nucleic acids
• Post-embedding	↗	→	
• Pre-embedding	↗	→	
• Frozen tissue	↘	→	
• Semithin sections	↗	→	
4. Permeabilization:			⇒ Increases the accessibility of targets
• Post-embedding	↘	→	
• Pre-embedding	↗	→	
• Frozen tissue	↘	→	
• Semithin sections	↘	→	
5. Acetylation:	↘	↘	⇒ Blocks functional NH_2
6. Prehybridization:			⇒ Blocks nonspecific binding sites
• Post-embedding	→	↘	⇒ Little effect
• Pre-embedding	→	↘	⇒ Blocks specific and nonspecific binding sites
• Frozen tissue	→	↘	⇒ Blocks specific and nonspecific binding sites
• Semithin sections	→	↘	
7. Denaturation:			
• 0	0	↗	⇒ If the target is DNA, the signal may not be present.
• +	→	→	⇒ Accessibility of double-stranded targets

8. Storage:

	Signal	Background	Comment
• Post-embedding	→	→	⇒ If possible in resin blocks
• Pre-embedding	↘	→	⇒ Store with care.
• Frozen tissue	→	→	⇒ At 4°C under a drop of saturated sucrose solution
• Semithin sections	→	→	⇒ At 4°C under a drop of saturated sucrose solution

8.1.4   Hybridization

Criteria

Criteria	Signal	Background	Comment
1. Probe concentration:			
• +	↘	↘	⇒ Targets are not saturated.
• ++	↗	→	⇒ Saturation of targets
• +++	→	↗	⇒ Increase in nonspecific binding
2. NaCl concentration: ≈ 600 mM	↗	↗	⇒ Facilitates hybridization.
3. Dextran sulfate concentration:			⇒ Concentration of probe
• Post-embedding	↗		⇒ Concentration of probe
• Pre-embedding	0		⇒ Does not penetrate vibratome sections.
• Frozen tissue	0		⇒ Deterioration in cellular structure.
• Semithin sections	↗	→	
4. tRNA, DNA concentration:	↘	↘	⇒ Saturation of nonspecific binding sites
5. Detergent:		↘	⇒ Reduction in hydrophobic regions
			⇒ Deterioration in cellular structure (frozen tissue)
• Post-embedding	→	↘	⇒ Lower nonspecific binding
• Pre-embedding	↗	↘	⇒ ↗ enables the penetration of solutions in vibratome sections.
• Frozen tissue	0	0	⇒ Deterioration in cellular structure
• Semithin sections	0	0	
6. DTT:	→	↘	⇒ Preserves ^{35}S functions
7. Formamide concentration:			⇒ Increases the specificity of the hybridization
• 10%	↘	↗	
• 30%	↗	↘	
• 50%	↘	↘	
8. Temperature:			
• RT	↗ or ↘	↗ or →	⇒ Nonspecific hybridization is possible.
• 37°C	↗ or →	→ or ↘	⇒ Reduction in nonspecific hybridization.
• Higher than 40°C	↘	↘	⇒ Denaturation is possible.
9. Time:			
• Post-embedding			
– Less than 1 h	→	↘	
– 1–3 h	↗	→	
– More than 5 h	→	↗	

- Pre-embedding
  - Less than 3 h   ↘ ↘
  - 5–6 h   ↗ →
  - More than 8 h   → ↗
- Frozen tissue
  - Less than 3 h   ↘ ↘
  - 3 h   ↗ →
  - More than 3 h   → ↗
- Semithin sections
  - Less than 3 h   ↘ ↘
  - 5 h   ↗ →
  - More than 6 h   → ↗

8.1.5 Washes

Criteria

Criteria	●	◯	
1. NaCl concentration:			
• +++	→	↗	⇨ Stability of nonspecific hybrids
• +	↘	↘	⇨ Denaturation of nonspecific hybrids
2. Temperature:			
• RT	↗	↗	⇨ Washes away nonspecific hybrids.
• 37°C	↘	↘	⇨ Hybrids denature
3. Time:	↘	↘	⇨ Hybrids become labile

8.1.6 Visualization

Criteria

Criteria	●	◯	
1. Autoradiography:			
• Post-embedding	↗	↗	
• Pre-embedding	↗	↗	
• Frozen tissue	→	↗	
• Semithin sections	↗	↗	
2. Immunocytochemistry:	↘	↘	⇨ Threshold for detection
• Direct method	↘	↗	⇨ Nonspecific adsorption
• Indirect method	↗	→	⇨ Increases the sensitivity
• Post-embedding			⇨ No penetration of solutions into the section
– Enzymatic labels	↘	↗	⇨ Nonspecific adsorption
– Particulate labels	→	↘	⇨ Colloidal gold
• Pre-embedding			⇨ Problem of penetration of solutions into vibratome sections
– Pre-embedding	→	↗	⇨ Enzymatic visualization on the section surface (1–3 μm)
– Ultrathin sections	↘	→	⇨ Loss of hybrids during embedding ⇨ Particulate label

- Frozen tissue → → ⇔ Comparable with autoradiography
  - Direct method ↘ ↗ ⇔ Little used
  - Indirect method ↗ ↘ ⇔ Particulate label
- Semithin sections
  - Direct method ↘ ↗
  - Indirect method ↗ ↘
  - Amplification ↗↗ ↗
3. Immunohistochemistry: ↘ ↘ ⇔ Frozen tissue technique
  - Direct method ↘ ↗ ⇔ Nonspecific adsorption
  - Indirect method ↗ → ⇔ Increases the sensitivity
  - Amplification ↗↗ ↗ ⇔ Lowers the threshold of detection

8.1.7 Staining

Criteria

- Post-embedding → →
- Pre-embedding → →
- Frozen tissue → →
- Semithin sections → ↘

8.1.8 Observation

Criteria

- Post-embedding
- Pre-embedding
- Frozen tissue
  - Electron microscopy → →
- Semithin sections
  - Bright field → →
  - Epipolarization ↗ ↘

8.2 SENSITIVITY/SPECIFICITY

8.2.1 Probes

Criteria

1. Nature:
  - cDNA ↘ → ⇔ Rehybridization of complementary strands
  - Single-stranded DNA → →

	sensitivity	specificity	
• Oligonucleotide	→	→	
• Oligonucleotides	↗	→	↝ Increases the possibility of hybridization
• cRNA	↗	→	↝ High specific activity and stability of hybrids
2. Homology:			
• 100%	→	→	↝ Optimal conditions
• Less than 100%	↘↘	↘↘	↝ Do not use.
3. Label:			
• Radioactive	↗	→	↝ No threshold of detection
			↝ Difficult to visualize and analyze statistically
• Antigenic	↘	↘	↝ Threshold of detection; endogenous antigens
4. Specific activity:			↝ Verify the probe label.
• +	↘	→	↝ Threshold of detection
• ++	→	→	
• +++	↗	→	↝ ↗ antigenic detection is possible
5. Purification:	↗	↗	↝ Lowers nonspecific binding

8.2.2 Samples

Criteria

	sensitivity	specificity	
1. Sample condition:	↗	↗	↝ Limits contamination
2. Preservation of nucleic acids:			↝ Limits breakdown
• Post-embedding	↘	→	
• Pre-embedding	↗↗	↗	
• Frozen tissue	↗↗	↗	
• Semithin sections	→	↗	

8.2.3 Pretreatments

Criteria

	sensitivity	specificity	
1. Fixation:			
• +	↗	↘	↝ Preservation of targets, but diffusion may occur.
• +++	↘	→	↝ Blocks nucleic acid/protein complexes
2. Preparation of tissue:			
• Embedding in resin			
– Acrylic resin	↘	→	↝ Extraction of RNA is greater than DNA.
– Epoxy resin	0		↝ Loss of hybridization potential
• Pre-embedding	↗	↗	↝ Preservation of targets

	sensitivity	specificity	
• Frozen tissue	↗	↗	⇨ Avoids dehydration and embedding
• Semithin sections			
– Freezing	→	→	⇨ Extraction of RNA is greater than DNA
– Epoxy resin	↘	↘	⇨ Pre-embedding technique after visualization
3. Deproteinization:			⇨ Increases accessibility of nucleic acid targets
• Post-embedding	↗	↗	
• Pre-embedding	↗	↗	
• Frozen tissue	↘	→	
• Semithin sections	↗	→	
4. Permeabilization:			⇨ Blocks functional NH_2
• Post-embedding	↘	↘	⇨ Does not penetrate into the section.
• Pre-embedding	↗	↗	⇨ ↗ of accessibility.
• Frozen tissue	↘	↘	⇨ Deterioration of cellular structures.
• Semithin sections	↘	↘	
5. Acetylation:	↘	→	⇨ Blocks functional NH_2
6. Prehybridization:			⇨ Blocks nonspecific sites
• Post-embedding	→	→ or ↗	⇨ Little effect
• Pre-embedding	→	↗	⇨ Blocks specific and nonspecific sites
• Frozen tissue	→	↗	⇨ Blocks specific and nonspecific sites
• Semithin sections	→	↗	
7. Denaturation:			
• 0	0	0	⇨ Indispensable if DNA is the target
• +	→	→	⇨ Accessibility of the target
8. Storage:			
• Post-embedding	→	→	⇨ If possible in resin blocks.
• Pre-embedding	↘	↘	⇨ Store with care.
• Frozen tissue	→	→	⇨ At 4°C under a drop of saturated sucrose solution
• Semithin sections	→ or ↘	→	⇨ At 4°C under a drop of saturated sucrose solution

8.2.4   Hybridization

Criteria

	sensitivity	specificity	
1. Probe concentration:			
• +	↘	↗	⇨ Not all of the targets are hybridized
• ++	↗	→	⇨ All of the targets are hybridized
• +++	→ or ↗	↘	⇨ Nonspecific binding rises
2. NaCl concentration ≈ 600 mM:	↗	→	⇨ Stabilization of nucleic acid/nucleic acid hybrids
3. Dextran sulfate concentration:			⇨ Probe concentration
• Post-embedding	↗	→	
• Pre-embedding	0	0	
• Frozen tissue	0	0	
• Semithin sections	↗	→	

Criteria	sensitivity	specificity	effect
4. tRNA, DNA concentration:	→	↗	⇔ Nonspecific sites are saturated.
5. Detergent:			
• Post-embedding	→ or ↘	↗	⇔ Not useful
• Pre-embedding	↗	↗	⇔ Necessary
• Frozen tissue	0 or ↘	↘	⇔ Disastrous
• Semithin sections	0 or ↘	0 or ↘	⇔ Not useful
6. DTT:	↗	↗	⇔ Protection of ^{35}S label
7. Formamide:			⇔ ↗ of its concentration
• Method			
– Post-embedding	→	↗	⇔ Little importance
– Pre-embedding	→	↗	⇔ Useful
– Frozen tissue	→	→	⇔ Little importance
– Semithin sections	→	↗	⇔ Useful
• Concentration			⇔ Increases the specificity of the hybridization
– 10%	↘	↘	
– 30%	↗	→	
– 50%	→	↗	
8. Temperature:			
• RT	↗	↘ or →	⇔ Facilitates the hybridization
• 37°C	↘ or →	→ or ↗	⇔ Optimal hybridization
• Higher than 40°C	↘	↗	⇔ Denaturation of hybrids
9. Time:			
• Post-embedding			
– Less than 3 h	→	→ or ↗	⇔ Generally sufficient
– 3 h	→ or ↗	→	⇔ Optimal duration
– More than 3 h	↗	↘	⇔ Nonspecific as well as specific hybridization
• Pre-embedding			
– Less than 3 h	↘	→ or ↗	⇔ Only specific hybridization
– 3–6 h	→	↗	⇔ Specific hybridization
– More than 8 h	↗	↘	⇔ Nonspecific as well as specific hybridization
• Frozen tissue			
– Less than 3 h	→ or ↘	↗	⇔ Generally sufficient
– 3 h	→ or ↗	↗	⇔ Optimal duration
– More than 3 h	↗	↘	⇔ Nonspecific as well as specific hybridization
• Semithin sections			
– Less than 3 h	↘	↗	⇔ Generally sufficient
– 5–6 h	→	↗	⇔ Optimal duration
– 16 h	↗	↘	⇔ Can be used

8.2.5 Washes

Criteria

Criteria	sensitivity	specificity	effect
1. NaCl concentration:			
• +++	→	↗	⇔ Stabilization of nucleic acid/nucleic acid hybrids
• +	↘	↘	⇔ Instability of hybrids

2. Temperature:
   - RT   ↗   ↘   ⇨ Weak denaturation of nonspecific hybrids.
   - Higher than 40°C   ↘   ↗   ⇨ Denaturation of nonspecific hybrids is greater than specific hybrids.
3. Time: ↘   ↘   ⇨ Hydrolyze hybrids

8.2.6   Visualization

Criteria

	sensitivity	specificity	
1. Autoradiography:	↗	↘	⇨ No threshold of detection, but background can be a problem
• Post-embedding	↗	↘	⇨ Little use
• Pre-embedding	↗	↘	⇨ Useful
• Frozen tissue	→ or ↗	↘	⇨ Little use
• Semithin sections	↗	↘	⇨ Useful
2. Immunocytochemistry:	↘	↗	⇨ Threshold of detection
• Direct method	↘	↘	⇨ Threshold of detection is important; background can be a problem
• Indirect method	↗	↗	⇨ Lower threshold of detection
• Post-embedding	→	→	⇨ Useful
– Direct method	↘	↘	
– Indirect method	↗ or ↘	↗	
• Pre-embedding	→	↘	⇨ Less important
– Direct method	↘	↘	
– Indirect method	↗	↗	
• Frozen tissue	→	→	⇨ Important
– Direct method	↘	↘	
– Indirect method	↗	↗	
• Semithin sections	→	→	⇨ Useful
			⇨ Amplification is possible, but background may occur.
– Direct method	↘	↘	
– Indirect method	↗	↗	
– Amplification	↗	↘	

8.2.7   Staining

Criteria

	sensitivity	specificity
• Post-embedding	→	→
• Pre-embedding	→	→
• Frozen tissue	→	→
• Semithin sections	→	→

8.2.8 Observation

Criteria

sensitivity | specificity

- Post-embedding
- Pre-embedding
- Frozen tissue
    - Electron microscopy → →
- Semithin sections
    - Bright field → ↘
    - Epipolarization ↗ ↗

8.3 CONTROLS

8.3.1 Probes

Criteria

1. Search for homologous sequences in gene bank

⇌ There is a possibility of the existence of sequences with high enough levels of homology to form nonspecific hybrids with the probe

2. Hybridization with a sense probe ✔

⇌ For oligonucleotides and RNA probes

3. Hybridization with a heterologous probe ✔

⇌ For all types of probes

4. Competition ✔

⇌ Between the labeled probe and a high concentration of unlabeled probe: negative hybridization

5. No denaturation of cDNA probe ✔

⇌ Verification of the reactivity of the tissue

8.3.2 Labeling

Criteria

1. Radioactive

⇌ Determination of the specific activity

2. Antigenic

⇌ Dot blot

8.3.3   Tissue

Criteria

1. Positive control tissue   ✔   → Tissue or cells known to express the target nucleic acid: positive hybridization.

2. Negative control tissue   ✔   → Tissue or cells know not to express the target nucleic acid: negative hybridization.

3. Destruction of the target by enzymatic treatment
   - DNase   ✔   → Degradation: hybridization is less than 0.
   - RNase   ✔   → Fragmentation of target RNA (hybridization is less than 0) that may remain after several washes

4. Destruction of non-target nucleic acids by enzymatic treatment
   - DNase   ✔   → Degradation of DNA
   - RNase   ✔   → Degradation of RNA that may remain after several washes

8.3.4   Hybridization/Washes

Criteria

1. Hybridization with a heterologous probe of the same type   ✔   → Verification of the reactivity of the tissue

2. Hybridization in the absence of the probe   ✔   → Verification of the absence of nonspecific binding of the probe label or of the visualization system in the tissue (e.g., biotin, enzyme, etc.)

3. Variation of the stringency   → Variation of the concentration of salt or the temperature of the hybridization and/or the washes: modification of the signal

8.3.5   Visualization

Criteria

1. Autoradiography   ✔   → Determination of the background
2. Immunohistochemistry   ✔   → Check endogenous activity:
   - Inhibition of endogenous activity   ✔
     - Endogenous biotin
     - Endogenous peroxidase
     - Endogenous alkaline phosphatase

8.3.6   Variation in Signal

Criteria

1. Experimental variation in the expression of the target nucleic acid
2. Quantification of the signal

↪ Determination of the background
↪ Increase in the signal after stimulation of expression
↪ Physiological controls or on the section

8.3.7   Verification by Other Techniques

Criteria

1. Dot blot, Southern blot, and Northern blot
2. PCR

3. *In situ* PCR
4. Immunocytochemistry

↪ Demonstration of the presence of the target nucleic acid in the tissue
↪ Determination of the presence or the absence of the target nucleic acid
↪ Amplification of the signal
↪ Demonstration of the translated protein (the product of the target nucleic acid)

8.4   SUMMARY TABLE

Criteria	Signal	Background	Sensitivity	Specificity
Probes				
Type				
• cDNA	→	→	↘	→
• Single-stranded DNA	↗	→	→	→
• Oligonucleotides	x oligo		↗	→
• cRNA	↗	↘	↗	→
Homology				
• 100%	→	→	→	→
• < 100%	↘	↗	↘↘	↘↘
Label				
• Radioactive	↗	↗	↗	→
• Antigenic	↘	↘	↘	↘
Specific activity				
• +	↘	↘	↘	→
• ++	→	→	→	→
• +++	↗	↗	↗	→
Purification	↗	↘	↗	↗

Criteria	Signal	Background	Sensitivity	Specificity
Samples				
Sample condition	↗		↗	↗
Preservation of nucleic acids				
• Post-embedding	↘		↘	→
• Pre-embedding	↗		↗↗	↗
• Frozen tissue	↗↗		↗↗	↗
• Semithin sections	↗		→	↗
Pretreatments				
Fixation				
• +	↗	↘	↗	↘
• +++	↘ or ↗	↗ or ↘	↘	→
Preparation of tissue				
• Post-embedding				
– Acrylic resin	↘	↗	↘	→
– Epoxy resin	0	↗	0	
• Pre-embedding	↗	→	↗	↗
• Frozen tissue	↗	↘ or →	↗	↗
• Semithin sections				
– Freezing	→	→	→	→
– Epoxy resin	↘	↘	↘	↘
Deproteinization				
• Post-embedding	↗	→	↗	↗
• Pre-embedding	↗	→	↗	↗
• Frozen tissue	↘	→	↘	→
• Semithin sections	↗	→	↗	→
Permeabilization				
• Post-embedding	↘	→	↘	↘
• Pre-embedding	↗	→	↗	↗
• Frozen tissue	↘	→	↘	↘
• Semithin sections	↘	→	↘	↘
Acetylation	↘	↘	↘	→
Prehybridization				
• Post-embedding	→	↘	→	→ or ↗
• Pre-embedding	→	↘	→	↗
• Frozen tissue	→	↘	→	↗
• Semithin sections	→	↘	→	↗
Denaturation				
• 0	0	↗	0	0
• +	→	→	→	→
Storage				
• Post-embedding	→	→	→	→
• Pre-embedding	↘	→	↘	↘
• Frozen tissue	→	→	→	→
• Semithin sections	→	→	→ or ↘	→

Criteria	Signal	Background	Sensitivity	Specificity
Hybridization				
Probe concentration				
• +	↘	↘	↘	↗
• ++	↗	→	↗	→
• +++	→	↗	→ or ↘	↘
NaCl concentration (≈ 600 mM)	↗	↗	↗	→
Dextran sulfate concentration				
• Post-embedding	↗		↗	→
• Pre-embedding	0		0	0
• Frozen tissue	0		0	0
• Semithin sections	↗	→	↗	→
tRNA, DNA concentrations	↘	↘	→	↗
Detergent				
• Post-embedding	→	↘	→ or ↘	↗
• Pre-embedding	↗	↘	↗	↗
• Frozen tissue	0	0	0 or ↘	0 or ↘
• Semithin sections	0	0	0 or ↘	0 or ↘
DTT	→	↘	↗	↗
Formamide				
• 10%	↘	↗	↘	↘
• 30%	↗	↘	↗	→
• 50%	↘	↘	→	↗
Temperature				
• RT	↗	→	↗	→ or ↘
• 37°C	→	↘	→ or ↘	→ or ↘
• > 40°C	↘	↘	↘	↗
Time				
• Post-embedding				
– 1 h	→	↘	→	↗ or →
– 3 h	↗	→	→ or ↗	→
– > 5 h	→	↗	↗	↘
• Pre-embedding				
– 3 h	↘	↘	↘	↗ or →
– 5–6 h	↗	→	→	↗
– > 8 h	→	↗	↗	↘
• Frozen tissue				
– < 3 h	↘	↘	→ or ↘	↗
– 3 h	↗	→	→ or ↘	↗
– > 3 h	→	↗	↗	↘
• Semithin sections				
– < 3 h	↘	↘	↘	↗
– 5–6 h	↗	→	→	↗
– 16 h	→	↗	↗	↘

Criteria	Signal	Background	Sensitivity	Specificity
Washes				
NaCl concentration				
• +++	→	↗	→	↗
• +	↘	↘	↘	↘
Temperature				
• RT	↗	↗	↗	↘
• ≥ 40°C	↘	↘	↘	↗
Time	↘	↘	↘	↘
Visualization				
Autoradiography				
• Post-embedding	↗	↗	↗	↘
• Pre-embedding	↗	↗	↗	↘
• Frozen tissue	→	↗	↗ or →	↘
• Semithin sections	↗	↗	↗	↘
Immunocytochemistry				
• Post-embedding				
– Direct method	↘	↗	↘	↘
– Indirect method	→	↘	↗	↗
• Pre-embedding				
– Direct method	→	↗	↘	↘
– Indirect method	↘	↘	↗	↗
• Frozen tissue				
– Direct method	↘	↗	↘	↘
– Indirect method	↗	↘	↗	↗
• Semithin sections				
– Direct method	↘	↗	↘	↘
– Indirect method	↗	↘	↗	↗
– Amplification	↗↗	↗	↗	↘
Staining				
Post-embedding	→	→	→	→
Pre-embedding	→	→	→	→
Frozen tissue	→	→	→	→
Semithin sections	→	↘	→	→
Observation				
Post-embedding				
Pre-embedding				
Frozen tissue				
• Electron microscopy	→	→	→	→
Semithin sections				
• Bright field	→	→	→	↘
• Epipolarization	↗	↘	↗	↗

Appendices

CONTENTS

——————— *A – EQUIPMENT* ———————

——————— *B – REAGENTS* ———————

This chapter documents the procedures for preparing materials for *in situ* hybridization to obtain the best morphology and preservation of nucleic acids.

—————— *A – EQUIPMENT* ——————

A1   PRACTICAL PRECAUTIONS

A1.1   RNase-Free Conditions

❏ *Working areas*

- Areas close to sterility

  ↝ Normally, for electron microscopy, samples are taken under conditions close to those considered RNase free.

  ↝ Reagents are available that destroy RNases.

❏ *Probe*

- Use of cRNA probes

  ↝ RNase activity is inhibited in the buffer (presence of RNasin).

  ↝ Commercial solutions for inhibiting RNases are available.

- Storage

  – DEPC-treated water
  – Choice of TE buffer (TE/NaCl)

  ↝ This buffer, which is very stable, inhibits the action of enzymes that break down DNA.

❏ *Tissue*

- Sampling in sterile conditions
- Storage of semithin sections

  ↝ Instruments must be sterile.

  ↝ Slides must be dried at RT prior to storage. RNases and DNases function only in the presence of water. Storage in the absence of water is the best inhibitor of RNases.

  – Semithin sections of tissue embedded in hydrophilic resin
  – Semithin frozen sections

  ↝ Place the dry slides in an airtight box containing desiccant (silica gel).

  ↝ At 4°C under a drop of sucrose used to pick up the sections

  ↝ Limited storage: 1–2 weeks

❏ *Equipment/Reagents/Solutions*

- Eppendorf tubes
- Sterilized pipette tips
- Gloves

  ↝ Disposable equipment is preferable to sterilized equipment.

  ↝ Do not touch the equipment or the reagents without gloves.

- New reagents

  ↬ Reserved exclusively for *in situ* hybridization.

- Recipients

  ↬ Reserved for *in situ* hybridization and sterilized immediately after use (otherwise, disposable equipment)

- Water treated with DEPC or containing RNase inhibitors

  ↬ The risk of contamination increases with time in relation to the frequency of opening.

❑ *Conclusion*

Avoiding RNase contamination is easier than destroying the RNase itself.

A1.2 Chemical Risks

Reagents/Resins

❑ *Reagents*

For electron microscopy, a number of products are used that are dangerous by contact and inhalation:

- Formaldehyde
- Glutaraldehyde
- Osmium tetroxide

Some organic solvents must be used in an area that vents to the outside.

Other products, such as:

- Formamide
- Lead nitrate
- Sodium cacodylate
- Uranyl acetate

are highly toxic if they are inhaled or ingested. They must be used with great care.

❑ *Resins*

Epoxy and acrylic resins are carcinogenic and highly toxic. Skin contact is dangerous.

Storage/Waste Products

Glassware, waste chemicals, and biological material for incineration are collected in special waste containers.

↬ In accordance with the manufacturer's instructions.

↬ **All experiments must be carried out under the fume hood.**

↬ Toxic vapors (double container)

↬ Toxic vapors (isoamyl acetate, xylene, acetone, etc.)

↬ α radiations

↬ Any trace of these products on the skin must be carefully washed off in running water.

↬ **Wear a mask.**

↬ Avoid all contact with the skin. In case of a spill, wash hands thoroughly with soapy water.

↬ Never mouth pipette these products.

A1.3   Radioactive Risks

Radioactive hazards are of different origins:

- Gloves

- Protection/control

- Storage of radioactive sources
- Radioprotection training courses

↝ Contact the person responsible.

↝ Change regularly to prevent contamination; gloves are not a barrier to radiation.
↝ Regularly check surfaces, screens, hands and equipment for contamination (^{35}S, ^{33}P).
↝ In containers stored in a special room
↝ Must be familiar with safety precautions

Table A1   Summary of Precautions to be Taken During the Use of Radioisotopes

Isotopes	Wear a Film Badge	Special Equipment	Risks, Controls
^{35}S	No	Screens: • Either 0.2 mm glass • Or 0.3 mm Plexiglas	No irradiation
^{33}P	No	2 pairs of gloves to be worn	Contamination

❏ *Precautions*
Soak contaminated material in a diluted solution of decontaminant and then wash in running water.

↝ Decontamination

❏ *Control/Wastes*
After radioactive decay (> 10 half-lives) waste can be disposed of as nonradioactive waste.

↝ Disposal must follow health and safety protocols.

A2   STERILIZATION

The sterilization of solutions and of small equipment is indispensable for *in situ* hybridization.

↝ Maintenance of clean conditions and the use of solutions exclusively for *in situ* hybridization

❏ *Equipment*

- Aluminum foil
- Autoclave
- Autoclave tape

- Oven

❏ *Minor equipment*
- Glass staining trays

↝ The surface pattern changes after sterilization.

↝ Glass equipment is autoclaved or sterilized in an oven at 50°C.

❑ *Protocol*

Treatment of equipment, including magnetic stirrers in:

• Oven	**2 h**	
	at 180°C	
• Autoclave	**2 h**	
	at 105°C	
	or	
	30 min	
	at 125°C	

↝ RNases are almost completely destroyed by this treatment.

↝ 2 bars; allow the equipment to cool in the oven (risk of breakage if it is placed on a cold surface).

A3   PRETREATMENT OF SLIDES/COVERSLIPS

A3.1   Pretreatment of Slides

❑ *Equipment*

- Metal forceps
- Oven or autoclave
- Trays, slide holders
- Slides

❑ *Reagents*

- Acetone
- Alcohol 95%
- 3-amino-propyl-tri-ethoxy-silane
- Hydrochloric acid 10 N
- Sterile distilled water

↝ *See* Appendix B1.1.

❑ *Solutions*

- Cleaning:
  Alcohol/HCl (5 mL HCl for 1 L 95% alcohol)

↝ Stable solution, which can be stored

- Treatment:
  3-amino-propyl-tri-ethoxy-silane at 2% in acetone

↝ The solution is unstable; prepare just before use.

❑ *Precaution*

Gloves should be worn throughout slide preparation.

❑ *Protocol*

1. Wash:
   - Alcohol/HCl   **overnight**
   - Running water   **1 h**
   - Distilled water   **1 min**
2. Dry the slides in an oven.   **180°C or 42°C**
                                15–60 min

↝ Sterilization at 180°C is optional.

3. Allow to cool.

4. Immerse in treatment solution.   **5–15 sec**

5. Wash:
   - Acetone   **2 × 1 min**
   - Sterile water   **1 min**

6. Dry   **overnight at 42°C**

❏ *Storage*

Up to a year at room temperature in dust-free conditions

A3.2   Slides for the Exposure of Ultrastructural Autoradiography

❏ *Materials*

- Diamond marker
- Glass slides   ⇝ Clean and grease free

❏ *Products*

- Superglue   ⇝ Water resistant
- Protective film   ⇝ Double-sided cellophane tape

❏ *Protocol*   ⇝ *See* Figure A1.

1. Cut the slide into 0.5-cm-wide strips   ⇝ With a diamond marker

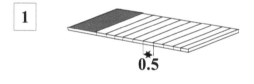

1 and 2 = Cut the slide into 0.5-cm-wide strips

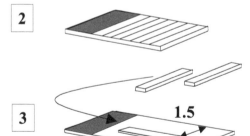

3 = Cover with double-sided cellophane tape
- **1.5 cm wide**

4 = Stick down the emulsion-coated grid

Figure A1   Preparation of slides.

2. Stick the glass strips to a glass slide

↪ Use superglue.

3. Dry.

4. Cover the glass strips with double-sided cellophane tape.

↪ Double-sided cellophane tape with a protective plastic film is best.
↪ Keep covered with the protective plastic film until use.

A3.3   Siliconized Glass Coverslips

1. Place the reaction mixture on the semithin sections:
   • Reaction mixture   **30 µL**

↪ Depends on the area covered by the sections

2. Cover with a sterile coverslip.

↪ To avoid bubbles position the coverslip carefully.

3. Seal the coverslip with rubber cement.

↪ **Optional**

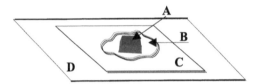

(A) Thick section
(B) Resin
(C) Siliconized coverslip
(D) Slide

Figure A2   Sealing coverslips.

4. Place the coverslips flat in a humid chamber.

A4   SUPPORT FILMS FOR GRIDS

A4.1   Aims

To maintain the stability of ultrathin sections under the electron beam, the grids may be coated in a nitrocellulose membrane (formvar or collodion). This support film should be as thin as possible.

The quality of a carbon membrane is limited by the following:

• Thickness
• Homogeneity
• Cleanliness
• Evenly fixed to the support film

↪ ≈ 20 to 30 nm in thickness
↪ Without holes
↪ Dirt interferes with observation of tissue.
↪ Prevents tearing and loss of sections

A4.2   Types of Films

Collodion Film

Collodion is a purified form of nitrocellulose:

• Formula

Its weak electric conductivity makes it fragile.

❑ *Protocol*

❑ *Advantages*

• Homogeneous film
• Easy to make

❑ *Disadvantages*

• Fragile
• Soluble in alcohol

Formvar Film

Formvar is usually used because of its superior strength. It can be used without carbon coating.

• Formula:

$$-CH_2-CH\,(OH)-CH_2-CH\,(OH)-$$

$$+ \;\; HCHO \;\; \Longrightarrow \;\; --CH_2-CH \quad CH-$$

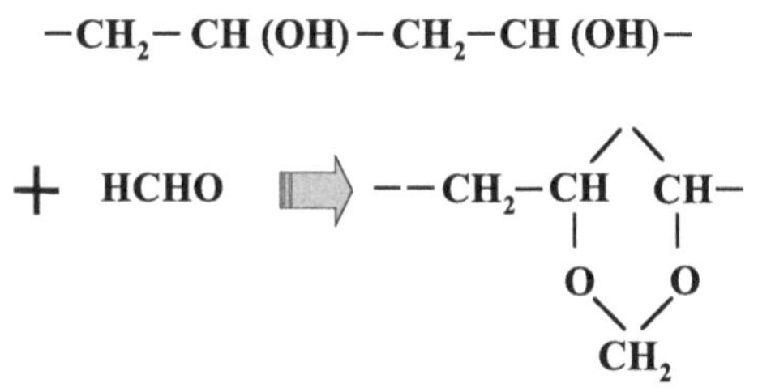

Figure A3 Formula of formvar.

❑ *Protocol*

❑ *Advantages*

• Stronger
• Not soluble in solvents

❑ *Disadvantages*

• Texture important

• Difficult to make

Carbon Coating

Carbon coating is used to strengthen an organic film. Carbon is evaporated under a vacuum (10^{-6} Torr) onto a collodion or formvar-coated grid.

❑ *Protocol*

❑ *Advantage*

• Stabilizes organic membrane

↪ Cellulose di-, -tri-, and tetranitrate

↪ Cellulose nitrate

↪ Carbon coating is recommended.

↪ *See* Appendix A4.3.3.

↪ Carbon coating is recommended.

↪ Stronger than collodion film

↪ Polyvinyl formula

↪ *See* Appendix A4.4.3.

↪ The texture of the membrane affects the carbon film.

↪ Carbon film

↪ *See* Appendix A4.5.3.

❏ *Disadvantages*
- Need specific equipment
- Extra stage in preparing grids

➴ Carbon coater

Table A2 Summary of the Different Characteristics of Support Films

	Organic Films		
Criteria	**Collodion**	**Formvar**	**Carbon**
Strength/stability	Fragile	Stronger	
Homogeneity	Very	Not at all	
Carbon coating	Yes	No	
Solubility	Soluble in alcohol	Insoluble	
Preparation	Easy	Difficult	Long

A4.3 Preparation of Collodion Films

A4.3.1 Summary of Different Stages

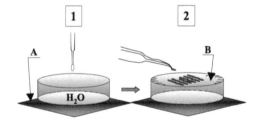

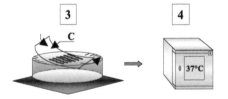

1 = **Place a drop of collodion onto the water surface in a wide bath.**
Wait several seconds
(**A**) Black background
2 = **Place grids onto film.**
(**B**) Collodion film

3 = **Pick up the collodion-coated grids.**
(**C**) Acetate film
4 = **Dry** (37°C).

❏ *Next stage:*
Carbon coating (*see* Figure A6)

Figure A4 Preparation of a collodion film.

A4.3.2 Materials/Solutions

1. Materials:
   - Acetate film or filter paper

➴ Acetate film is better since there are no loose fibers on its surface.
➴ Rapid filter paper is best to easily pick up collodion-coated grids.

- Fine forceps
- Nickel or gold grids ≥200 mesh

↪ For picking up grids

↪ The grids should be degreased in alcohol or acetone/ether ($^v/_v$); this ensures good adhesion between the film and the metal.

↪ The choice of grid depends on the type of protocol:

- Copper can react chemically, so nickel or gold is recommended.
- If the stability of the sample is important a fine grid is recommended (300 mesh or less). If a support film is not used, a finer mesh has to be used (> 300), which limits the surface that can be observed.

- Pasteur pipettes
- Petri dish or glass evaporating dish with a diameter of up to 10 cm

↪ The drop of collodion (or formvar) must spread without touching the edges of the container.

2. Solutions

- 2% collodion in isoamyl acetate

↪ Solution commercially available
↪ Solution for use is 0.5%.
↪ Collodion dissolved in ether is not suitable for making membranes.
↪ The solutions must be kept in tightly closed containers.

- Distilled water
- Isoamyl acetate

↪ Untreated or DEPC
↪ AR quality

A4.3.3 Protocol

↪ Drop technique (most widely used method) (*see* Figure A4)
↪ Advantage: rapid preparation

1. Fill an evaporating dish with distilled water.
2. Drop a single drop of 0.5% solution in the center of the water.

↪ The collodion forms a layer of equal thickness.

3. Wait a few seconds.

↪ The solvent evaporates and the collodion polymerizes.
↪ Any folded areas must be avoided.

4. Place the grids delicately on the surface of the film.

↪ The grids are placed shiny side down on the film.
↪ The edge of the forceps must not touch the film.

5. Pick up the collodion film with an acetate film

↪ Acetate film is better than filter paper because there are no loose fibers on its surface.

6. Dry in a dust-free atmosphere.   **37°C**

↪ A temperature higher than 37°C will damage the film.

7. Carbon coating

↪ **Necessary** (*see* Figure A6)

❏ *Storage*
- At room temperature
- In a dry, dust-free atmosphere

⇝ In a Petri dish
⇝ It is better to carbon coat the grids immediately after drying (*see* Appendix A4.5.3).

A4.4 Preparation of Formvar Film

A4.4.1 Summary of Different Stages

A formvar film is made on a glass slide and then
floated off on the surface of water. Grids are
then placed on the floating film.

⇝ Difficult technique that requires practice

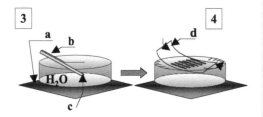

1 = **Dipping the slide in a formvar solution**
2 = **Drying** (at RT)

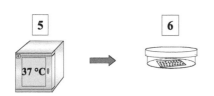

3 = **Floating off the film**
   Angle of the slide 30–40°
   (a) Black background
   (b) Slide
   (c) Formvar film
4 = **Picking up formvar-coated grids**
   (d) Filter paper

5 = **Drying** (37°C)
6 = **Storage:**
   - In a dry, dust-free atmosphere
   - At room temperature

Figure A5 Preparation of a formvar film.

A4.4.2 Materials/Products/Solutions

1. Materials:
   - Acetate film or filter paper

   - Deep glass container

   - Filter paper
   - Fine forceps
   - Glass tube
   - Nickel or gold grids ≥200 mesh
   - Scalpel
   - Slides for histology

⇝ Acetate film is better than filter paper
because it has no loose fibers on its surface.
⇝ Staining dish or evaporating dish diameter
is larger than 10 cm.
⇝ Rapid filter paper
⇝ For picking up grids
⇝ Clean and degreased
⇝ Copper is oxidized during hybridization.

⇝ Clean and degreased

2. Products:
   - Formvar powder

3. Solutions:

   - Distilled water
   - 0.15–0.25% formvar
   - Solvents

     – Dichloroethane
     – Dioxane

↝ Untreated or DEPC
↝ Solution for use
↝ Formvar is soluble in dichloroethane and dioxane.

A4.4.3   Protocol

1. Dilute formvar in dioxane or dichloroethane to make a 0.15 to 0.25% solution.
2. Filter the solution.
3. Clean a glass slide and grease lightly.

4. Dip the slide in the glass tube containing the formvar solution.

5. Remove and dry the slide vertically.

6. When the film is perfectly dry:
   - Delicately scrape the borders of the slide with a scalpel blade.
   - Remove the film by floating on the surface of clean water, slowly submerging the slide while holding it at an angle of 30 to 40°.
7. Place the grids on the film.

8. Lower a rectangle of acetate gently to pick up the grids. Lift the grids and formvar membrane gently from the surface of the water.
9. Dry.   **37°C**

↝ *See* Figure A5.

↝ The more concentrated the solution, the thicker the film.

↝ To eliminate all dust
↝ Allows the formation of the film

↝ The quality of the surface of the film depends largely on the surface of the glass slide.

↝ In the formation of the formvar film, it is important to avoid any variations in the thickness.

↝ To allow the film to be removed from the slide

↝ Place black paper under the glass dish to allow the film to be seen; film should be smooth and without wrinkles.

↝ The grids are placed shiny side down on the film.

↝ Any sudden movement will break or crease the film.

↝ A temperature higher than 37°C will damage the film.
↝ The grids can be used when they are perfectly dry.

❑ *Storage*

- At room temperature
- In a dry, dust-free atmosphere

A4.5 Preparation of a Carbon Coating

A4.5.1 Summary of Different Stages

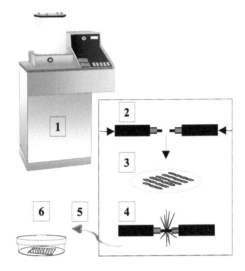

1 = **Carbon coater**

❑ *Parameters for evaporation:*
2 = **Size and position of the electrodes**
3 = **Distance between the electrodes and the grids:**
     ≈ 10 cm
4 = **Evaporation**
    • Intensity ≈ 100 A
    • Vacuum ≈ 10^{-6} Torr
5 = **Time of evaporation**
6 = **Storage**
    • At room temperature
    • In a dust-free atmosphere

Figure A6 **Carbon coating collodion-coated grids.**

A4.5.2 Materials/Products

• Carbon electrodes

↝ Single or multiple use; for the latter, the size must be regulated carefully. After adjusting the size, evaporate the carbon without coating grids to ensure an even coating the next time.

• Carbon coater

↝ Chamber under a vacuum in which the carbon electrodes are heated until evaporation.

• Sharpener for carbon electrodes

↝ Automatic sharpeners are available.

A4.5.3 Protocol

↝ *See* Figure A6.

1. Place the sharpened electrodes one opposite the other with the points just lightly touching.

↝ The electrodes are kept in permanent contact due to spring-loaded holders.

2. Place the collodion-coated grids in the carbon coater.

↝ The grids should be placed ≈ 10 cm from the electrodes.

3. Make a vacuum of 10^{-6} Torr.

↝ The vacuum is important in the evaporation of the carbon so that the coating is regular and consists of small particles.

4. Pass a current between the electrodes.   ≈ **100 A**

5. Evaporate.   **Some seconds**

↝ The time of evaporation for the carbon should be judged by placing a piece of white paper by the side of the grids and judging the thickness of the coating from the change in color.

⤳ The following formulas apply:
- $U = RI$
- $R = \rho \times L/S$ (L = length)
  - The intensity (I) is related to the size of the electrode.
  - For a given intensity, the smaller the section (S) of the electrode, the greater the resistance (R), the higher the temperature, and the better the evaporation of the carbon.

6. Wait before using the grids.　　　**1 h**

❑ *Storage*

- At room temperature
- In a dry, dust-free atmosphere

❑ *Usage*

The carbon film can be ionized just prior to use to allow ultrathin sections to be mounted on it easily.

A5   KNIFE-MAKING

A5.1   Apparatus

Glass knives for cutting frozen or resin (epoxy or acrylic) embedded tissues are made with a knife-maker.

⤳ A diamond knife may be used for cryo-ultramicrotomy and ultramicrotomy.

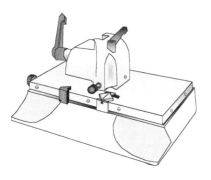

Figure A7   Knife-maker.

A5.2   Protocol

The knives are made from a block of glass 25 mm wide and 4.5 mm thick. After cleaning well with detergent, the glass is cut transversely in the knife-maker (*see* Figure A7) and broken into squares (*see* Figure A8).

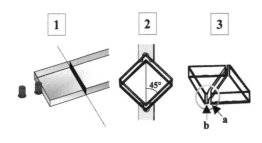

1 = Breaking the glass block into squares
2 = Making two knives by cutting the squares at 45°
3 = (a) a knife and its other corresponding half (b).

Figure A8 Making glass knives.

The squares are placed in the knife-maker and cut diagonally before being broken into two pieces, one of which will possess a straight sharp edge (*see* Figure A8).

☞ Tissue embedded in methacrylate and epoxy resin requires a knife angle of less than 50°. Routinely, a 45° angle is used.
☞ A knife angle larger than 45° is recommended for cutting very hard materials.

Small plastic boats are fixed onto the knife with nail polish (*see* Figure A9).

☞ The boats are filled with water onto which the freshly cut sections of tissue embedded in resin float.

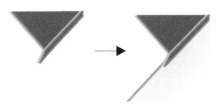

Figure A9 Mounting a plastic boat onto a glass knife.

The most important quality of a knife is that it have an edge as long as possible with no knife marks.
Straight convex breaks are best.

☞ Or microfractures

☞ Straight edges allow ribbons of sections to be cut.

Knives should be made just before use for cutting ultrathin sections.
Store away from dust.

☞ To ensure the best quality knife

☞ Fix the base of the knife into plasticine in the bottom of a Petri dish.

A5.3 Use

The tissue block is trimmed using the right-hand side of the razor edge (*see* Figure A10).
Cutting of ultrathin sections is carried out on the right-hand side of the cutting edge (*see* Figure A10).
Sections are cut moving progressively toward the right of the cutting edge.

☞ The part of the knife with the most knife marks
☞ Best part of the cutting edge

☞ When small knife marks appear on ultrathin sections, the block is moved progressively to the right of the knife (*see* Figure A10)

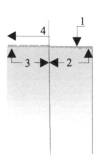

1 = **Cutting edge**
2 = **Part with most knife marks**
  • For smoothing the block
  • For cutting semithin sections
3 = **Part with least knife marks**
  • For cutting ultrathin sections
4 = **Cutting the first ultrathin sections** and the different positions possible on the best part of the cutting edge.

Figure A10   Different parts of the cutting edge.

A6   SMALL PIECES OF EQUIPMENT

A6.1   Making Comets

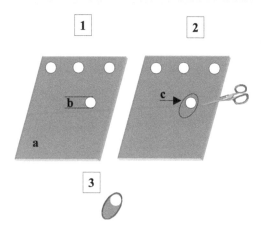

1 = **Making a hole for picking up sections**
  (a) Sheet of polypropylene 12/100 in thickness
  (b) Diameter of hole is 4 mm
2 = **Cut the comet with fine scissors.**
  (c) Final shape

3 = **Oval comet**

Figure A11   Making comets.

A6.2   Preparing Materials for Ultrastructural Autoradiography

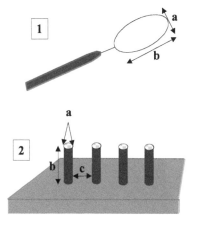

1 = **Tungsten ring (0.5 mm in diameter)**
  (a) 1 cm
  (b) 2 cm
2 = **Grid support made from metal, Plexiglas, or wood**
  (a) Diameter of grid is 3 mm
  (b) Height of support is 2 cm
  (c) Space between supports is 1 cm

Figure A12   Equipment used for ultrastructural autoradiography.

B – REAGENTS

B1   WATER

B1.1   Sterile

❑ *Reagents*
Distilled water

❑ *Precaution*
Do not keep after opening

❑ *Protocol*
Sterilize in an autoclave   **2 h**
                                        at 105°C

❑ *Storage*
At room temperature

↝ Sufficient for the detection of double-stranded DNA (e.g., viruses).

B1.2   Diethylpyrocarbonate (DEPC)

❑ *Reagents/Solutions*

• DEPC
• Distilled water

❑ *Precaution*
DEPC is dangerous.

❑ *Protocol*

1. Mix

• DEPC   **0.5–1 mL**
• Water   **1000 mL**

2. Shake under a fume hood   **overnight**
3. Sterilize in an autoclave   **30 min**
                              at 105°C

❑ *Storage*
At room temperature

↝ **Indispensable** with cRNA probes

↝ $C_6H_{10}O_5$

↝ MW = 162.1

B2 SOLUTIONS

B2.1 7.5 *M* Ammonium Acetate

⇨ Stock solution

❑ *Reagents/Solutions*

- Ammonium acetate
- 10 N hydrochloric acid
- Sterile water

⇨ $CH_3 CO_2NH_4$

⇨ *See* Appendix B1.1.

❑ *Precautions*

- Check to make sure the powder is dry.
- Volatile reagent

❑ *Protocol*

- Ammonium acetate   **57.81 g**
- Sterile water   **80 mL**

⇨ MW = 77.08

  1. Mix
  2. Adjust the pH to 5.5 with HCl.

- Sterile water   **to 100 mL**

❑ *Sterilization*
Sterilize in an autoclave.   **2 h**
          at 105°C

❑ *Storage*

- At –20°C in aliquots of 50 to 100 μL
- At room temperature for some months

B2.2 3 *M* Sodium Acetate

⇨ Stock solution

❑ *Reagents/Solutions*

⇨ AR quality, to be used only for *in situ* hybridization

- Glacial acetic acid
- Sodium acetate
- Sterile water

⇨ $CH_3 CO_2H$
⇨ $CH_3 COONa$
⇨ *See* Appendix B1.1.

❑ *Precautions*
None

❑ *Protocol*

- Sodium acetate   **24.6 g**
- Sterile water   **80 mL**

⇨ MW = 82.03

  1. Mix with magnetic stirrer.
  2. Adjust the pH to 4.8 with acetic acid.

- Sterile water   **to 100 mL**

❑ *Sterilization*

Sterilize in an autoclave.   **2 h**
                                          at 105°C

❑ *Storage*

Store at room temperature for some months, in aliquots of 50 to 100 µL.

B2.3   DNA

↝ Stock solution; 10 mg/mL sterile water

❑ *Reagents/Solutions*

↝ Use commercially available preparations.

- Herring sperm DNA
- Salmon sperm DNA
- Sterile water

↝ *See* Appendix B1.1.

❑ *Precaution*
Risk of bacterial contamination

❑ *Protocol*

- DNA   **10 mg**
- Sterile water   **to 1 mL**

Sonicate for 10 min.   ↝ Maximum power

❑ *Storage*

Store at –20°C.   ↝ May be thawed and refrozen

B2.4   2% Agarose

❑ *Products/Solutions*

- Agarose   ↝ Molecular biology quality
- Fixation buffer   ↝ PBS buffer, phosphate or cacodylate
                                         ↝ *See* Appendix B3.1 and 3.4.

❑ *Precaution*

Cover the container when the agarose is being dissolved to avoid evaporation.

❑ *Protocol*

1. Dissolve the agarose in the fixation buffer: ↝ Use a water bath

   - Agarose   **2 g**
   - Fixation buffer   **100 mL**

2. After dissolving the agarose, keep the solution in an oven.

❑ *Storage*

Store in an oven.   **56°C**   ↝ Keep at this temperature until use.

B2.5   RNA

❑ *Reagents/Solutions*

• DEPC-treated water
• Yeast tRNA

❑ *Precaution*
Risk of hydrolysis

❑ *Protocol*

• RNA	**10 mg**
• DEPC-treated water	**to 1 mL**

Sonicate for 10 min.

❑ *Storage*
Store at –20°C.

↝ Stock solution; 10 mg/mL sterile water

↝ *See* Appendix B1.2.

↝ Maximum power

↝ May be thawed and refrozen

B2.6   Calcium Chloride

B2.6.1   1 *M* Calcium Chloride

❑ *Reagents/Solutions*

• Calcium chloride
• Sterile water

❑ *Precautions*
None

❑ *Protocol*

• Calcium chloride	**14.7 g**
• Sterile water	**to 100 mL**

❑ *Sterilization*

Sterilize in an autoclave.	**2 h**
	at 105°C

❑ *Storage*
Store at room temperature.

↝ Stock solution

↝ AR quality, to be used only for *in situ* hybridization
↝ $CaCl_2 \cdot 2H_2O$
↝ *See* Appendix B1.1.

↝ MW = 147.02

B2.6.2   Calcium Chloride/Cobalt Chloride

❑ *Reagents/Solutions*

• Calcium chloride
• Cobalt chloride
• Sterile water

❑ *Precautions*
None

↝ Stock solution; 2 m*M* $CaCl_2$/2 m*M* $CoCl_2$

↝ $CaCl_2 \cdot 2H_2O$
↝ $CoCl_2$
↝ *See* Appendix B1.1.

❑ *Protocol*

• Calcium chloride	**0.029 g**	➥ MW = 147.02
• Cobalt chloride	**0.025 g**	➥ MW = 129.83
• Sterile water	**to 100 mL**	

❑ *Sterilization*

Sterilize in an autoclave   **2 h**
at 105°C

❑ *Storage*
Store at room temperature.

B2.7   4 *M* Lithium Chloride   ➥ Stock solution

❑ *Reagents/Solutions*   ➥ AR quality; to be used only for *in situ* hybridization

• Lithium chloride	➥ LiCl
• Sterile water	➥ *See* Appendix B1.1.

❑ *Precaution*
Risk of bacterial contamination

❑ *Protocol*

• Lithium chloride	**8.48 g**	➥ MW = 42.39
• Sterile water	**to 50 mL**	

1. Mix
2. Filter on a 0.22 μm filter

❑ *Storage*
Store at 4°C for a few weeks.

B2.8   1 *M* Magnesium Chloride   ➥ Stock solution

❑ *Reagents/Solutions*   ➥ AR quality; to be used only for *in situ* hybridization

• Magnesium chloride	➥ MgCl$_2$ · 6 H$_2$O
• Sterile water	➥ *See* Appendix B1.1.

❑ *Precautions*
None

❑ *Protocol*

• Magnesium chloride	**20.3 g**	➥ MW = 203.30
• Sterile water	**to 100 mL**	

❑ *Sterilization*

Sterilize in an autoclave.   **2 h**
at 105°C

❑ *Storage*
Store at room temperature.

B2.9 5 *M* Sodium Chloride

⇨ Stock solution

❑ *Reagents/Solutions*

- Sodium chloride
- Sterile water

⇨ AR or molecular biology quality, to be used only for *in situ* hybridization
⇨ NaCl
⇨ *See* Appendix B1.1.

❑ *Precautions*
None

❑ *Protocol*

- Sodium chloride **14.6 g**
- Sterile water **to 50 mL**

⇨ MW = 58.44

❑ *Sterilization*
Sterilize in an autoclave. **2 h at 105°C**

❑ *Storage*
Store at room temperature.

B2.10 0.1% Collagenase in PBS

⇨ Solution for use

❑ *Products/Solutions*

- Collagenase

⇨ Each new batch must be assessed before use.
⇨ 0.25% trypsin in PBS may be used.

- PBS

⇨ *See* Appendix B3.4.3.

❑ *Precautions*
None

❑ *Protocol (immediate)*

⇨ Start with a fresh aliquot.

- Collagenase **1 mg**
- PBS **to 1 mL**

⇨ Start with a dry aliquot.

❑ *Storage*
Store 1 mg dry at –80°C.

⇨ Dry powder

B2.11 1 *M* Dithiothreitol (DTT)

⇨ Stock solution

❑ *Reagents/Solutions*

⇨ Molecular biology quality

- DTT
- Sterile water

⇨ $C_4H_{10}O_2S_2$
⇨ *See* Appendix B1.1.

❑ *Precautions*

- Very volatile reagent; unstable at room temperature
- Does not resist denaturation

⇨ Is denatured quickly in the hybridization buffer

❏ *Protocol*

1. In a sterile Eppendorf tube, place:

• Dithiothreitol	**1.54 g**	⇝ MW = 154.24
• Sterile water	**10 mL**	

2. Add

❏ *Storage*
Store at –20°C in aliquots of 100 μL.   ⇝ Must never be refrozen; the strong smell indicates that the product is in good condition.

B2.12 DNase I

⇝ Stock solution; 1 mg/mL

❏ *Reagents/Solutions*

⇝ Molecular biology quality

• DNase I   ⇝ Check the activity of the enzyme.
• Sterile water   ⇝ *See* Appendix B1.1.

❏ *Precaution*
Risk of RNase contamination

❏ *Protocol*
Dissolve:

• DNase I	**1 mg**	⇝ To be expressed in units.
• Sterile water	**1 mL**	

❏ *Storage*
Store at –20°C in aliquots of 100 μL.

B2.13 500 m*M* Ethylene Diamine Tetra-acetic Acid (EDTA)

⇝ Stock solution

❏ *Reagents/Solutions*

⇝ AR quality; to be used only for *in situ* hybridization

• EDTA   ⇝ $C_{10}H_{14}N_2O_8Na_2 \cdot 2H_2O$ or Titriplex III
• 10 N Sodium hydroxide   ⇝ *See* Appendix B2.20.
• Sterile water   ⇝ *See* Appendix B1.1.

❏ *Precaution*
Toxic reagent

❏ *Protocol*

1. Dissolve:

• EDTA	**186 g**	⇝ MW = 372.24
• Sterile water	**to 1 L**	

2. Adjust the pH to 8.0 with sodium hydroxide.

❑ *Sterilization*
Sterilize in an autoclave. **2 h**
at 105°C

❑ *Storage*
Store at room temperature, in aliquots.

B2.14 Deionized Formamide

❑ *Equipment*

- Sterile flask
- Whatman filter N° 1M

❑ *Reagents/Solutions*

- Amberlite resin, 20–50 mesh
- Formamide

❑ *Precaution*
Avoid contact

❑ *Protocol*

- Amberlite **5 g**
- Formamide filter **50 mL**

  1. Shake gently. **30 min**
  2. Filter on Whatman paper.

❑ *Storage*
Store at –20°C in sterile 1 mL tubes in light-free conditions.

⇝ Molecular biology quality

⇝ CH_3NO

⇝ Commercially available solution

⇝ MW = 45.04

⇝ Do not thaw. If it is liquid at –20°C, it must not be used.

B2.15 Proteinase K

⇝ Stock solution; 10 mg/mL

❑ *Reagents/Solutions*

- Proteinase K
- Sterile water

❑ *Precaution*
Weak proteolysis at room temperature

❑ *Protocol*

1. In a sterile Eppendorf tube, place:

   - Proteinase K **10 mg**
   - Sterile water **1 mL**

2. Mix.

❑ *Storage*
Store at –20°C in aliquots of 100 µL.

⇝ *See* Appendix B1.1.

⇝ Use commercially available solutions.

⇝ Dilute in Tris–HCl/CaCl$_2$ buffer (*see* Appendix B3.7.2).

B2.16 RNase A

⇒ Stock solution; 100 µg/mL

❑ *Reagents/Solutions*

⇒ Molecular biology quality

- RNase A ⇒ Bovine pancreas ribonuclease
- Sterile water ⇒ *See* Appendix B1.1.
- TE/NaCl buffer ⇒ *See* Appendix B3.6.2.

❑ *Precaution*

Risk of DNase contamination

❑ *Protocol*

• RNase A	**100 µg**
• Sterile water	**1 mL**

❑ *Storage*

Store at –20°C in aliquots of 40 µL.

B2.17 Sucrose

⇒ Solution for use: 0.4 to 2.3 *M*

❑ *Reagents/Solutions*

- Phosphate buffer or PBS ⇒ *See* Appendix B3.4.
- Sucrose ⇒ $C_{12}H_{22}O_{11}$

❑ *Precaution*

Risk of contamination

❑ *Protocol*

Sucrose 0.4 *M*

• Sucrose	**13.70 g**	⇒ MW = 342.34
• Buffer	**to 100 mL**	

Sucrose 2.3 *M*

• Sucrose	**78.73 g**	⇒ MW = 342.34
• Buffer	**to 100 mL**	

Stir regularly. ⇒ Dilute slowly over several days.

❑ *Storage*

It is best to store the solution at 4°C.

B2.18 Sarcosyl

⇒ Stock solution 20% buffered

❑ *Reagents*

- Phosphate buffer or PBS ⇒ *See* Appendix B3.4.
- Sarcosyl ⇒ 0.2% saponin may be used.

❑ *Precaution*

Toxic

❏ *Protocol*

• Sarcosyl	**20 mL**
• Sterile water	**80 mL**

❏ *Storage*
Store at 4°C for some weeks.

B2.19   50X Denhardt's Solution

⇨ Stock solution

❏ *Reagents/Solutions*

- Bovine serum albumin (fraction V)
- Ficoll 400
- Polyvinylpyrrolidone
- Sterile water

⇨ BSA, crystallized 5 times

⇨ Synthetic polymer of sucrose
⇨ *See* Appendix B1.1.

❏ *Precaution*
Risk of bacterial contamination

❏ *Protocol*

1. Add:

• BSA	**1 g**
• Ficoll 400	**1 g**
• Polyvinylpyrrolidone	**1 g**
• Sterile water	**to 100 mL**

2. Leave the mixture to hydrate overnight before shaking.
3. Shake the mixture gently and intermittently for some days.

❏ *Storage*
Store at –20°C in aliquots.

⇨ Can be thawed and refrozen

B2.20   Sodium Hydroxide

⇨ Stock solution: 10 N

❏ *Reagents/Solutions*

- Sodium hydroxide
- Sterile water

⇨ AR quality

⇨ NaOH
⇨ *See* Appendix B1.1.

❏ *Precaution*
Extremely corrosive reagent

❏ *Protocol*

• Sodium hydroxide	**40 g**
• Sterile water	**to 100 mL**

⇨ MW = 40.00

Mix slowly.

❏ *Storage*
Store at room temperature.

B2.21   Dextran Sulfate

↪ Stock solution: 50%

❑ *Reagents/Solutions*

↪ Molecular biology quality

- Dextran sulfate
- Sterile water

↪ *See* Appendix B1.1.

❑ *Precaution*

Reagent is very difficult to pipette. Prepare only the quantity necessary.

❑ *Protocol*

1. In a sterile Eppendorf tube, place:
   - Dextran sulfate   **50 mg**   ↪ MW = 500.00
   - Sterile water   **65 µL**
2. Mix.
3. Microcentrifuge for a few seconds.

❑ *Storage*

Store at –20°C, in aliquots of 100 µL.

↪ It is recommended that it be stored in aliquots to which the other components of the hybridization mixture should be added.

B2.22   1 *M* Tris

↪ Stock solution

❑ *Reagents/Solutions*

- 10 N hydrochloric acid
- Sterile water   ↪ *See* Appendix B1.1.
- Tris hydroxymethyl aminomethane   ↪ $C_4H_{11}NO_3$; Tris base in powder

❑ *Precautions*

None

❑ *Protocol*

Dissolve:

- Tris base   **121 g**   ↪ MW = 121.16
- Sterile water   **to 1 L**

❑ *Storage*

Store at room temperature.

↪ Some weeks

B2.23   Trypsin ± EDTA

↪ Stock solution; 0.25% trypsin ± 0.05% EDTA in PBS

❑ *Products/Solutions*

↪ AR quality

- PBS   ↪ *See* Appendix B3.4.3.
- EDTA   ↪ $C_{10}H_{14}N_2O_8Na_2 \cdot 2H_2O$ (ethylene diamine tetra-acetic acid, or Titriplex III) or 500 m*M* solution stock (*see* Appendix B2.13).

- Trypsin
- Sterile water

➭ *See* Appendix B1.1.

❏ *Precaution*
Avoid contact with the skin

❏ *Protocol*
Dissolve:

• Trypsin	**25 mg**
• ± EDTA	**5 mg**
• PBS	**to 10 mL**

➭ Not indispensable

❏ *Storage*
Store in ready-to-use solution at –20°C.

➭ May be prepared just prior to use

B3   BUFFERS

B3.1   Cacodylate Buffer

➭ 200 mM cacodylate buffer; pH 7.4

❏ *Products/Solutions*

- N hydrochloric acid
- Sodium cacodylate
- Sterile water

➭ AR quality

➭ Na(CH$_3$)$_2$AsO$_2$ · 3H$_2$O
➭ *See* Appendix B1.1.

❏ *Precautions*
None

❏ *Protocol*

1. Dissolve:

• Sodium cacodylate	**4.28 g**
• Sterile water	**to 100 mL**

➭ MW = 214.05

2. Check to make sure the pH is 7.4.

❏ *Sterilization*
Sterilize in an autoclave.   **2 h**
                            at 105°C

❏ *Storage*
Store at 4°C.

B3.2   100 mM Acetylation Buffer

➭ 100 mM of a triethanolamine buffer; pH 8.0

❏ *Reagents/Solutions*

- 10 N hydrochloric acid
- Sterile water
- Triethanolamine

➭ AR quality

➭ *See* Appendix B1.1.
➭ C$_6$H$_{15}$NO$_3$ [Tris (hydroxy-2-ethyl)-amine]

❑ *Precaution*
Toxic

❑ *Protocol*

• Triethanolamine	**3.32 mL**
• Sterile water	**200 mL**

⇝ MW = 149.19

Adjust: **pH 8.0**

• HCl	**≈ 1 mL**
• Sterile water	**to 250 mL**

❑ *Storage*
Store at 4°C.

B3.3   DNase I Buffer 10X

⇝ Stock solution

❑ *Reagents/Solutions*

⇝ AR quality

• Bovine serum albumin
• 10 N hydrochloric acid
• 500 m*M* magnesium chloride
• Sterile water
• 1 *M* Tris

⇝ BSA 5X, crystallized

⇝ *See* Appendix B2.8.
⇝ *See* Appendix B1.1.
⇝ *See* Appendix B2.22.

❑ *Precaution*
Risk of RNase contamination; prepare under sterile conditions

❑ *Protocol*

• Tris–HCl; pH 7.6	**500 m*M***
• MgCl₂	**100 m*M***
• BSA	**0.5 mg /mL**

⇝ *See* Appendix B3.7.1.

❑ *Storage*
Store at –20°C in aliquots.

B3.4   Phosphate Buffer

B3.4.1   1 *M* Phosphate

⇝ Stock solution

❑ *Reagents/Solutions*

⇝ AR quality

• Disodium phosphate

⇝ Na₂HPO₄. Hydrated reagents can be used, taking into account variations in molar mass.

• Monosodium phosphate
• Sterile water

⇝ NaH₂PO₄ · H₂O
⇝ *See* Appendix B1.1.

❑ *Precaution*
Preparations obtained in sachets should be transferred to sterile containers, with sterile water.

❏ *Protocol*

1. Dissolve:
   - Disodium phosphate   **14.19 g**   ⇨ MW = 141.96
   - Monosodium phosphate   **13.8 g**   ⇨ MW = 138.00
   - Sterile water   **to 100 mL**
2. Check.   **pH 7.4**

❏ *Sterilization*
Sterilize in an autoclave.   **2 h**
    at 105°C

❏ *Storage*
Store at room temperature or 4°C.

B3.4.2   Phosphate/NaCl

⇨ 100 mM phosphate buffer/300 mM NaCl; pH 7.4

❏ *Products/Solutions*

⇨ AR quality

- 1 M phosphate buffer   ⇨ *See* Appendix B3.4.1.
- 5 M sodium chloride   ⇨ *See* Appendix B2.9.
- Sterile water   ⇨ *See* Appendix B1.1.

❏ *Precautions*
None

❏ *Protocol*

1. Add:
   - 5 M sodium chloride   **6 mL**
   - 1 M phosphate buffer   **10 mL**
   - Sterile water   **to 100 mL**
2. Check.   **pH 7.4**

❏ *Sterilization*
Sterilize in an autoclave.   **2 h**
    at 105°C

❏ *Storage*
Store at room temperature or 4°C.

B3.4.3   PBS

⇨ Phosphate buffered saline
⇨ 10 mM mono-di-phosphate buffer/150 mM NaCl/10 mM KCl; pH 7.4
⇨ Can be prepared in stock solution, 10X

❏ *Reagents/Solutions*

⇨ AR quality; to be used only for *in situ* hybridization

- Disodium phosphate   ⇨ $Na_2HPO_4 \cdot 12\ H_2O$
- Monopotassium phosphate   ⇨ $KH_2PO_4 \cdot 3\ H_2O$
- Potassium chloride   ⇨ KCl
- Sodium chloride   ⇨ NaCl
    ⇨ *See* Appendix B2.9.

- 10 N sodium hydroxide
- Sterile water

⤳ *See* Appendix B2.20.
⤳ *See* Appendix B1.1.

❑ *Precaution*
Risks of contamination

❑ *Protocol*

1. Add:
   - Sodium chloride   **8.76 g**   ⤳ MW = 58.44
   - Potassium chloride   **0.74 g**   ⤳ MW =74.56
   - Disodium phosphate   **3.58 g**   ⤳ MW = 358.14
   - Monopotassium phosphate **1.90 g**   ⤳ MW = 190.15
   - Sterile water   **to 1 L**
2. Adjust the pH with   **7.4**
   sodium hydroxide.

❑ *Sterilization*
Sterilize in an autoclave.   **2 h**
                              at 105°C

❑ *Storage*
Store at room temperature until opened, then at 4°C.

B3.5   SSC Buffer 20X

⤳ Standard saline citrate
⤳ Stock solution

⤳ 3 *M* sodium chloride; 300 m*M* sodium citrate; pH 7.0
⤳ AR quality

❑ *Reagents/Solutions*

- 10 N sodium hydroxide
- Sodium chloride
- Sodium citrate
- Sterile water

⤳ *See* Appendix B2.20
⤳ NaCl
⤳ $C_6H_5Na_3O_7 \cdot 2H_2O$
⤳ *See* Appendix B1.1.

❑ *Precautions*
None

❑ *Protocol*

1. Add:
   - Sodium chloride   **175.3 g**   ⤳ MW = 58.44
   - Sodium citrate   **88.2 g**   ⤳ MW = 294.10
   - Sterile water   **to 1 L**
2. Adjust the pH with   **7.0**
   sodium hydroxide.

❑ *Sterilization*
Sterilize in an autoclave.   **2 h**
                              at 105°C

❑ *Storage*
Store at room temperature prior to opening, then at 4°C.

B3.6 TE (Tris–EDTA) Buffer

❏ *Reagents*

- EDTA

- Tris (hydroxymethyl–aminomethane)

↝ Tris/EDTA

↝ AR quality; to be used only for *in situ* hybridization

↝ $C_{10}H_{14}N_2O_8Na_2 \cdot 2H_2O$ (ethylene diamine tetra-acetic acid, or Titriplex III)

↝ $C_4H_{11}NO_3$

B3.6.1 TE Buffer 10X

❏ *Reagents/Solutions*

- EDTA

- 10 N hydrochloric acid
- Sterile water
- Tris

↝ Stock solution
↝ 100 mM Tris–HCl/10 mM EDTA; pH 7.6

↝ $C_{10}H_{14}N_2O_8Na_2 \cdot 2H_2O$ or 500 mM stock solution (*see* Appendix B2.13)

↝ *See* Appendix B1.1.
↝ $C_4H_{11}NO_3$ or 1 M stock solution (*see* Appendix B2.22)

❏ *Precaution*
Toxicity of EDTA

❏ *Protocol*

1. Add:
   - Tris   **12.1 g**
   - EDTA   **3.7 g**
2. Adjust the pH with HCl: **7.6**

   - Sterile water   **to 1 L**

↝ MW = 121.16
↝ MW = 372.24

❏ *Sterilization*
Sterilize in an autoclave.   **2 h**
                            at 105°C

❏ *Storage*
Store at room temperature prior to opening, then at 4°C

B3.6.2 TE–NaCl Buffer 10X

❏ *Products/Solutions*

- 10 N hydrochloric acid
- Sodium chloride
- Sterile water
- 10X TE buffer

❏ *Precaution*
Toxicity of EDTA

↝ 100 mM Tris–HCl/10 mM EDTA/5 M NaCl; pH 7.6

↝ AR quality

↝ NaCl
↝ *See* Appendix B1.1.
↝ *See* Appendix B3.6.1.

❑ *Protocol*

1. Add:
   - Sodium chloride   **29.2 g**   ↪ MW = 58.44
   - 10X TE buffer   **to 100 mL**
2. Adjust the pH with HCl.   **7.6**

❑ *Sterilization*
Store in an autoclave.   **2 h**
                      at 105°C

❑ *Storage*
Store at room temperature until opened, then at 4°C.

B3.7   Tris–HCl Buffer

B3.7.1   100 m*M* Tris–HCl Buffer

❑ *Reagents/Solutions*   ↪ AR quality

- 10 N hydrochloric acid
- Sterile water   ↪ *See* Appendix B1.1.
- 1 *M* Tris   ↪ *See* Appendix B2.22.

❑ *Precaution*
Risk of contamination by exogenous organisms

❑ *Protocol*

1. Add:
   - Tris   **1 vol**
   - Sterile water   **9 vol**
2. Adjust the pH with HCl:
   - pH   **7.6**
   - pH   **8.5**

❑ *Sterilization*   ↪ Very difficult, if not impossible

❑ *Storage*
Store at 4°C.

B3.7.2   Tris–HCl/CaCl₂ Buffer   ↪ 20 m*M* Tris–HCl/2 m*M* CaCl$_2$; pH 7.6

❑ *Reagents/Solutions*   ↪ AR quality

- 1 *M* calcium chloride   ↪ *See* Appendix B2.6.
- 10 N hydrochloric acid
- Sterile water   ↪ *See* Appendix B1.1.
- 1 *M* Tris   ↪ *See* Appendix B2.22.

❑ *Precaution*
Risk of contamination

❏ *Protocol*

1. Add:
   - Tris   **20 mL**
   - Calcium chloride   **2 mL**
2. Adjust the pH with HCl: **7.6**
   - Sterile water   **to 1 L**

❏ *Sterilization*

↝ Very difficult, if not impossible

❏ *Storage*
Store at 4°C.

B3.7.3   Tris–HCl/Glycine Buffer

↝ 50 mM Tris–HCl/50 mM glycine; pH 7.4

❏ *Reagents/Solutions*

↝ AR quality

- Glycine
- 10 N hydrochloric acid
- Sterile water
- 1 M Tris

↝ $C_2H_5NO_2$

↝ *See* Appendix B1.1.
↝ *See* Appendix B2.22.

❏ *Precaution*
Risk of bacterial contamination

❏ *Protocol*

1. Add:
   - Glycine   **0.37 g**
   - Tris   **5 mL**
   - Sterile water   **to 100 mL**
2. Adjust the pH with HCl.   **7.4**

↝ MW = 75.07

❏ *Sterilization*

↝ Very difficult, if not impossible

❏ *Storage*

- At –20°C in aliquots
- At 4°C

B3.7.4   Tris–HCl/MgCl$_2$ Buffer

↝ 10 mM Tris–HCl/5 mM MgCl$_2$; pH 7.3 (dilution buffer for DNase)

❏ *Reagents/Solutions*

↝ AR quality

- 10 N hydrochloric acid
- 1 M magnesium chloride
- Sterile water
- 1 M Tris

↝ *See* Appendix B2.8.
↝ *See* Appendix B1.1.
↝ *See* Appendix B2.22.

❏ *Precaution*
Risk of bacterial contamination

❏ *Protocol*

1. Add:
   - 1 M Tris   **1 mL**
   - 1 M MgCl$_2$ **500 µL**
2. Adjust the pH with HCl.   **7.3**
   - Sterile water   **to 100 mL**

❑ *Storage*

- At –20°C in aliquots
- At 4°C

B3.7.5   Tris–HCl/NaCl Buffer

↝ 20 mM Tris–HCl/300 mM NaCl; pH 7.6
↝ Visualization buffer

❑ *Reagents/Solutions*

↝ AR quality

- 10 N hydrochloric acid
- 5 M sodium chloride
- Sterile water
- 1 M Tris

↝ *See* Appendix B2.9.
↝ *See* Appendix B1.1.
↝ *See* Appendix B2.22.

❑ *Precaution*
Risk of bacterial contamination

❑ *Protocol*

1. Add:
   - 5 M sodium chloride   **6 mL**
   - 1 M Tris   **2 mL**
   - Sterile water   **to 100 mL**
2. Adjust the pH with HCl.   **7.6**

❑ *Storage*
Prepare just before use.

B3.7.6   Tris–HCl/NaCl/MgCl$_2$ Buffer

↝ 20 mM Tris–HCl/300 mM NaCl/50 mM MgCl$_2$

❑ *Reagents/Solutions*

↝ AR quality

- 10 N hydrochloric acid
- 1 M magnesium chloride
- 5 M sodium chloride
- Sterile water
- 1 M Tris

↝ *See* Appendix B2.8.
↝ *See* Appendix B2.9.
↝ *See* Appendix B1.1.
↝ *See* Appendix B2.22.

❑ *Precaution*
Risk of bacterial contamination

❑ *Protocol*
1. Add:
   - 1 M Tris   **20 mL**
   - 5 M NaCl   **60 mL**
   - 1 M MgCl$_2$ **50 mL**
2. Adjust the pH with HCl:

   - pH   **7.6**   ↝ Visualization of peroxidase
   - pH   **9.5**   ↝ Visualization of alkaline phosphatase
   - Sterile water   **to 1 L**

❑ *Storage*
Prepare extemporaneously.

B4   FIXATIVES

B4.1   2% Osmium Tetroxide

❑ *Products/Solutions*

- Osmium tetroxide

- Sterile water

❑ *Precautions*

- Toxic
- Volatile and dangerous chemical

❑ *Protocol*
1. In a tightly sealed glass container, place:
   - Sterile water   **25 mL**
2. Clean the ampoule thoroughly using detergent; rinse carefully with distilled water and dry.
3. Break the ampoule, placing the two halves carefully in the container:
   - Osmic acid   **0.5 g**
4. Close the glass container and place inside a tightly closed tin; leave to dissolve at RT.
5. Treat the container with care.

❑ *Storage*

- Double container
- At room temperature under fume hood.

⇝ Stock solution

⇝ Electron microscope quality

⇝ OsO_4; improperly called osmic acid
⇝ Commercially available in breakable sealed ampoules or in a 2% solution
⇝ *See* Appendix B1.1.

⇝ Use under a fume hood.

⇝ Use a paper towel that does not shed fibers.

⇝ It is better to break the ampoule inside the glass container.
⇝ MW = 254.20
⇝ It dissolves more rapidly at 37°C.

B4.2   2.5% Glutaraldehyde

❑ *Products/Solutions*

- 25–70% glutaraldehyde

- 200 mM phosphate buffer

❑ *Precaution*
Volatile and dangerous chemical

❑ *Protocol*
Mix:

- 25% glutaraldehyde   **1 mL**

⇝ Solution should be prepared just before use.

⇝ Electron microscope quality

⇝ $OCH(CH_2)_3CHO$
⇝ The higher the concentration, the more stable the solution.
⇝ *See* Appendix B3.4.1.

⇝ Use under a fume hood.

⇝ Depending on the concentration of the original solution

- Phosphate buffer   **5 mL**
- Distilled water   **to 10 mL**

❑ *Storage*
Use immediately after preparation.   ↬ A couple of hours at 4°C

B4.3   Paraformaldehyde

❑ *Equipment*

- Hood
- Magnetic stirrer, with heat

❑ *Reagents/Solutions*   ↬ AR quality

- Distilled water   ↬ *See* Appendix B1.1.
- Paraformaldehyde in powder   ↬ $HO(CH_2O)_nH$
                                                     ↬ In the form of nonhydrated powder
- 1 *M* phosphate buffer   ↬ *See* Appendix B3.4.1.
- 10 N sodium hydroxide   ↬ *See* Appendix B2.20.

B4.3.1   4% Paraformaldehyde

❑ *Precaution*
Manipulate under fume hood

❑ *Protocol*

1. In a glass flask, place:
   - Paraformaldehyde   **4 g**
   - Distilled water   **50 mL**
2. Stir while heating.   **60°C**
3. Allow to cool, and add:
   - 1 *M* phosphate buffer   **10 mL**   ↬ *See* Appendix B3.4.1.
4. Adjust the pH with NaOH: **pH 7.4**
   - Distilled water   **to 100 mL**
5. Filter and allow to cool.

❑ *Storage*

- To be used immediately
- At –20°C   ↬ Do not refreeze.

B4.3.2   4% Paraformaldehyde/0.05% Gluta-   ↬ Working solution
raldehyde

❑ *Reagents/Solutions*

- Glutaraldehyde 25–75%   ↬ $OCH(CH_2)_3CHO$; microscopy quality; the higher the concentration, the more stable the solution.
- 4% paraformaldehyde (PF) in 100 m*M* phosphate buffer; pH 7.4   ↬ *See* Appendix B4.3.1.

❑ *Precaution*

Manipulate under hood

❑ *Protocol*

- 4% buffered paraformaldehyde **200 mL**
- Glutaraldehyde **133–400 µL** ⮫ According to the glutaraldehyde concentration

❑ *Storage*

Must be used immediately

B5 EMBEDDING MEDIA

B5.1 Materials

- Disposable pipettes
- Fume hood
- Glass containers
- Gloves
- Magnetic stirrer ⮫ Or electric stirrer
- Mask
- Propipette

B5.2 Epoxy Resins

B5.2.1 Epon–Araldite

❑ *Products/Solutions*

- Araldite 502
- Dodecenyl succinic anhydride ⮫ DDSA, hardener
- Hexahydrophthalic anhydride ⮫ Epox 812 resin
- 2,4,6 Tris-dimethylamine methyl phenol ⮫ DMP 30 (accelerator); very dangerous, irritant chemical

❑ *Precautions*

- Toxic products ⮫ **Risk of developing allergies**
  ⮫ **Wear a mask.**
- Avoid all contact with skin. ⮫ **Wear gloves.** All traces of resin on the skin must be removed with soapy water.

❑ *Protocol*

1. Mix all the following products in a glass container: ⮫ The mixture is very viscous.

   - Araldite 502 **20 mL**
   - DDSA **60 mL**
   - Epox 812 **25 mL**

2. Add the hardener to the mixture:

• Epon-araldite mixture	**20 mL**	↪ 300 µL of resin is required to fill a small gelatin capsule (size 00).
• DMP 30	**300 µL**	↪ The amount of the product required can rise by 20% depending on how often the container is opened.

❏ *Storage*

Prepare just before use and keep away from light.

B5.2.2   Epon

❏ *Products/Solutions*

- Dodecenyl succinic anhydride → DDSA, hardener
- Epox 812 resin → Aliphatic carbon chain (mixture of glycidyl ethers and mono- trisubstituted glycerol)

- Methyl nadic anhydride → MNA, hardener
- 2,4,6 -Tris-dimethylamine methyl phenol → DMP 30, accelerator; very dangerous, irritant chemical

❏ *Precautions*

- Toxic products → Risk of developing allergies
  → **Wear a mask**
- Avoid all skin contact. → **Gloves must be worn.**

❏ *Protocol*

1. Stock solution:

   - Thoroughly mix all the following products in a glass container: ↪ The mixture is very viscous. A glass rod fixed to an electric mixer may be used to prepare a large quantity.

– DDSA	**50 mL**
– Epox 812	**81 mL**
– MNA	**44.5 mL**

2. Solution for use:

   - Add the hardener to the mixture ↪ To polymerize the resin

– Epon mixture	**20 mL**
– DMP 30	**300 µL**

   ↪ The amount of the product required can rise by 20% depending on how often the container is opened.

❏ *Storage*

- The mixture without the hardener can be kept at room temperature for 1 month
- The mixture with the hardener can be kept at 4°C for several hours or overnight. ↪ Time taken to impregnate the tissue

B5.3   Acrylic Resins

B5.3.1   Lowicryl K4M

❏ *Products/Solutions*

- Lowicryl K4M
  - Solution A (cross linker)
  - Solution B (monomer)
  - Initiator C

❏ *Precautions*

- Toxic products
- Very toxic resin

- Avoid all contact with skin.
- Avoid inhaling vapors.

❏ *Protocol*

1. Thoroughly mix the following products in a glass container:
   - Initiator C   **0.1 g**
   - Solution B   **17.30 mL**
2. Add
   - Solution A   **2.70 mL**
3. Embed the samples

❏ *Storage*
- The different products of the kit are kept at RT in a dry, dust-free place away from heat.
- The mixture can be kept for up to two days at –30°C.

❏ *Wastes*
Keep waste in special containers.

B5.3.2   LR White Medium

❏ *Products/Solutions*
- LR White medium

- Accelerator

❏ *Precautions*
- Weakly toxic and irritant resin

➥ Methacrylate resin

➥ 2-hydroxyethyl-acrylate; resin sold as a kit

➥ Clear liquid, distinctive smell

➥ Accelerates polymerization

➥ **Risk of developing allergies**
➥ An automatic embedding system is available (*see* Chapter 4, Section 4.3.7).
➥ Methacrylate is irritant to the skin, eyes, and respiratory system. Gloves and a mask must be worn. All traces of resin on the skin must be removed with running water.
➥ Pipette using a propipette.

➥ Use a fume hood.
➥ Quantity to prepare:
  - 1 mL per tube for impregnation
  - 500 μL per gelatin capsule for embedding

➥ The mixture is very runny.

➥ **Wear a mask** during weighing.
➥ Pipette using an automatic pipette.

➥ Gentle stirring (to avoid bubbles)
➥ *See* Chapter 4, Section 4.3.4.4.

➥ 1–2 years.

➥ The storage depends on the time taken to impregnate the samples.

➥ Do not throw down the sink.
➥ May be destroyed by incineration

➥ Or London Resin medium hard

➥ Acrylic polyhydroxyaromatic resin
➥ A single stock solution of methacrylate

➥ Made up of monomers used in medicine and dentistry
➥ Less allergenic than Lowicryl K4M

- Avoid contact with skin.
- Avoid inhaling vapors.

❏ *Protocol*

↬ **Recommended to wear a mask and gloves;** pipette with a propipette.

↬ **Use in a fume hood.**
↬ Quantity to prepare:
  - 1 mL per tube for impregnation
  - 0.5 mL per gelatin capsule for embedding

1. Take out the products just before making the mixture.
2. Place in a glass container:
   - LR White   **5 mL**
   - Accelerator   **5 μL**
3. Mix immediately.   **1–2 min**
4. Embed the samples very quickly.

↬ **Use in a fume hood.**
↬ Use a propipette.
↬ ≈ 1 drop (optional)
↬ Very quickly with a vortex
↬ *See* Chapter 4, Section 4.3.5.4.
↬ Polymerization is very fast: less than 15 min with an accelerator and several hours without an accelerator.

❏ *Storage*
Store at 4°C for several months.

↬ Very sensitive to heat

❏ *Wastes*
Keep the waste in glass containers.

↬ Do not throw down the sink.

B5.3.3   Unicryl

❏ *Product*

- Unicryl

↬ Stock solution ready to use.

↬ The commercially available solution is ready to use.

❏ *Precautions*

- Toxic resin
- Avoid contact with skin
- Avoid inhaling vapors

❏ *Protocol*

↬ **Always wear gloves.**
↬ **Always wear a mask;** pipette using a propipette under a fume hood.

↬ Use a fume hood.
↬ Quantity to prepare:
  - 1 mL per tube for impregnation
  - 0.5 mL per gelatin capsule for embedding

1. Take out the stock solution just before impregnation.
2. Embed biological samples at a low temperature:
   Samples are dehydrated progressively at a low temperature (from 0°C down to −20°C) without a substitution stage. After the last change of solvent, the stock solution is directly impregnated at −20°C for several hours.

↬ *See* Chapter 4, Section 4.3.6.4.

↬ Rapid protocol

↬ Easy to use, and the resin is ready to use, which cuts down the risk of handling.

❑ *Storage*
  - At −30°C for several months
  - At 4°C for one year

B6   IMMUNOCYTOCHEMISTRY

B6.1   Blocking Solutions

B6.1.1   Nonspecific Sites

❑ *Reagents/Solutions*
  - Blocking agents
    – Bovine serum albumin (BSA)
    – Dried skim milk
    – Fish gelatin
    – Goat serum
    – Ovalbumin
    – 0.01%Triton X-100
  - Buffers
    – 100 mM phosphate buffer; pH 7.4   ⇒ *See* Appendix B3.4.1.
    – 50 mM Tris–HCl buffer/300 mM NaCl;   ⇒ *See* Appendix B3.7.5.
      pH 7.6

❑ *Precaution*
Do not leave at room temperature.

❑ *Working solution*
Dilute:

  - Blocking agents   **1–2%**
  - Buffer   **to 100 mL**

❑ *Storage*
Store at −20°C.

B6.1.2   Endogenous Alkaline Phosphatases

⇒ Endogenous enzyme activity may be inhibited by levamisole or by heat treatment.

B6.1.2.1   Levamisole

❑ *Reagents/Solutions*

  - Levamisole   ⇒ $C_{11}H_{12}N_2S$
    (2,3,5,6-tetrahydro-6-phenylimidazole)   (2,3,5,6-tetrahydro-6-phenylimidazole)
  - 50 mM Tris–HCl buffer/100 mM NaCl/50   ⇒ *See* Appendix B3.7.6.
    mM MgCl$_2$; pH 9.5

❑ *Precautions*
None

❏ *1 M stock solution*
Dissolve:

- 100 m*M* levamisole   **24 mg**   ↪ MW = 240.80
- Visualization buffer   **1 mL**

❏ *1 mM working solution*

- Stock solution   **1 µL**
- Visualization buffer   **990 µL**

❏ *Storage*
Use immediately after preparation.

B6.1.3   Endogenous Peroxidases

B6.1.3.1   HYDROGEN PEROXIDE/BUFFER

❏ *Reagents/Solutions*

- 30% hydrogen peroxide   ↪ 110 vol
- 50 m*M* Tris–HCl/300 m*M* NaCl; pH 7.6   ↪ *See* Appendix B3.7.5; or phosphate buffer (*see* Appendix B3.4.1)

❏ *Precaution*
Avoid contact with skin.   ↪ Wash in running water.

❏ *Solution for use*
Mix:

- Hydrogen peroxide   **3 mL**
- Buffer   **100 mL**

❏ *Storage*
Use immediately after preparation.

B6.1.3.2   HYDROGEN PEROXIDE/METHANOL

❏ *Reagents/Solutions*

- 30% hydrogen peroxide   ↪ 110 vol
- 100% methanol

❏ *Precaution*
Contact dangerous

❏ *Working solution*
Mix:

- Hydrogen peroxide   **3 mL**
- Methanol   **100 mL**

❏ *Storage*
Use immediately after preparation.

B6.2 Chromogens

B6.2.1 Alkaline Phosphatase

B6.2.1.1 NBT-BCIP

↝ There are other chromogen substrates that give different color precipitates (i.e., Fast Red, etc.).
↝ Use commercial solution.

❑ *Reagents/Solutions*
- Dimethylformamide
- Substrates
  - NBT (nitro blue tetrazolium)
  - BCIP
    (5-bromo-4-chloro-3-indolyl phosphate)
- Tris–HCl/NaCl/MgCl$_2$ buffer; pH 9.5

↝ (CH$_3$)$_2$NOCH

↝ C$_{40}$H$_{30}$Cl$_2$N$_{10}$O$_6$
↝ C$_8$H$_6$NO$_4$BrCIP × C$_7$H$_9$N

↝ *See* Appendix B3.7.6.

❑ *Precaution*
- Avoid contact with skin.
- Use immediately after preparation and keep out of the light.

❑ *Working solution*
Mix:

• NBT	**0.34 mg or**	↝ MW = 817.70
	4.5 µL	↝ 75 mg/mL dimethylformamide
• BCIP	**0.18 mg or**	↝ MW = 433.60
	3.5 µL	↝ 50 mg/mL dimethylformamide
• Visualization buffer	**to 1 mL**	

❑ *Storage*
Reagents in commercial solutions are stored at 4°C.

B6.2.2 Peroxidase

↝ Inhibition of peroxidase activities

B6.2.2.1 Diaminobenzidine Tetrahydrochloride

↝ There are other chromogen substrates that give different color precipitates (i.e., 4-chloro-1-naphthol or 3-amino-9-ethylcarbazole).

❑ *Reagents/Solutions*
- DAB
- 30% hydrogen peroxide
- Tris–HCl/NaCl buffer; pH 7.6

↝ C$_{12}$H$_{14}$N$_4$ · 4HCl
↝ 110 vol
↝ *See* Appendix B3.7.5.

❑ *Precaution*
Reagent dangerous if inhaled

↝ Tablet form is preferable.

❑ *Working solution*

↝ Prepare in darkness just before use.

1. Dissolve:

• DAB or 1 tablet	**10 mg**	↝ MW = 360.12
• Visualization buffer	**10 mL**	↝ DAB tablet should be dissolved in water.

2. Filter the solution.

3. To the filtrate, add:
   - 3% hydrogen peroxide/H_2O **10 µL**

❑ *Storage*
Store at −20°C, in aliquots, without hydrogen peroxide.

⇒ Commercial solutions (DAB and H_2O_2) can be stored at 4°C.

B7   STAINING/COATING

B7.1   Light Microscopy

B7.1.1   Toluidine Blue

⇒ 1% aqueous toluidine blue

❑ *Reagents/Solutions*

- Distilled water
- Sodium tetraborate
- Toluidine blue

⇒ $Na_2B_4O_7 \cdot 12H_2O$
⇒ $C_{15}H_{16}N_3 \cdot SCl$

❑ *Precaution*
Filter the solution before use.

❑ *Stock solution*

1. Mix and dissolve at room temperature:
   - Toluidine blue   **1 g**
   - Sodium tetraborate   **1 g**
   - Distilled water   **100 mL**
2. Filter.

⇒ MW = 305.80
⇒ MW = 381.37

❑ *Working solution*
Dilute:

- Stock solution   **1 vol**
- Distilled water   **50 vol**

❑ *Storage*
Store at room temperature for some months.

⇒ Filter before use.

B7.1.2   Methyl Green

❑ *Reagents/Solutions*

- Distilled water
- Methyl green

⇒ $C_{27}H_{35}N_3BrCl \cdot ZnCl_2$

❑ *Precaution*
Filter before use.

❑ *Working solution*
Dissolve:

- Methyl green   **1 g**
- Distilled water   **100 mL**

⇒ MW = 653.2

❑ *Storage*
Store at room temperature.

B7.2   Electron Microscopy

B7.2.1   Uranyl Acetate

❑ *Products*

- Ammonia
- 95% ethanol
- Oxalic acid
- Sterile distilled water
- Uranyl acetate

❑ *Precautions*

- Uranium is radioactive
- Avoid all contact with skin
- Avoid inhaling uranyl powder.

⇝ Electron microscope quality (EM)

⇝ NH_4OH (10% solution)

⇝ $C_2O_4H_2 \cdot 2H_2O$
⇝ *See* Appendix B1.1.
⇝ $(CH_3COO)_2\ UO_2 \cdot 2H_2O$

⇝ **Use a fume hood.**

⇝ Emits α radiation

⇝ **Wear a mask while weighing out the uranyl acetate.**

B7.2.1.1   2.5% ALCOHOLIC URANYL ACETATE

❑ *Protocol*

1. Prepare just before use in a brown glass container:

   - 5% aqueous uranyl acetate   **1 vol**
   - 95% alcohol   **1 vol**

2. Filter in the dark.
3. Stain in the dark.

❑ *Storage*
Do not store.

⇝ **In a dark room with a safe light**

⇝ Saturated solution (see Appendix B7.2.1.2)

⇝ Alcoholic uranyl is very sensitive to light.

B7.2.1.2   2–5 % AQUEOUS URANYL ACETATE

❑ *Protocol*

- Uranyl acetate   **0.5–1.25 g**

- Distilled water   **25 mL**

1. Mix.
2. Filter using a Millipore.   **0.22 μm**

❑ *Storage*
Store at 4°C for 1 month away from light.

⇝ Stock solution

⇝ Dissolve at room temperature in the dark on a magnetic stirrer for 1 to 2 hours.

B7.2.1.3   4% Neutral Uranyl Acetate

❏ *Solutions*

- 4% aqueous uranyl acetate
- 10% ammonia
- 300 m*M* oxalic acid

❏ *Precaution*
Away from light

❏ *Protocol*

1. Mix:
   - 4% aqueous uranyl acetate   **1 vol**
   - 300 m*M* oxalic acid   **1 vol**
2. Adjust the pH
   - 10% ammonia   **7–7.5**

3. Filter using a Millipore.   **0.22 μm**

❏ *Storage*
Store at 4°C for 1 month.

➥ 4% aqueous uranyl acetate/300 m*M* oxalic acid; pH 7.5
➥ Electron microscope quality

➥ *See* Appendix B7.2.1.2.

➥ Use pH paper; uranyl salts cause deterioration of the electrodes.

B7.2.2   Lead Citrate

➥ According to Reynolds
➥ Electron microscope quality

❏ *Products*

- Lead nitrate
- Sodium citrate
- Sodium hydroxide

➥ Pb(NO3)$_2$
➥ C$_6$H$_5$Na$_3$O$_7$ · 2 H$_2$O
➥ NaOH

❏ *Solutions*

- 1 N sodium hydroxide
- Distilled water

➥ *See* Appendix B2.20.
➥ *See* Appendix B1.1.

❏ *Storage*
Store at 4°C for 1 month.

❏ *Protocol*

- Sodium citrate   **0.882 g**

- Lead nitrate   **0.662 g**

➥ MW = 294.10. 1% sodium tartrate can also be used.
➥ MW = 331.23

1. Dissolve the two products separately in distilled water.
2. Mix the two solutions thoroughly.
3. Add:
   - 1 N sodium hydroxide   **4 mL**
4. Mix until the precipitate dissolves completely.
5. Add:
   - Distilled water   **to 25 mL**
6. Filter with a 0.22 μm Millipore filter

➥ A homogeneous precipitate forms.

➥ MW = 40.00
➥ The precipitate dissolves at a basic pH (≈ 10).

❑ *Storage*
Store at 4°C for 1 month in aliquots.

➥ Fill the aliquots to the top to avoid contact with carbon dioxide in the air.

B7.2.3 Methylcellulose

B7.2.3.1 2% METHYLCELLULOSE

❑ *Product*
• Methylcellulose (Tylose)

➥ Stock solution

➥ Electron microscope quality
➥ Tylose is the commercial name for methyl cellulose.

❑ *Solution*
• Distilled water

➥ *See* Appendix B1.1.

❑ *Precaution*
Do not stir. Take the quantity needed from the top of the tube.

❑ *Protocol*

• Methylcellulose	**2 g**
• Distilled water	**100 mL**
1. Heat.	**95°C**
	15 min
2. Stir.	**4°C**
	overnight
3. Leave without stirring until the tylose has dissolved.	**15–30 days**
	at 4°C
4. Centrifuge.	**100,000 *g***
	60 min

➥ Make up stocks and keep at 4°C.

➥ The solution forms very slowly (longer than 1 month).
➥ **Not indispensable**

❑ *Storage*

• At 4°C for several weeks
• At –20°C for several months

➥ Solution in use
➥ Reserve solution

B7.2.3.2 0.8% METHYLCELLULOSE, 0.2% NEUTRAL URANYL ACETATE

➥ A staining solution can also be used for coating sections of frozen tissue.

❑ *Solutions*

• Distilled water
• 2% neutral uranyl acetate
• 2% methylcellulose

➥ *See* Appendix B1.1.
➥ *See* Appendix B7.2.3.2.
➥ *See* Appendix B7.2.3.1.

❑ *Precautions*
None

❑ *Protocol*

1. Add the methylcellulose and the neutral uranyl acetate to give a final concentration of 0.2%:
   • Neutral uranyl acetate   **200 µL**

• Tylose	**800 μL**
• Distilled water	**100 mL**

⇨ Or more, if necessary

2. Stir.
3. Centrifuge.

⇨ Eliminates bubbles

❑ *Storage*

* At 4°C for several weeks
* At –20°C for several months

⇨ Solution for use

B7.2.4   0.5 % Sodium Silicotungstate

❑ *Solutions*

* Distilled water
* Sodium silicotungstate

⇨ *See* Appendix B1.1.
⇨ AR quality

❑ *Precaution*
Avoid all contact with skin.

❑ *Protocol*
Dissolve in distilled water:

• Sodium silicotungstate	**0.5 g**
• Distilled water	**100 μL**

❑ *Storage*
Store at 4°C for several weeks.

B8   MOUNTING MEDIA

B8.1   Aqueous

Aqueous mountants are used where the reaction precipitate is soluble in alcohol:

* Buffered glycerine
* Commercial media
* Moviol

⇨ *See* Appendix B8.1.1.

⇨ *See* Appendix B8.1.2.

B8.1.1   Buffered Glycerine

❑ *Reagents/Solutions*

* Glycerol
* 100 m*M* phosphate buffer, or PBS

⇨ *See* Appendix B3.4.1 and 3.4.3.

❑ *Precautions*
None

❑ *Preparation*

• Glycerol	**1 vol**
• Buffer	**1 vol**

❏ *Storage*
Store at 4°C for some days.

B8.1.2   Moviol

❏ *Reagents/Solutions*

- Distilled water
- Glycerol anhydride
- Moviol 4–88
- 100 mM tris–HCl buffer; pH 8.5

➥ AR quality

➥ $C_3H_8O_3$

➥ *See* Appendix B3.7.1.

❏ *Precautions*
None

❏ *Preparation*

1. Mix:
   - Moviol 4–88   **2.4 g**
   - Glycerol   **4.8 mL**
2. Add:
   - Distilled water   **6 mL**
3. Mix and leave.   **2–48 h**
                                     at 60°C
4. Add:
   - Buffer   **12 mL**
5. Incubate.   **overnight**
                                     at 60°C
6. Mix, centrifuge.   **5000 g**
                                     15 min

➥ MW = 92.10

❏ *Storage*
Store at –20°C in aliquots.

B8.2   Permanent

Permanent mounting media are used after dehydration and passage through xylene (or a similar chemical).

❏ *Reagent*
Commercial media

❏ *Precautions*

- Toxic on contact or by inhalation
- Polymerization on drying

➥ Aromatic solvents, or substitutes

❏ Storage
Store at room temperature in tightly closed containers.

➥ Xylene can be used to make the solution more fluid.

Examples of Results

Figure 1

1. Description

    ↪ Detection of mRNA encoding growth hormone in the rat pituitary

2. Method

    ↪ Semithin sections of frozen tissue

    ↪ *See* Chapter 7.

3. Tissue

    ↪ Rat pituitary

4. Fixation

    ↪ 4% paraformaldehyde

5. Pretreatment
    • Proteinase K

6. Probes
    • **(A)** α^{35}S dATP
    • **(B)** dUTP 11 digoxigenin

    ↪ Anti-sense oligonucleotide (30 mers) for mRNA encoding growth hormone labeled by 3′ incorporation

7. Visualization
    • **(A)** Autoradiography
    • **(B)** Immunocytology

    ↪ Peroxidase/DAB

8. Bar

    ↪ 10 μm

9. Comments

    ↪ mRNA encoding growth hormone is visualized in the somatotrophs of the anterior pituitary:
    • **(A)** Autoradiography visualizes the mRNA by black silver grains overlying some cells.
    • **(B)** The brown precipitate DAB is present in the cytoplasm of the same cells.

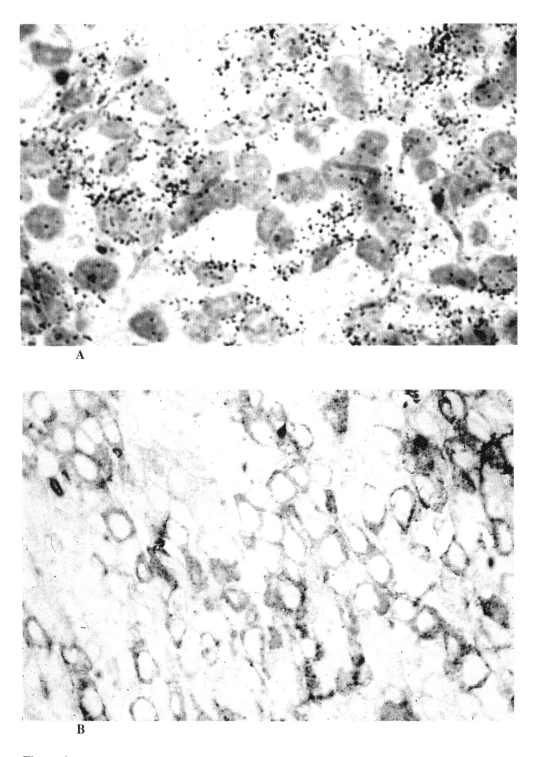

A

B

Figure 1

Figure 2

1. Description

    ⇨ Detection of mRNA encoding oxytocin or growth hormone on semithin sections

2. Method

    ⇨ Semithin sections are derived by the pre-embedding technique.
    ⇨ *See* Chapter 5.

3. Tissue
    - **A**
    - **B**

    ⇨ Rat
    ⇨ Supraoptic nuclei
    ⇨ Anterior pituitary

4. Fixation

    ⇨ 4% paraformaldehyde

5. Pretreatment

    ⇨ None

6. Probes
    - **A**

    ⇨ Anti-sense oligonucleotide (25 mers) for mRNA encoding oxytocin labeled by 3′ incorporation of tritiated ATP

    - **B**

    ⇨ Anti-sense oligonucleotide (30 mers) for mRNA encoding growth hormone labeled by 3′ incorporation with dUTP 21 biotin

7. Visualization
    - **(A)** Autoradiography
    - **(B)** Immunocytology

    ⇨ Exposure time is 5 months.
    ⇨ Avidin–biotin–peroxidase complex/DAB

8. Embedding

    ⇨ Epon

9. Bar

    ⇨ 10 μm

10. Comments

    ⇨ Probes:
    - **(A)** Transverse section (T) of a 50 μm vibratome section. The signal is present throughout the thickness of the section. A number of magnocellular neurones are labeled (arrowed) at the center of the vibratome section demonstrating that the labeled probe has diffused throughout the section (Trembleau, Calas, and Fevre-Montange, Ultrastructural localization of oxytocin mRNA in the rat hypothalamus by *in situ* hybridization using a synthetic oligo-nucleotide, *Mol. Brain Res.*, 8, 37, 1990, with permission).
    - **(B)** The reaction precipitate is observed in 40% of cells of the anterior pituitary corresponding to the percentage of somatotrophs present (for details, see Figure 24) (Le Guellec, Trembleau, Pechoux, Gossard, and Morel, Ultrastructural non-radioactive *in situ* hybridization of GH mRNA in rat pituitary glands: pre-embedding vs. ultrathin frozen sections vs. post-embedding, in *J. Histochem. Cytochem.*, 40, 979, 1992, with permission).

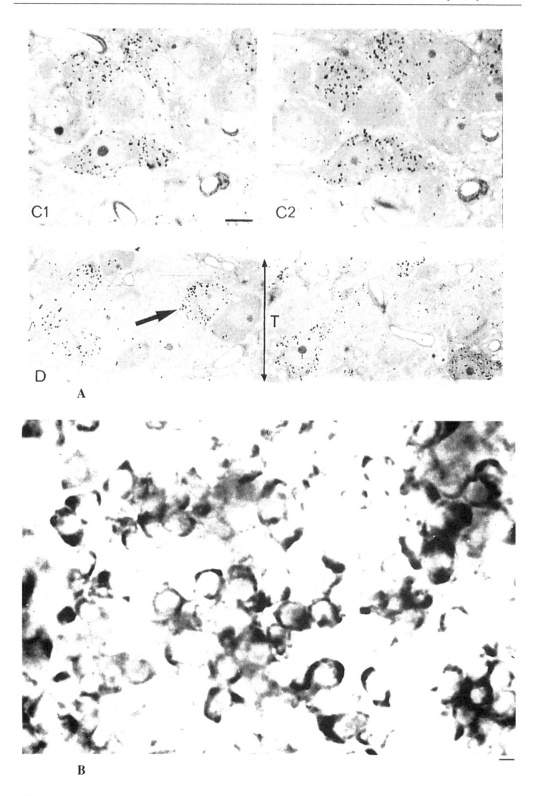

Figure 2

Figure 3

1. Description	↪ Detection of the *Herpes simplex* virus, type 1 DNA
2. Method	↪ Post-embedding technique
	↪ *See* Chapter 4.
3. Tissue	↪ Infected HeLa cells
4. Fixation	↪ 4% formaldehyde
5. Embedding	↪ Lowicryl K4M
6. Pretreatments	
• RNase	↪ Destruction of RNA
• NaOH	↪ Denaturation
7. Probe	↪ Biotinylated viral DNA
8. Visualization	
• Immunocytology	↪ Colloidal gold (10 nm)
9. Bar	↪ 0.1 μm
10. Comments	↪ The signal is limited to the viral genome accessible at the surface of the section. The viral genome present within the cell is not labeled:

- **(A)** The particles of colloidal gold are localized over the viral nucleoid, no labeling can be seen over the capsomere, the tegument, or the envelope (arrowed).
- **(B)** In this part of the nucleus, the signal is restricted to the viral genome in some of the capsomeres (arrowed). The empty capsomeres are not labeled (arrowhead).

↪ From Puvion-Dutilleul, F., with permission.

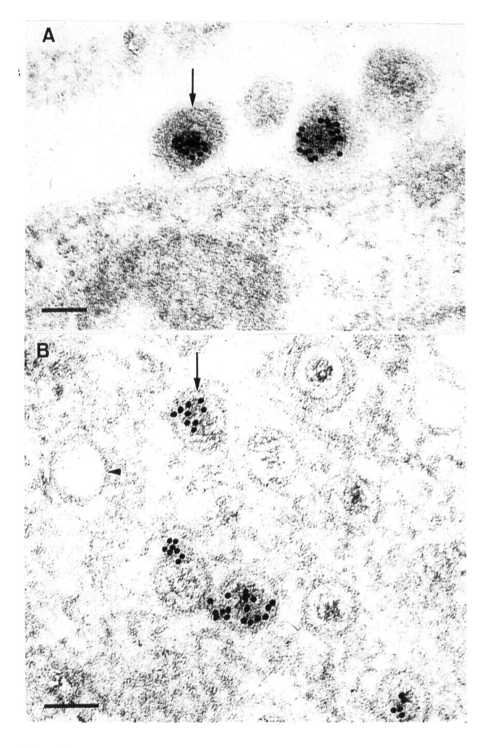

Figure 3

Figure 4

1. Description	➯ Localization of ribosomal RNA
2. Method	➯ Post-embedding technique
	➯ *See* Chapter 4.
3. Tissue	➯ Ehrlich tumor cells (mouse)
4. Fixation	➯ 4% paraformaldehyde, 0.2% glutaraldehyde
5. Embedding	➯ Lowicryl K4M
6. Pretreatment	➯ None
7. Probe	➯ Biotinylated mouse nuclear ribosomal DNA
8. Visualization	
• Immunocytology	➯ Colloidal gold (5 nm)
9. Bar	➯ 0.5 μm
10. Comments	➯ The colloidal gold particles (arrowheads) are present in the nucleolus and the cytoplasm. (N) nucleus, (n) nucleolus.
	➯ From Fournier, J-G., with permission.

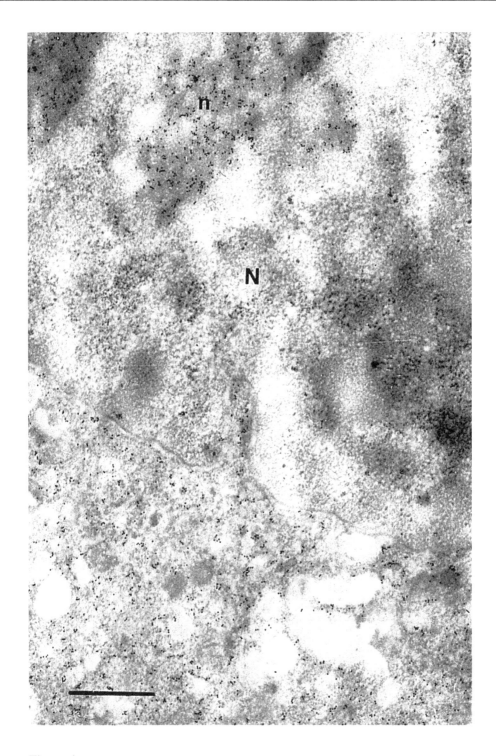

Figure 4

Figure 5

1. Description	↪ Detection of mitochondrial ribosomal RNA
2. Method	↪ Post-embedding technique
	↪ *See* Chapter 4.
3. Tissue	↪ CEM cells
4. Fixation	↪ 4% paraformaldehyde, 0.1% glutaraldehyde
5. Embedding	↪ Lowicryl K4M
6. Pretreatment	↪ None
7. Probe	↪ Biotinylated RNA
8. Visualization	
• Immunocytology	↪ Colloidal gold (10 nm)
9. Bar	↪ 0.5 μm
10. Comments	↪ The particles of colloidal gold are present exclusively over the mitochondrial matrix (m). None are present over the cytoplasm and the nucleus (N).
	↪ From Fournier, J-G., with permission.

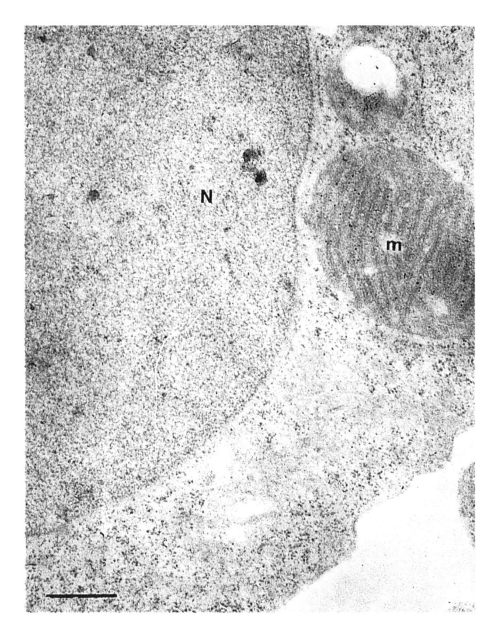

Figure 5

Figure 6

1. Description	➾ Simultaneous detection of cellular and mitochondrial ribosomal RNA
2. Method	➾ Post-embedding technique
	➾ *See* Chapter 4.
3. Tissue	➾ CEM cells
4. Fixation	➾ 4% paraformaldehyde, 0.1% glutaraldehyde
5. Embedding	➾ Lowicryl K4M
6. Pretreatment	➾ None
7. Probe	➾ Sulfonated nuclear ribosomal DNA and biotinylated mitochondrial DNA
8. Visualization	
• Dual immunocytochemistry	➾ Colloidal gold (5 and 10 nm)
9. Bar	➾ 0.5 µm
10. Comments	➾ The particles of 10 nm colloidal gold represent mitochondrial DNA present exclusively over the mitochondria (m), while the particles of 5 nm colloidal gold represent ribosomal nucleic acids and are found over the cytoplasm.
	➾ From Fournier, J-G., with permission.

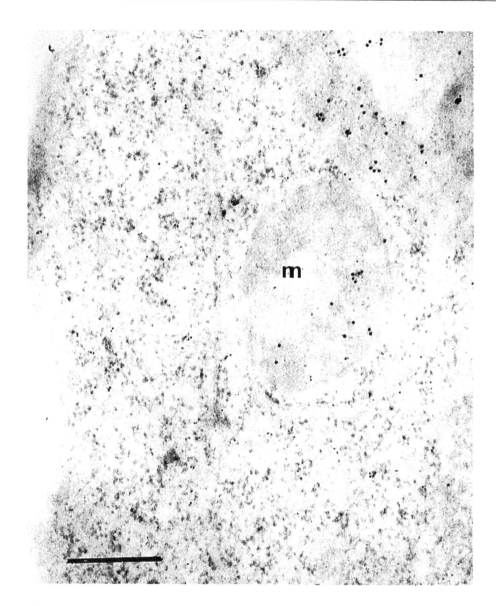

Figure 6

Figure 7

1. Description
2. Method

3. Tissue
4. Fixation
5. Embedding
6. Pretreatment
7. Probe

8. Visualization
   • Autoradiography
9. Bar
10. Comments

⇀ Detection of mRNA encoding fibronectin
⇀ Post-embedding technique.
⇀ *See* Chapter 4.
⇀ Chicken fibroblast
⇀ 4% paraformaldehyde
⇀ Lowicryl K4M
⇀ None
⇀ cDNA labeled with α[^{35}S]dATP by random priming

⇀ 0.2 μm
⇀ RNA is visualized by the silver grains (twisted strands) present over both the cytoplasm and the nucleus. Cytoplasmic RNA is found close to the endoplasmic reticulum (Morel, Le Guellec, Mertani, and Trembleau, *In situ* hybridization for electron microscopy, in *Visualization of Nucleic Acids*, CRC Press, Boca Raton, 1995, 229, with permission).

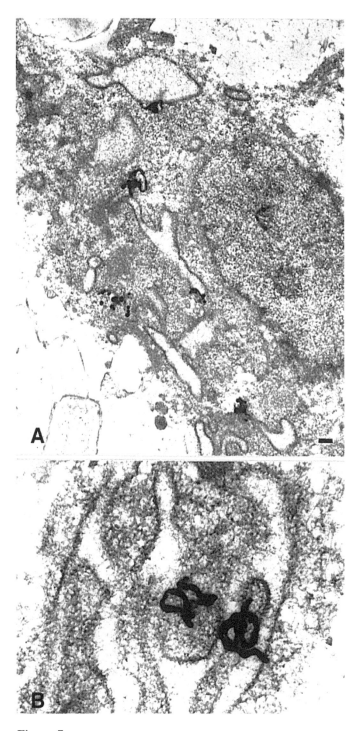

Figure 7

Figure 8

1. Description

2. Method

3. Tissue

4. Fixation
5. Embedding
6. Pretreatment
7. Probe

8. Visualization
   • Immunocytology
9. Bar
10. Comments

↪ Detection of mRNA encoding, type I collagen
↪ Post-embedding technique
↪ *See* Chapter 4.
↪ Cells from a fish scale in the process of regeneration (6 days).
↪ 4% paraformaldehyde
↪ Lowicryl K4M
↪ None
↪ cRNA labeled with UTP 16 biotin by *in vitro* transcription

↪ Colloidal gold (10 nm)
↪ 0.2 μm
↪ RNA encoding Type 1 collagen is visualized by grains of colloidal gold over the endoplasmic reticulum and the external nuclear membrane (N) [Le Guellec and Zylbergerg, Expression of type I and type V collagen in mRNA in the elasmoid scales of a teleost fish as revealed by *in situ* hybridization, *Connective Tissue Res.*, 39, 257, 1998, with permission).

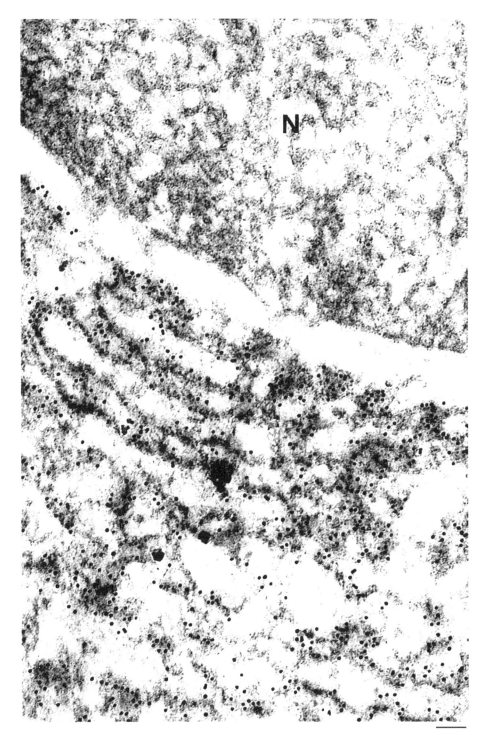

Figure 8

Figure 9

1. Description	↪ Localization of mRNA encoding the interleukin Type I receptor in the testis
2. Method	↪ Post-embedding technique
	↪ *See* Chapter 4.
3. Tissue	↪ Mouse testis
4. Fixation	↪ 4% paraformaldehyde, 0.5% glutaraldehyde
5. Embedding	↪ Lowicryl K4M
6. Pretreatment	↪ None
7. Probe	↪ cDNA labeled with dUTP 11-digoxigenin by random priming
8. Visualization	
• Immunocytology	↪ Colloidal gold (10 nm)
9. Bar	↪ 0.5 µm
10. Comments	↪ The messenger RNA encoding the interleukin Type I receptor is expressed in the Sertoli (S) cells and in the peritubular myoid cells (P). The particles of colloidal gold (arrowheads) are detected in the cytoplasm of these two cell types. (m) mitochondria, (b) basal membrane. (Gomez, Morel, Cavalier, Liénard, Haour, Courtens, and Jegou, Type I and type II interleukin-1 receptor expression in rat, mouse, and human tests, *Biol. Reprod.*, 56, 1513, 1997, with permission).

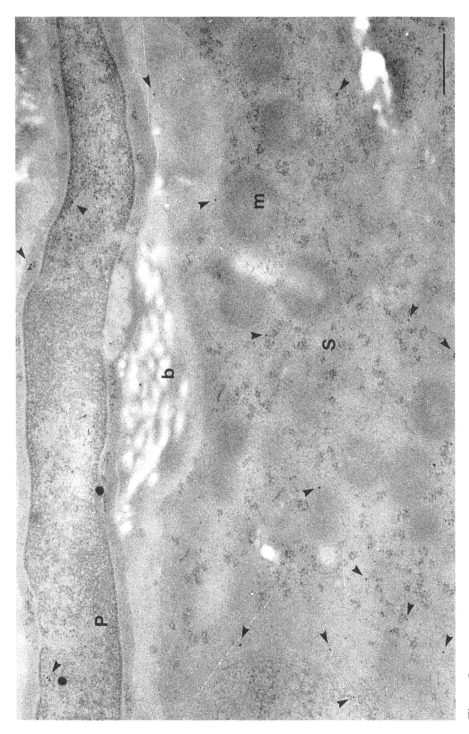

Figure 9

Figure 10

1. Description	➯ Detection of different types of Type 5 adenovirus nucleic acids (*see* Figures 10 to 13): • Detection of single and double-stranded viral DNA
2. Method	➯ Post-embedding technique ➯ *See* Chapter 4.
3. Tissue	➯ Infected HeLa cells
4. Fixation	➯ Formaldehyde
5. Embedding	➯ Lowicryl K4M
6. Pretreatments	
• Protease	➯ Removal of proteins is indispensable for a good hybridization with this virus.
• RNase	➯ Destruction of RNA.
• NaOH	➯ Denaturation.
7. Probe	➯ Biotinylated viral DNA.
8. Visualization	
• Immunocytology	➯ Colloidal gold (10 nm)
9. Bar	➯ 0.5 µm
10. Comments	➯ There is a strong signal over the two nuclear compartments of the fibrogranular network (F) made up of viral nucleic acids actively engaged in replication and transcription, and the pale annular (star) structure made up of single-stranded DNA. CH: condensed chromatin of the host cell. ➯ From Puvion-Dutilleul, F., with permission.

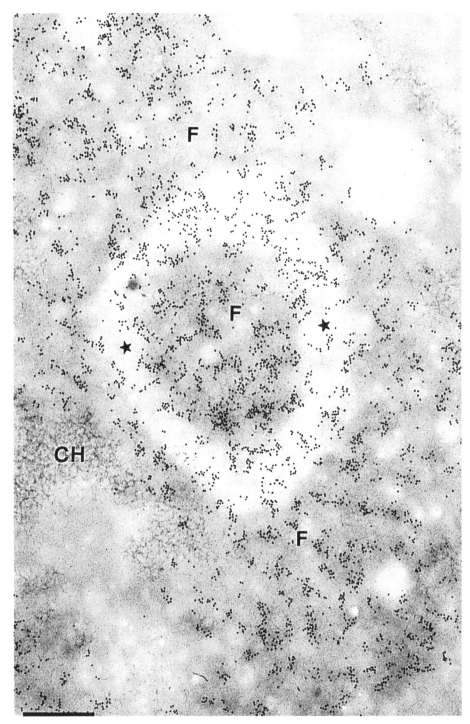

Figure 10

Figure 11

1. Description

⇨ Detection of different types of type 5 adenovirus nucleic acid (*see* Figures 10 to 13)
  • Detection of double-stranded viral DNA

2. Method

⇨ Post-embedding technique
⇨ *See* Chapter 4.

3. Tissue

⇨ Infected HeLa cells.

4. Fixation

⇨ Formaldehyde

5. Embedding

⇨ Lowicryl K4M

6. Pretreatments
  • Protease

⇨ Proteins must be removed for good hybridization with this virus.

  • Nuclease S1

⇨ Destruction of single-stranded DNA

  • NaOH

⇨ Denaturation

7. Probe

⇨ Biotinylated viral DNA

8. Visualization
  • Immunocytology

⇨ Colloidal gold (10 nm)

9. Bar

⇨ 0.5 µm

10. Comments

⇨ The colloidal gold particles are present over the fibrogranular network (F), but are completely absent over the pale annular (star) structure.
⇨ From Puvion-Dutilleul, F., with permission.

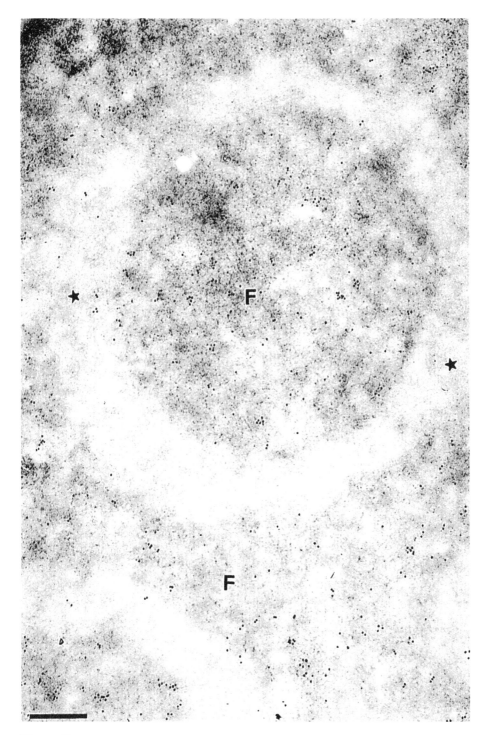

Figure 11

Figure 12

1. Description ⟿ Detection of different types of Type 5 adenovirus nucleic acid (*see* Figures 10 to 13)
   • Detection of single-stranded viral DNA

2. Method ⟿ Post-embedding technique
   ⟿ *See* Chapter 4.

3. Tissue ⟿ Infected HeLa cells

4. Fixation ⟿ Formaldehyde

5. Embedding ⟿ Lowicryl K4M

6. Pretreatments
   • Protease ⟿ Proteins must be removed for good hybridization with this virus.
   • RNase ⟿ Destruction of RNA

7. Probe ⟿ Biotinylated viral DNA

8. Visualization
   • Immunocytology ⟿ Colloidal gold (10 nm)

9. Bar ⟿ 0.5 μm

10. Comments ⟿ In the absence of denaturation (pretreatment with sodium hydroxide), colloidal gold particles are mainly found over the annular structure (star) and more rarely over the fibrogranular network (F). (CH) condensed host chromatin.
    ⟿ From Puvion-Dutilleul, F., with permission.

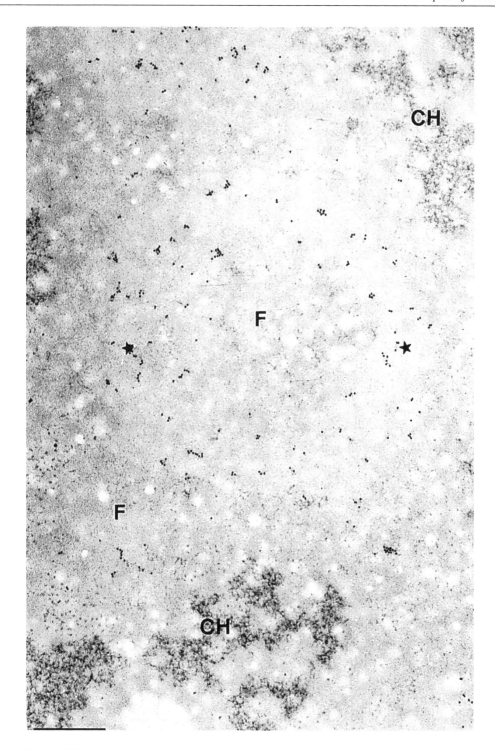

Figure 12

Figure 13

1. Description

    → Detection of different types of Type 5 adenovirus nucleic acid (*see* Figures 10 to 13)
       • Detection of viral RNA

2. Method
    → Post-embedding technique
    → *See* Chapter 4.

3. Tissue
    → Infected HeLa cells

4. Fixation
    → Formaldehyde

5. Embedding
    → Lowicryl K4M

6. Pretreatments
  • Protease
    → Proteins must be removed for good hybridization with this virus.

  • DNase 1
    → Destruction of DNA

7. Probe
    → Biotinylated viral DNA

8. Visualization
  • Immunocytology
    → Colloidal gold (10 nm)

9. Bar
    → 0.5 μm

10. Comments
    → In the absence of pretreatment with RNase, Nuclease S1, and denaturation with sodium hydroxide, the probe cannot hybridize with RNA. The colloidal gold particles are present over the fibrogranular network (F) that contains the viral genome, as in Figure 10, but they are absent over the pale annular (star) structure that corresponds to an accumulation of single stranded viral DNA (*see* Figure 12).
    → From Puvion-Dutilleul, F., with permission.

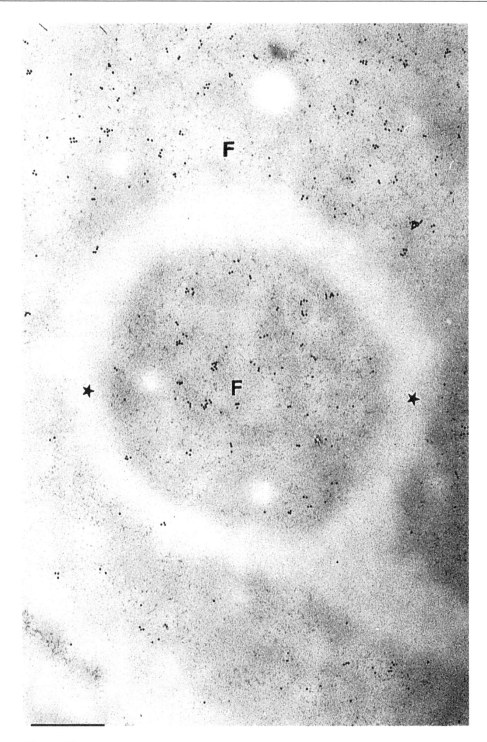

Figure 13

Figure 14

1. Description

2. Method

3. Tissue
4. Fixation
5. Pretreatment
   • Permeabilization
6. Probe

7. Visualization
   • Autoradiography
8. Embedding
9. Bar
10. Comments

⇨ Detection of RNA encoding oxytocin in the neurones of the suprachiasmatic nuclei
⇨ Pre-embedding technique
⇨ *See* Chapter 5.
⇨ Rat suprachiasmatic nuclei
⇨ Perfusion with 4% formaldehyde

⇨ Detergent
⇨ Oligonucleotide labeled with [^{3}H]dATP by 3′ extension

⇨ Epon
⇨ 0.5 μm
⇨ The hybrids are visualized by silver grains (twisted strands) that are present over the endoplasmic reticulum, the cytoplasmic matrix, and the nucleus of the magnocellular neuron (*see* Figure 2 for the semithin section) (Trembleau, Calas, and Fevre-Montange, Ultrastructural localization of oxytocin mRNA in the rat hypothalamus by *in situ* hybridization using a synthetic oligonucleotide, *Mol. Brain Res.*, 8, 37, 1990, with permission).

Figure 14

Figure 15

1. Description	⇨ Localization of mRNA encoding oxytocin in the hypothalamus
2. Method	⇨ Pre-embedding technique
	⇨ *See* Chapter 5.
3. Tissue	⇨ Rat hypothalamus
4. Fixation	⇨ Perfusion with 4% paraformaldehyde
5. Pretreatment	
• Permeabilization	⇨ Detergent (Triton X100) or a freeze/thaw cycle
6. Probe	⇨ Oligonucleotide (37 mers) labeled with dUTP-16 digoxigenin by 3' extension
7. Visualization	
• Immunocytology	⇨ Peroxidase/DAB
8. Embedding	⇨ Epon
9. Bar	⇨ 1 μm
10. Comments	⇨ The reaction precipitate (DAB brown precipitate made electron opaque by osmium postfixation) is present in the endoplasmic reticulum (arrowed). No signal is found in the nucleus (N), the mitochondria or the Golgi apparatus (Morel, Le Guellec, Mertani, and Trembleau, *In situ hybridization for electron microscopy*, in *Visualization of Nucleic Acids*, CRC Press, Boca Raton, 1995, 229, with permission).

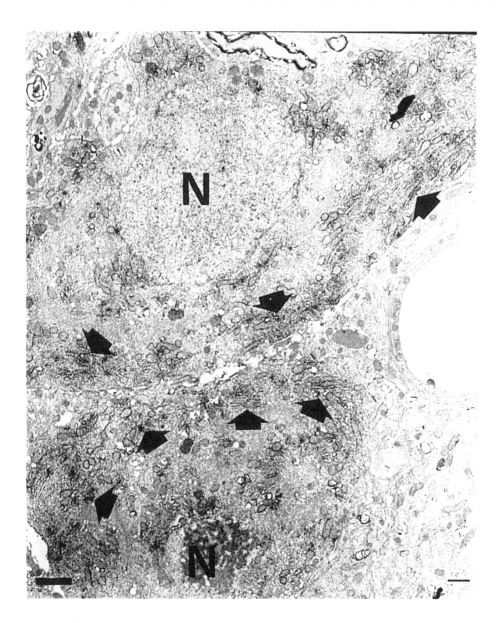

Figure 15

Figure 16

1. Description

    ⮑ Localization of LSU sub-units in a tobacco leaf

2. Method

    ⮑ Pre-embedding technique
    ⮑ *See* Chapter 5.

3. Tissue

    ⮑ Tobacco leaf (*Nicotinia tabacum*)

4. Fixation

    ⮑ 4% paraformaldehyde in sucrose (50 m*M*)

5. Pretreatment
    • Deproteinization

    ⮑ Pronase (25 µg/mL)

6. Probe

    ⮑ cDNA labeled with dUTP 11-biotin by nick translation

7. Visualization
    • Immunocytology

    ⮑ Avidin–ferritin

8. Embedding

    ⮑ Spurr

9. Bar

    ⮑ 1 µm

10. Comments

    ⮑ Ferritin (arrowed) reveals the presence of sequences encoding LSU (large subunit of ribulose bisphosphate carboxylase) localized exclusively in the chloroplast stroma. (N) nucleus, (cy) cytoplasm, (cw) cellular membrane (Brangeon and Sossountzov, Electron microscopic *in situ* hybridization to RNA or DNA in plants cells, in *Hybridization Techniques for Electron Microscopy*, CRC Press, Boca Raton, 1993, 301, with permission).

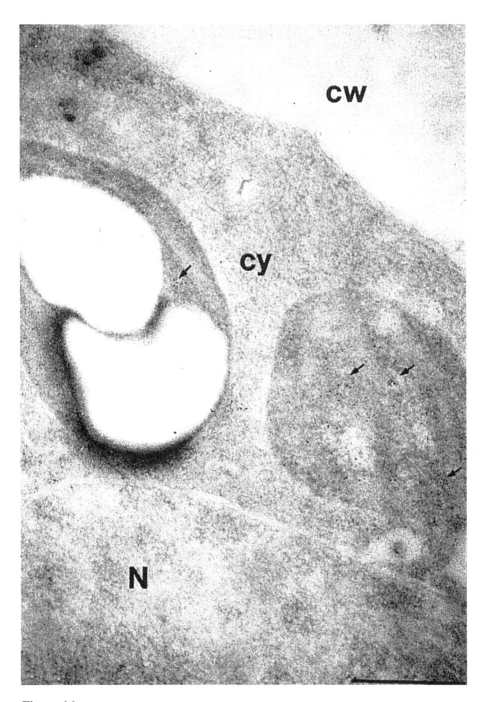

Figure 16

Figure 17

1. Description	⇀ Simultaneous localization of mRNA encoding prolactin and the protein itself respectively using *in situ* hybridization and immunocytochemistry
2. Method	⇀ Pre-embedding technique
	⇀ *See* Chapter 5.
3. Tissue	⇀ Rat anterior pituitary
4. Fixation	⇀ 4% paraformaldehyde
5. Embedding	⇀ Epon
6. Pretreatment	⇀ None
7. Probe	⇀ cDNA encoding prolactin labeled with $\alpha[^{35}S]$ dCTP by random priming
8. Visualization	
• Autoradiography	⇀ For *in situ* hybridization
• Immunocytochemistry	⇀ For protein (protein A/colloidal gold)
9. Bar	⇀ 1 μm
10. Comments	⇀ The silver grains (twisted threads) are found over the endoplasmic reticulum (RER) and the colloidal gold particles are over the secretion granules where the protein is stored (arrow). No signal is detectable over the somatotroph (GH). (N) nucleus. (Tong, Zhao, Simard, Labrie, and Pelletier, Electron microscopic autoradiographic localization of prolactin mRNA in rat pituitary, *J. Histochem. Cytochem.*, 37, 567, 1989, with permission).

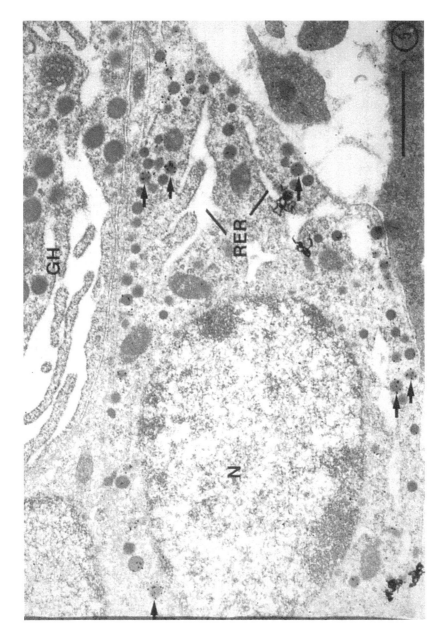

Figure 17

Figure 18

1. Description	➥ Visualization of mRNA encoding prolactin in a lactotroph
2. Method	➥ Frozen tissue technique
	➥ *See* Chapter 6.
3. Tissue	➥ Rat anterior pituitary
4. Fixation	➥ 4% paraformaldehyde
5. Embedding	➥ None
6. Pretreatment	➥ None
7. Probe	➥ Anti-sense oligonucleotide (30 mers) for mRNA encoding prolactin labeled with dUTP 16-digoxigenin by 3′ extension
8. Visualization	
• Immunocytology	➥ Colloidal gold (10 nm)
9. Bar	➥ 0.5 μm
10. Comments	➥ The particles of colloidal gold are detected in lactotrophs characterized by the rows of parallel endoplasmic reticulum and the large polymorphous secretion granules (≈ 600 nm in diameter). The colloidal gold particles are localized in the cytoplasmic matrix next to the endoplasmic reticulum and over the nucleus (Morel, *In situ* hybridization on ultrathin frozen sections, in *Hybridization Techniques for Electron Microscopy*, CRC Press, Boca Raton, 1993, 163, with permission).

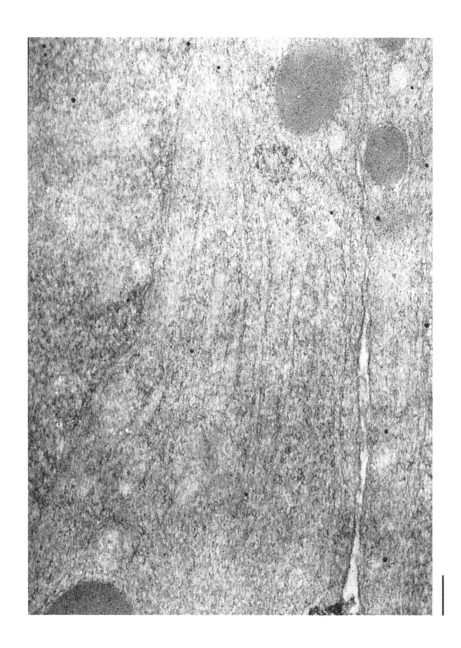

Figure 18

Figure 19

1. Description

2. Method

3. Tissue
4. Fixation
5. Embedding
6. Pretreatment
7. Probe

8. Visualization
   • Autoradiography
9. Bar
10. Comments

↬ Visualization by ultrastructural autoradiography of mRNA encoding growth hormone in a somatotroph

↬ Frozen tissue technique
↬ *See* Chapter 6.

↬ Rat anterior pituitary
↬ 2% paraformaldehyde
↬ None
↬ None
↬ cDNA labeled with $\alpha[^{35}S]dCTP$ by nick translation

↬ 1 µm
↬ The silver grains (twisted threads) are found over cells characteristic of somatotrophs. Other cells are not labeled (i.e., gonadotrophs: LH-FSH). The signal is present over the cytoplasm and the nucleus of somatotrophs (Morel, Dihl, and Gossard, Ultrastructural distribution of GH mRNA and GH intron I sequence in rat pituitary gland: effects of GH-releasing factor and somatostatin, *Mol. Cell. Endocrinol.*, 65, 81, 1989, with permission).

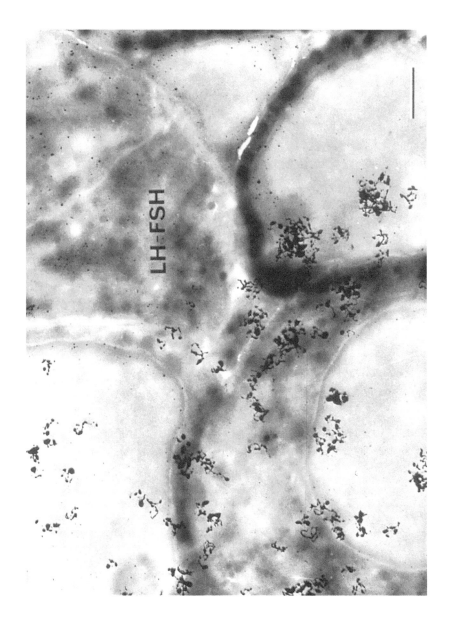

Figure 19

Figure 20

1. Description

   ⇝ Detection of mRNA encoding the prolactin receptor in an hepatocyte

2. Method

   ⇝ Frozen tissue technique

   ⇝ *See* Chapter 6.

3. Tissue

   ⇝ Rat liver

4. Fixation

   ⇝ 4% paraformaldehyde

5. Embedding

   ⇝ None

6. Pretreatment

   ⇝ None

7. Probe

   ⇝ Anti-sense oligonucleotide (30 mers) encoding the prolactin receptor labeled with dUTP 16-digoxigenin by 3′ incorporation

8. Visualization
   • Immunocytology

   ⇝ Colloidal gold (10 nm)

9. Bar

   ⇝ 0.5 μm

10. Comments

   ⇝ The colloidal gold particles (arrowheads) are localized in the cytoplasmic matrix close to the endoplasmic reticulum in a hepatocyte. (M) mitochondria. (Ouhtit, Ronsin, Kelly, and Morel, Ultrastructural expression of prolactin receptor in rat liver, *Biol. Cell*, 82, 169, 1994, with permission).

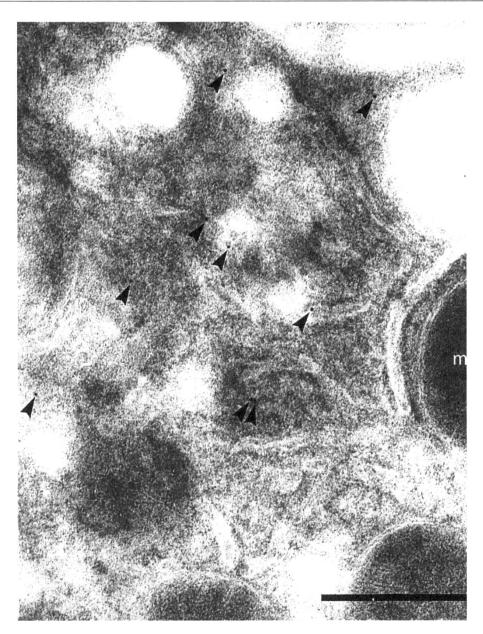

Figure 20

Figure 21

1. Description	↪ Detection of mRNA encoding the GnRH receptor and growth hormone (GH) receptor
2. Method	↪ Frozen tissue technique
	↪ *See* Chapter 6.
3. Tissue	↪ Rat anterior pituitary
4. Fixation	↪ 4% paraformaldehyde
5. Embedding	↪ None
6. Pretreatment	↪ None
7. Probes	↪ Oligonucleotide (30 mers) labeled with dUTP 16-digoxigenin by 3′ incorporation
• Probe A	↪ Anti-sense mRNA encoding the GH receptor
• Probe B	↪ Anti-sense mRNA encoding the GH receptor
8. Visualization	
• Immunocytology	↪ Colloidal gold (10 nm)
9. Bar	↪ 0.5 μm
10. Comments	↪ Probes:

- **(A)** The particles of colloidal gold (arrowheads) revealing the presence of mRNA encoding the GnRH receptor are localized in the cytoplasm close to the endoplasmic reticulum in a gonadotroph (Mertani, Testard, Ouhtit, Brisson, and Morel, Gonadotropin-releasing hormone receptor gene expression in rat anterior pituitary, *Endocrine*, 4, 1, 1996, with permission).
- **(B)** The mRNA encoding the GH receptor is found in three cell populations in the anterior pituitary (somatotrophs, lactotrophs, gonadotrophs). In somatotrophs, particles of colloidal gold (arrowheads) are found in the cytoplasm (cytoplasmic matrix, endoplasmic reticulum (er)) and the nucleus (N), but not the mitochondria (m) or the secretion granules (sg) (Mertani, Waters, Jambou, Gossard, and Morel, Growth hormone receptor/binding protein in rat anterior pituitary, *Neuroendocrinology*, 59, 483, 1994, with permission.)

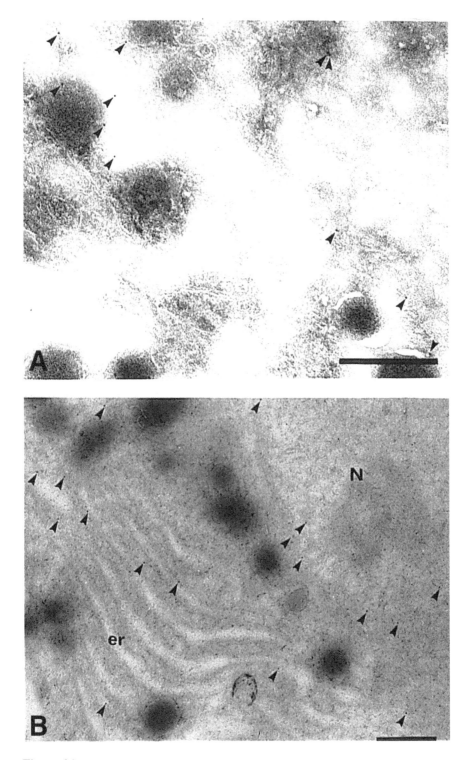

Figure 21

Figure 22

1. Description

      ⇝ Simultaneous localization of mRNA coding the atrial natriuretic factor (ANF) receptor (A) and the GnRH receptor (B) and detection of prolactin by immunocytochemistry to characterize lactotrophs

2. Method

      ⇝ Frozen tissue technique

      ⇝ *See* Chapter 6.

3. Tissue

      ⇝ Rat anterior pituitary

4. Fixation

      ⇝ 4% paraformaldehyde

5. Embedding

      ⇝ None

6. Pretreatment

      ⇝ None

7. Probes

      ⇝ Oligonucleotide (30 mers) labeled with dUTP 16-Digoxigenin by 3′ incorporation

   • Probe A

      ⇝ Anti-sense for mRNA encoding the type A ANF receptor

   • Probe B

      ⇝ Anti-sense for mRNA encoding the GnRH receptor

8. Visualization

   • Immunocytology

      ⇝ Colloidal gold:
   - Particles of 15 nm for the detection of the hybrid
   - Particles of 5 nm for the detection of the protein.

9. Bar

      ⇝ 0.5 μm

10. Comments

      ⇝ The lactotroph is characterized by the presence of secretion granules (g) where the protein is stored (prolactin). The 5 nm particles of colloidal gold are localized over the secretion granules while the 15 nm particles of colloidal gold (localization of mRNA) are localized over the cytoplasmic matrix close to the endoplasmic reticulum (er). (N) nucleus (A: Grandclément, Cavalier, and Morel, Visualization of mRNA receptor by ultrastructural *in situ* hybridization, in *Visualization of Receptors*, CRC Press, Boca Raton, 1997, 321, with permission; B: Mertani, Testard, Ouhtit, Brisson, and Morel, Gonadotropin-releasing hormone receptor gene expression in rat anterior pituitary, *Endocrine*, 4, 1, 1996, with permission).

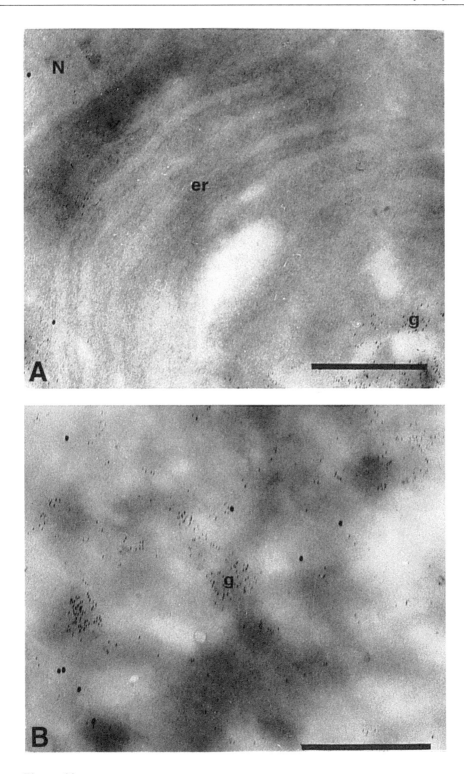

Figure 22

Figure 23

1. Description

⇝ Comparison of the three techniques of *in situ* hybridization to visualize mRNA encoding growth hormone in the anterior pituitary of the rat (*see* Figures 23 to 25)

2. Method

⇝ Post-embedding technique
⇝ *See* Chapter 4.

3. Tissue

⇝ Rat anterior pituitary

4. Fixation

⇝ 4% paraformaldehyde

5. Embedding

⇝ Lowicryl K4M

6. Pretreatment

⇝ None

7. Probe

⇝ Oligonucleotide (30 mers) anti-sense of mRNA for growth hormone labeled with dUTP 21 biotin by 3' incorporation

8. Visualization
   • Immunocytology

⇝ Colloidal gold (5 and 10 nm)

9. Bar

⇝ 0.5 µm

10. Comments

⇝ The colloidal gold particles 5 or 10 nm (insert) are localized close to the endoplasmic reticulum found in parallel rows characteristic of somatotrophs. The density of the signal is weaker with the 10 nm particles (Le Guellec, Trembleau, Pechoux, Gossard, and Morel, Ultrastructural non-radioactive *in situ* hybridization of GH mRNA in rat pituitary gland: pre-embedding vs. ultrathin frozen sections vs. post-embedding, in *J. Histochem. Cytochem.*, 40, 979, 1992, with permission).

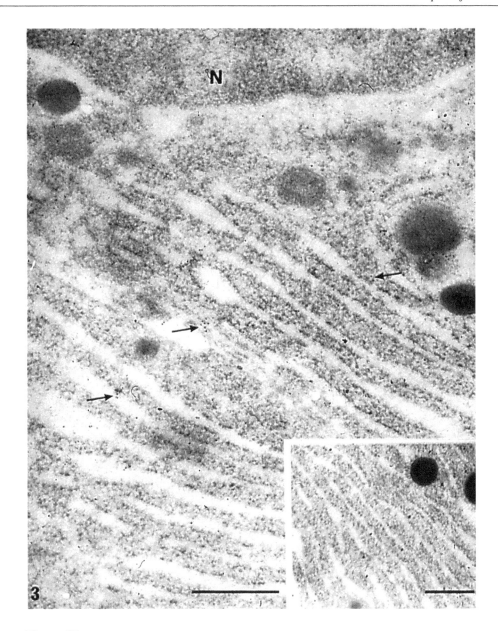

Figure 23

Figure 24

1. Description

⇨ Comparison of the three techniques of *in situ* hybridization to visualize mRNA encoding growth hormone in the anterior pituitary of the rat (*see* Figures 23 to 25)

2. Method

⇨ Pre-embedding technique
⇨ *See* Chapter 5.

3. Tissue ⇨ Rat anterior pituitary
4. Fixation ⇨ 4% paraformaldehyde
5. Embedding ⇨ Epon
6. Pretreatment ⇨ None
7. Probe

⇨ Anti-sense oligonucleotide (30 mers) to mRNA for growth hormone labeled with dUTP 21 biotin by 3′ incorporation

8. Visualization
   • Immunocytology

⇨ Avidin-biotin-peroxidase/DAB complex

9. Bar ⇨ 0.5 μm
10. Comments

⇨ The precipitate of the immunocytochemical reaction is found between the parallel rows of endoplasmic reticulum (er) characteristic of somatotrophs. The signal is absent over the mitochondria (m) and the nucleus (N) (Le Guellec, Trembleau, Pechoux, Gossard, and Morel, Ultrastructural non-radioactive *in situ* hybridization of GH mRNA in rat pituitary gland: pre-embedding vs. ultrathin frozen sections vs. post-embedding, *J. Histochem. Cytochem.*, 40, 979, 1992, with permission).

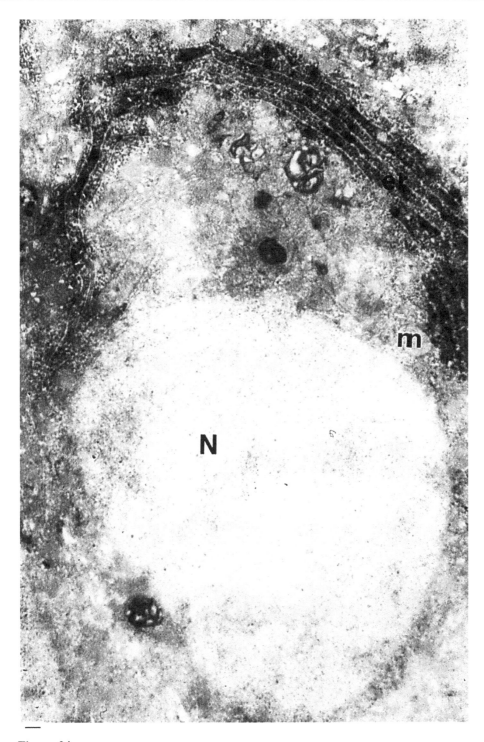

Figure 24

Figure 25

1. Description	➥ Comparison of the three techniques of *in situ* hybridization to visualize mRNA encoding growth hormone in the anterior pituitary of the rat (*see* Figures 23 to 25)
2. Method	➥ Frozen tissue technique
	➥ *See* Chapter 6.
3. Tissue	➥ Rat anterior pituitary
4. Fixation	➥ 4% paraformaldehyde
5. Embedding	➥ None
6. Pretreatment	➥ None
7. Probe	➥ Anti-sense oligonucleotide (30 mers) to mRNA for growth hormone labeled with dUTP 21 biotin by 3′ incorporation
8. Visualization	
• Immunocytology	➥ Colloidal gold (5 and 10 nm)
9. Bar	➥ 0.5 µm
10. Comments	➥ The 5 and 10 nm (insert) colloidal gold particles are localized close to the endoplasmic reticulum in parallel rows, characteristic of somatotrophs over the cytoplasmic matrix between the 350 nm secretion granules. The density of the signal is weaker with the 10 nm particles. The concentration of the probe used to give a signal of comparable intensity to Figure 24 is 25 times weaker (Le Guellec, Trembleau, Pechoux, Gossard, and Morel, Ultrastructural non-radioactive *in situ* hybridization of GH mRNA in rat pituitary gland: pre-embedding vs. ultrathin frozen sections vs. post-embedding, *J. Histochem. Cytochem.*, 40, 979, 1992, with permission).

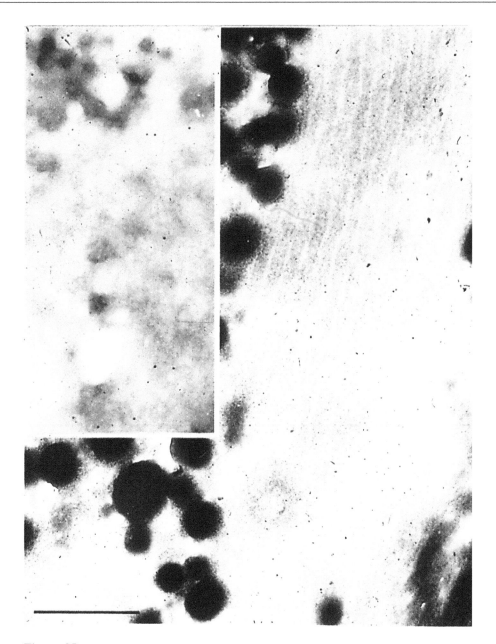

Figure 25

Glossary

A

Acetylation
⇨ Transformation of the amino group, $-NH_2$ (reactive), in a protein into a substituted amide group, $-NH-CO-CH_3$ (neutral), by the action of an acetic anhydride, $CH_3-CO-CH_3$, in a triethanolamine buffer with a pH of 8. This reaction is effective in reducing background noise levels, though it also acts on the nitrogenous bases of nucleic acids and can cause a reduction in the hybridization signal.

Acrylic
⇨ Resin (*see* Chapter 4, Section 4.3.2).

Adenine
⇨ A purine base which is found in nucleosides, nucleotides, coenzymes, and nucleic acids. It establishes two hydrogen bonds with thymine and deoxyuracil in double-stranded nucleic acids.

Alkaline phosphatase
⇨ A phosphatase extracted from calf intestine (MW = 80 kDa). Its optimum pH is alkaline (*see* Chapter 7, Section 7.2.5.2).

Amino acid
⇨ The elementary unit in proteins. Each one contains an amino group, $-NH_2$, and an organic acid group, $-COOH$. The amino acid chains that make up the proteins are called polypeptide chains.

Ångström
⇨ 1 Å = 10^{-1} nanometer (or 10^{-10} m).

Antibody
⇨ A glycoprotein produced as a response to the introduction of an antigen. It can combine with the antigen that induced its production.

monoclonal
⇨ An antibody that is specific to an epitope produced by a cell culture resulting from the fusion of a tumoral cell and a cell that produces antibodies.

polyclonal
⇨ All the immunoglobulins contained in the serum from an animal immunized with a given antigen. Polyclonal antibodies recognize several epitopes or antigenic determinants.

primary
⇨ The first antibody used in the specific immunological reaction of the antigen being sought.

secondary
⇨ Antibody used for indirect immunological reactions, directed against the species in which the primary antibody was raised. May or may not be conjugated.

Antigen
⇨ Any substance that is foreign to a receiving organism is said to be immunogenic when it induces an immune response (which can engender the production of antibodies).

Antigenic determinant	➩ *See* Epitope.
Antiserum	➩ Serum containing immunoglobulins.
Araldite	➩ Epoxy resin made up of chains with aromatic rings of polyarylethers of glycerol (*see* Chapter 5, Section 5.7.2.1.3).
Artifact	➩ Background noise, nonspecific reactions.
ATPase	➩ Adenosine triphosphatase (an enzyme which catalyzes the dephosphorylation of ATP).
Autoclave	➩ A device which is used to sterilize objects using pressurized steam.
Autoradiographic activity	➩ This expresses the number of disintegrations per second which take place in a radioactive source. The unit is the Becquerel, which corresponds to one disintegration per second; the previous unit was the Curie (1 Bq = 2.7×10^{-11} Curie).
Autoradiographic efficacy	➩ The number of silver grains obtained by β^- radiation.
Autoradiographic resolution	➩ Distance between the location of a silver grain and the source of the radioactive emission (a function of the energy of the isotope and the autoradiographic system).
Autoradiography/radioautography	➩ Method for visualizing hybrids containing isotopes.
Avidin	➩ A glycoprotein (MW = 16.4) produced in the uterus of the chicken. It contains four subunits, each of which binds a molecule of biotin with a very high affinity: Kd 10^{-15}.

B

Background	➩ Nonspecific signal (artefact).
Base pair	➩ This is a parameter which is used for expressing the length of a double strand of DNA. A base pair consists of an association of two nucleotides one on each of the two complementary strands of DNA.

C

Carbon coating	➩ Organic film evaporated onto collodion or formvar coated grids.
Cellular clone	➩ All the cells that are produced by the division of a given mother cell (genetically identical cells).

Chelator	➳ Agent (e.g., EDTA) that captures Ca^{++} ions. It blocks enzymatic action.
Chemography	➳ Chemical effect of biological preparation on photographic emulsion, which causes background.
Chromogen	➳ Substrate modified by an enzyme to produce a colored substance.
Chromosome	➳ The name comes from the Greek *khroma*, which means color. These structures contain most or all of the cellular DNA and, thus, the genetic information.
Cloning vector	➳ A DNA molecule which is capable of replicating itself (replicon). It is used to transport a piece of foreign DNA (e.g., a gene) into a target cell. This vector can be a plasmid, a phage, etc.
Codon	➳ A sequence of three bases (triplet) in the mRNA that codes for an amino acid. It can also signal the initiation or the end of the transcription process.
initiation	➳ Codon, AUG, in mRNA that starts protein synthesis. AUG also codes for the amino acid methionine.
termination	➳ Codons that do not code for an amino acid (e.g., UAA, UAG, UGA), but signal termination of protein synthesis.
Collodion	➳ Collodion is a purified form of nitrocellulose used as an organic film on an electron microscopic grid.
Complementary DNA (cDNA)	➳ A DNA copy of an RNA molecule, generally messenger RNA.
Counterstaining	➳ Secondary staining which makes it possible to obtain a different contrast from that of the hybridization signal.
Cryofixation	➳ Stabilization of tissue by frozen tissue technique.
Cryopolymerization	➳ The hardening of resin by UV irradiation at a low temperature.
Cryoprotector	➳ Molecule with the property of passing, in aqueous solution, from a liquid to a solid state via an amorphous phase (i.e., without crystallization).
Cryoprotection	➳ Incubation of samples in a solution that limits crystallization during freezing.
Cryosection	➳ Section cut from frozen tissue.
Cryosubstitution	➳ Infiltration of a fixed specimen, at low temperature, by mixtures of solvent–resin, with increasing concentrations of resin then by pure resin.

Cryosubstitution chamber	⇝ Automatic apparatus for « Automatic Freeze-Substitution-System » (AFS). Dehydrates, impregnates and polymerizes biological samples in hydrophilic resin at low temperature.
Cryo-ultramicrotome	⇝ Apparatus for cutting ultrathin and semithin sections of frozen tissue.
Cryo-ultramicrotomy	⇝ Technique for cutting ultrathin and semithin sections of frozen tissue.
Cytosine	⇝ A pyrimidine base which is a component of nucleosides, nucleotides and nucleic acids.

D

Dehydration	⇝ Progressive replacement of cell water by successive changes in increasing concentrations of solvents (e.g., ethanol, acetone…).
Denaturation	⇝ Separation of the two strands of the double helix of DNA, either by heat or through the action of other chemical agents (change of configuration of the nucleic acids). It can also be applied to certain RNAs.
Denhardt's solution	⇝ Mixture of polymers which reduces background.
Deoxynucleoside	⇝ Molecule consisting of a combination of a base and a deoxyribose.
Deoxynucleotide	⇝ Molecule consisting of a combination of a base, a deoxyribose, and one or more phosphates.
Deoxyribonucleic acid (DNA)	⇝ Nucleic acid which is found in the nucleus of the cell (and carries genetic information in all cells). The DNA molecule is a polynucleotide which is composed essentially of deoxyribonucleotides linked by phosphodiester bonds.
Deoxyribose	⇝ A ribose which has lost an atom of oxygen at $2'$.
Deproteinization	⇝ Enzymatic treatment which partially eliminates proteins, more particularly, those which are associated with nucleic acids.
Dextran sulfate	⇝ It increases the efficiency of the hybridization process by concentrating the probe. Useful in the case of cDNA probes.
Dissecting needle	⇝ Tool with fine point for positioning semithin or ultrathin sections without damage.

DNA ligase	⇝ An enzyme combining two fragments of DNA by the formation of a new phosphodiester bond.
DNA polymerase	⇝ An enzyme which synthesizes a new DNA molecule from a DNA matrix.
DNase	⇝ An enzyme which destroys DNA in a specific way.
Duplication	⇝ The DNA molecule is able to make an exact copy of itself by a process called duplication.

E

Electron microscope	⇝ Microscope for observing ultrathin sections of tissue. Sections are mounted on grids and an electron beam passes through them. Heavy atomic weight substances deflect the electrons, producing contrast.
ELISA test	⇝ Enzyme-linked immunosorbent assay: technique which is used to detect and quantify specific antibodies and antigens.
Embedding	⇝ Treatment (dehydration, impregnation, polymerization) results in hardened tissue and allows ultrathin sectioning of tissue.
Endogen	⇝ A molecule which is found in a biological sample.
Energy of emission	⇝ Energy liberated by the disintegration of an isotope (the energy is measured in MeV, i.e., mega-electron-volts).
Enrobage	⇝ A layer of sucrose or methylcellulose is placed on a tissue or section.
Enzyme	⇝ A protein which catalyzes a specific chemical reaction. The name of an enzyme generally specifies its function: the addition of the suffix -ase to the name of the type of molecule on which the enzyme acts means that it is broken down by this enzyme (e.g., ribonuclease cuts RNA molecules; proteases hydrolyze proteins).
Epipolarization microscope	⇝ Polarized light can be used to reinforce the contrast observed with certain transparent or opaque reflecting objects.

Epitope	➯ Also termed antigenic determinant. The part of the antigen which is recognized by an antibody (an epitope is a structure which can combine with a single antibody molecule). An antigen can have several epitopes, each one corresponding to a sequence of 3–6 contiguous or non-contiguous amino acids.
Epon	➯ Epoxy resin (*see* Chapter 5, Section 5.7.2.1.1).
Epon-Araldite	➯ Epoxy resin (*see* Chapter 5, Section 5.7.2.1.4).
Epoxy	➯ Hydrophobic resin (*see* Chapter 5, Section 5.7.2).
Exogen	➯ A substance introduced by contamination from outside (recipients, etc.).
Exon	➯ A translated (or coding) sequence of DNA.
Expression vector	➯ A particular cloning vector for expressing recombinant genes in host cells. The recombinant gene is transcribed, and the protein for which it codes is synthesized.

F

$F(ab')_2$	➯ Fragment obtained by digesting the antibody with the enzyme pepsin (MW = 100 kDa).
Fab	➯ Antigen-binding fragment of immunoglobulin (MW = 45 kDa) which is obtained by cutting an antibody with an enzyme (papain). The enzymatic cut is situated upstream from the disulfur bonds which link the two heavy chains of the immunoglobulin. The fragment represents 2/3 of the native molecule.
Fc fragment	➯ A fragment (MW = 50 kDa) which represents 1/3 of the native molecule. It has no antibody activity, but is easily crystallized; hence its name.
Formamide	➯ A molecule which favors the denaturation of nucleic acids. The addition of 1% formamide to the hybridization solution lowers the hybridization temperature by 0.65°C for DNA and by 0.72°C for RNA.
Formvar	➯ Organic film that supports and improves the adherence to grids of ultrathin sections.
Freezing	➯ The stopping of all biological processes and the solidification of the sample by low temperatures.
Freezing temperature	➯ Temperature of a liquid–solid mixture.
	➯ Temperature at which the sample becomes frozen.

G

Gene
⇨ A segment of DNA involved in producing a polypeptide chain. It includes regions preceding and following the coding region (exons) and introns.

Genetic engineering
⇨ In general terms, the techniques used to deliberately modify genetic information about an organism through changes in its genomic nucleic acid.

Genome
⇨ The entire set of genes to be found in an eucaryotic cell, bacterium, yeast, or virus.

Genotype
⇨ The genotype represents the totality of genetic information for all the genes of a given organism.

Glutaraldehyde
⇨ Fixative that forms bridges between cellular and tissue components (*see* Chapter 3, Section 3.4.1).

GMA
⇨ Glycol methacrylate is a very fluid hydrophilic acrylic resin made up of a mixture of several methacrylates and Luperco which polymerizes under UV at 4°C. This was one of the first resins used for electron microscopy.

Guanine
⇨ A purine base which is found in nucleosides, nucleotides, and nucleic acids.

H

Hapten
⇨ A nonimmunogenic molecule which is nonetheless able to induce the production of antibodies directed against itself when it is coupled to a macromolecular carrier.

Haptenization
⇨ Labeling of nuclear probes using haptens introduced either by a chemical reaction or by enzymatic action.

Histones
⇨ Proteins which are rich in arginine and lysine, associated with DNA in eucaryotic cells and chromosomes.

Horseradish peroxydase
⇨ An enzyme extracted from horseradish, a plant of the Cruciferae family (*see* Chapter 7, Section 7.2.5.2).

Hybridization of nucleic acids
⇨ The formation of hybrid molecules of double-stranded DNA, DNA–RNA, or RNA–RNA. Insofar as the sequences are complementary, stable hybrids will be formed.

| Hydrophilic | ⇝ This is said of a polar substance which has a strong affinity for water or which dissolves easily in water. |
| Hydrophobic | ⇝ This is said of a non-polar substance which has no affinity for water or which does not easily dissolve in water. |

I

Immunocytochemistry	⇝ Provides cytological evidence for antigenic constituents by means of an immunochemical reaction.
Immunocytology	⇝ Allows the localization of antigens by an immunological reaction.
Immunoglobulin	⇝ *See* Antibody
Immunoglobulin G (IgG)	⇝ A class of immunoglobulins which are predominant in serum (> 85% of the immunoglobulins in serum) [MW = 150 kDa].
Immunoglobulin M (IgM)	⇝ A class of serous immunoglobulins (MW = 900 kDa) formed of heavy chains and light chains such as IgGs. The molecules are in the shape of a 5-pointed star with a central ring.
Impregnation	⇝ Water in the tissue is gradually replaced by embedding media which is solidified by polymerization.
Intron	⇝ Segment of DNA that is transcribed but excised during splicing, leading to the maturation of mRNA.
Isotopes	⇝ Atoms of a given element which have different atomic masses, though their chemical properties are very similar.

K

| Knife-maker | ⇝ Apparatus for making glass knives for cutting frozen or resin embedded sections. |

L

| Lowicryls | ⇝ Composed of a mixture of acrylate and methacrylate monomers. These resins are very fluid at low temperatures.
⇝ Hydrophilic and/or hydrophobic resins (*see* Chapter 4, Section 4.3.4). |

hydrophilics	↪ Or polar.
	↪ These resins have the same viscosity.
Lowicryl K4 M	↪ Polymerization at −20°C–35°C.
Lowicryl K11 M	↪ Polymerization at −60°C.
hydrophobics	↪ Or non polar.
Lowicryl HM 20	↪ Polymerization at −40°C.
Lowicryl HM 23	↪ Polymerization at −80°C.
LR White (London Resin Gold)	↪ Very hydrophilic acrylic resin composed of a mixture of methacrylate and hardener (*see* Chapter 4, Section 4.3.5).
	↪ Polymerization may take place in two ways depending on the hardness required (4°C or 50°C).

M

Melting temperature (Tm)	↪ The temperature at which 50% of double-stranded DNA separates out into individual chains. It depends on the C and G content of the DNA, the Na^+ concentration, and the length of the strands (*see* Chapter 7, Section 7.2.2.1).
Mer	↪ Unit of nucleic acid: 1 mer corresponds to 1 nucleotide or 1 base-pair.
Methacrylate	↪ Acrylic resin made up of a monomer of aromatic polyhydroxyl.
Micrometer	↪ 1 µm = 10^{-3} millimeter (or 10^{-6} m).

N

Nanometer	↪ 1 nm = 10^{-3} micrometer (or 10^{-9} m).
Nonspecificity	↪ Non-homogeneous (probe–non-complementary nucleic acid), or heterogeneous (protein–nucleic acid) matching.
Northern blot technique	↪ Hybridization of RNA on a membrane.
Nucleic acids	↪ Macromolecules, the vehicles for genetic information.
Nucleic acids	↪ Macromolecules that carry the genetic information comprising Deoxyribonucleic acid (DNA) and ribonucleic acid (RNA).
targets	↪ Nucleic acids to be localized in the cells.
complimentary sequences	↪ Nucleic acids which act as probes.
genome	↪ Contains the genetic information of a cell
exogen	↪ Nucleic acids present in a cell but not part of expression of its genome (i.e., viral, bacterial, or fungal origin).
mitochondrial	↪ Nucleic acids present in the mitochondria.

Nucleoside	↪ A molecule which is a combination of a purine or pyrimidine base and a sugar (ribose).
Nucleosome	↪ A molecule of DNA associated with basic proteins (histones) in the eucaryotic nuclear matrix. Histones participate in the formation of chromatins, which are made up of flexible chains of repeated units, namely the nucleosomes.
Nucleotide	↪ A nucleoside plus 1 or more phosphates.
Nucleotide triphosphate	↪ A nucleotide triphosphate is made up of a purine or pyrimidine base plus a sugar (ribose or deoxyribose) and 3 phosphate groups.

O

Oligonucleotide	↪ From the Greek *oligoi*, meaning "few." Oligonucleotides are short fragments of DNA, with 15 to 60 nucleotides.
Osmium tetroxide	↪ Fixative sometimes incorrectly called osmic acid (OsO_4).

P

Palindrome	↪ DNA containing two complementary sequences which form an intrachain hybrid.
Peptide	↪ A short chain of amino acids.
Phenotype	↪ A set of characteristics of an organism resulting from interaction of the genome and the environment.
Phosphatase	↪ An enzyme which hydrolyzes phosphate groups in molecules.
Phosphorylation	↪ The addition of a phosphate group by the action of an enzyme.
Plasmid	↪ A molecule of DNA into which it is possible to introduce a DNA insert. It is used as a cloning vector.
PLT	↪ Progressive lowering temperature; a method of progressive dehydration and embedding at low temperatures using liquid nitrogen vapors (*see* Chapter 4, Section 4.3.7).
Poly (A)	↪ A repetitive polynucleotide sequence which is attached by the action of a ligase to the 3′ end of a transcribed mRNA molecule. This is characteristic of mRNA.
Polymerase	↪ A synthesizing enzyme for DNA or RNA.

Polymerase chain reaction (PCR)	⇝ A reaction that takes place in repeated cycles (denaturation, hybridization, elongation) which makes possible the amplification of a given sequence of DNA.
Polymerization	⇝ Polymerization results in very hard resin enabling sectioning. Hardening is made possible by the addition of a catalyst that starts a chain reaction. The reaction also takes place between chains for acrylic resins or in three dimensions with epoxy resins.
Polynucleotide	⇝ A chain formed by a combination of several nucleotides (RNA is a polynucleotide).
Polynucleotide kinase	⇝ The kinase enzyme (which is extracted from calf thymus) is used in the radioactive labeling of oligonucleotide probes by phosphorylating the 5′ end (phosphate group in γ position).
Post-fixation	⇝ A complementary fixation using another fixative.
Prehybridization	⇝ Incubation of sections in the hybridization solution without the probe.
Pre-messenger RNA	⇝ Single-stranded RNA.
Pretreatment	⇝ Stage before hybridization.
Primer	⇝ Oligonucleotide which acts as a template to start the synthesis of nucleic acid by DNA polymerase (e.g., random priming, PCR).
Probe	⇝ A fragment of nucleic acid (DNA or RNA) whose nucleotide sequence is complementary to that of the nucleic acid being sought, which is immobilized in the preparation (target).
anti-sense probe	⇝ A complementary sequence to that of the target; it hybridizes specifically with the complementary sequence which is immobilized *in situ* (target).
non-sense	⇝ A complementary but sense sequence to the target used for negative hybridization.
oligonucleotide probe	⇝ In cases where the sequence is known, it is possible to make a single-stranded probe using synthetic oligonucleotides (≈ 30 nucleotides) whose sequences are complementary to those of the nucleic acids being sought.
sense probe	⇝ A sense probe is a homologous copy to that of the target. It serves as a negative control.
Promoter	⇝ A sequence of a sense strand of DNA situated upstream from a gene, to which polymerase RNA is linked before the start of the transcription process.
Protein	⇝ A macromolecule composed of amino acids.

Proteinase	⮑ Enzyme which hydrolyzes proteins into amino acids.
Purine bases	⮑ The purine bases are guanine and adenine, (*see* Chapter 7, Section 7.2.1) whose chemical structures resemble those of the pyrimidines, combined with a pentagonal structure (pyrimidine cycle plus imidazol cycle).
Purine	⮑ A basic nitrogenous molecule composed of two aromatic nuclei, found in nucleic acids and other cell components. It is made up principally of the bases adenine and guanine.
Pyrimidine bases	⮑ The pyrimidine bases are cytosine and thymine (DNA) or uracil (RNA) [*see* Chapter 7, Section 7.2.1], whose chemical structure are hexagonal.
Pyrimidine	⮑ A basic nitrogenous molecule composed of an aromatic nucleus, found in nucleic acids and other cell components. It is made up principally of the bases cytosine, uracil, and thymine.

R

Radioactivity	⮑ A property possessed by certain elements of transforming themselves by disintegration into other elements, with the emission of different forms of radiation.
Radioautography	⮑ *See* Autoradiography.
Random priming	⮑ A method for random labeling by duplication of all DNA, starting with nonspecific primers (*see* Chapter 1, Section 1.3.1).
Recombinant DNA technology	⮑ Techniques used in genetic engineering for the identification and isolation of specific genes, their insertion into a vector such as a plasmid to form recombinant DNA, and the production of large quantities of the gene and its product.
Renaturation	⮑ Rematching of complementary nucleic acids.
Replication	⮑ Synthesis of two new strands of DNA by complementary matching of the nucleotides which line each of the two strands.
Resins	⮑ Epoxy and/or acrylic resins allow very thin sections of tissue to be cut.
acrylic	⮑ These resins are principally hydrophilic and are used almost exclusively for immunocytochemistry (e.g., Lowicryls, LR White, Unicryl, etc.).

	↪ These are made up of a mixture of acrylates–methacrylates, and are very fluid before polymerization, allowing rapid infiltration into the tissue.
epoxy	↪ These resins are principally hydrophobic resins used exclusively for morphology (e.g., Epon, Araldite, Spurr…).
hydrophilic	↪ Resins that will polymerize in the presence of a small amount of water (e.g., acrylic resins).
hydrophobic	↪ Resins that will not polymerize in the presence of water (e.g., epoxy resins).
Restriction enzyme	↪ A catalytic protein which cuts DNA at a specific site. Such enzymes are indispensable to the manipulation of DNA *in vitro*.
Restriction map	↪ DNA map that shows the sites recognized and cut by restriction enzymes.
Ribonuclease (RNase)	↪ An enzyme which breaks down RNA molecules.
Ribonucleic acid (RNA)	↪ A polynucleotide composed of ribonucleotides linked by phosphodiester bonds. There is messenger RNA, transfer RNA, ribosomic RNA, and viral RNA.
transfer ribonucleic acid (tRNA)	↪ A short RNA chain which transports amino acids during the synthesis of proteins.
messenger ribonucleic acid (mRNA)	↪ Single-stranded RNA synthesized from a DNA matrix during transcription. It links to ribosomes, and carries messages used for protein synthesis.
ribosomal ribonucleic acid (rRNA)	↪ Ribosomal RNA is a component of ribosomes. Several single-stranded rRNA molecules of different size contribute to the structure of ribosomes and are also directly involved in protein synthesis.
Ribose	↪ A ribonucleic acid (sugar in RNA).
RNA polymerase	↪ Enzyme which catalyzes the synthesis of RNA from a DNA matrix. RNA polymerase recognizes the errors that result from incorrect matching. Thanks to its exonucleasic activity, it removes erroneous nucleotides for which it substitutes the appropriate nucleotides.
RNase	↪ An enzyme which breaks down RNA.
RNase A	↪ This enzyme breaks down only single-stranded RNA. RNase treatment carried out after hybridization makes it possible to reduce background noise.
RNase H	↪ This enzyme destroys DNA–RNA hybrids. This RNase H treatment can be used as negative control after hybridization.

RNase-free conditions

⇨ Experimental conditions in which all contamination by exogenous ribonucleases is eliminated so as to preserve mRNA.

S

Saline concentration

⇨ Ionic strength, *see* Salinity

Salinity

⇨ The concentration of Na ions, which influences the stability of the hybrids. The hybridization speed increases with the concentration of salts.

Sections

⇨ *See* Ultramicrotomy.

  semithin

⇨ Sections (≤ 1 μm) mounted on a glass slide for light microscopy.

  ultrathin

⇨ Ultrathin sections (80–100 nm) mounted on nickel grids for electron microscopy.

Sensitivity

⇨ The smallest quantity of nucleic acid target that can be detected in a cell or tissue, or the number of molecules that can be detected by a given *in situ* hybridization reaction.

Serum

  immune

⇨ Serum after immunization.

  non immune

⇨ Serum from the same species—non immunized.

  pre-immune

⇨ Serum taken prior to immunization.

Signal

⇨ Visible *in situ* hybridization reaction product allowing localization of the hybrid.

Southern blot technique

⇨ Detection of specific fragments of DNA transferred to a membrane by hybridization with a labeled probe.

Specific activity

⇨ The specific activity of a probe results from its labeling; i.e., the number of isotopes or antigens incorporated, in comparison with the mass, or the concentration, of the probe.

Specificity

⇨ Total complementarity of matching between two molecules of nucleic acid.

Spliceosome

⇨ Site where the splicing of the mRNA precursor takes place.

Splicing of RNA

⇨ Process during which introns are cut out of the primary transcripts of RNA in the nucleus during the formation of mRNA.

Stability

⇨ Between two molecules of nucleic acid. It depends on their nature, and there are three types of duplex which can be formed: DNA–DNA, DNA–RNA, RNA–RNA, in increasing order of stability.

Stain

⇨ Heavy metal salts (i.e., uranium and lead salts).

Staining	⇒ Heavy metal salts are deposited onto ultrathin sections to provide contrast for electron microscopy.
positive	⇒ Stained structures appear dark on a light background.
negative	⇒ Stained structures appear light on a dark background.
Sterilization	⇒ A process whereby all living cells and viruses are either destroyed or eliminated from an object or a habitat.
Streptavidin	⇒ A protein of bacterial origin which has a high affinity for four biotin molecules and a very weak charge and which causes little background. Its characteristics are similar to those of avidin.
Stringency	⇒ A parameter which is used to express the efficacy of hybridization and washing conditions (depending on the concentrations of salt and formamide and the temperature). A low level of stringency favors nonspecific matchings, whereas too high a level gives rise to a specific signal of low intensity.
Substrate	⇒ A substance which is acted on by an enzyme.
Super-cooling temperature (Tm)	⇒ Nucleation point.
	⇒ The lowest temperature obtained in the liquid phase during freezing (*see* Figure 7.8).

T

Target	⇒ The nucleic sequence that is being sought within the cell.
TdT or terminal deoxytransferase	⇒ An enzyme which is used for labeling oligonucleotide probes by elongation of the 3′ end, provided that this is hydroxylated (free OH). It can polymerize NTP and dNTP.
Thymine	⇒ A pyrimidine base which is found in nucleosides, nucleotides, and DNA.
Transcriptase	⇒ Enzyme which catalyzes transcription. In RNA viruses it is an RNA-dependent RNA polymerase which is used to make mRNA from RNA genomes.
Transcription	⇒ A process in the course of which single-stranded RNA is synthesized using a DNA matrix.
Transgenic	⇒ An animal or plant whose genome contains new genetic information in a stable manner, due to the acquisition of foreign DNA.

Translation	↪ Process that occurs whereby information in mRNA is transformed into protein synthesis.

U

Ultramicrotome	↪ Apparatus for cutting resin embedded tissues of variable thickness: semithin (≤ 1 µm) and ultrathin (80 to 100 nm). The section thickness is controlled by a thermal advance mechanism.
Ultramicrotomy	↪ Cutting ultrathin sections of embedded tissue using an ultramicrotome.
Unicryl	↪ Very hydrophilic methacrylate resin with low viscosity allowing rapid infiltration of tissue.
Uracil	↪ A pyrimidine base which is found in nucleosides, nucleotides, and RNA.
Uridine	↪ Nucleotide formed by uracil and ribose.

V

Viral ribonucleic acid	↪ RNA found in viruses.
Viscosity	↪ Viscosity of a resin can be high or low. Low viscosity allows rapid infiltration of the resin into the tissue (i.e., Cps is 0.7 for acrylic resins). Viscosity is higher for epoxy resins (e.g., Araldite: 1300 to 1650 centipoises at 25°C).

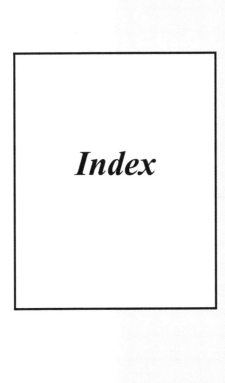

Index